Biochemistry of Non-Shivering Thermogenesis in Vertebrates

This comprehensive reference book covers non-shivering thermogenesis (NST), metabolic rates, and associated diseases. The book describes how the incidence of metabolic syndromes has reached epidemic levels worldwide due to dysregulation in energy balance. Various chapters deal with NST in brown adipose tissue (BAT) and in skeletal muscle and the implications these mechanisms have for understanding vertebrate evolution. It covers the role of NST in birds and mammals, as well as muscle adaptation during hibernation. The book is meant for researchers and students of biochemistry, molecular medicine, and physiology. It is also useful to physicians and health care professionals.

Biochemistry of Non-Shivering Thermogenesis in Vertebrates

Evolutionary Perspectives

Edited by
Naresh Chandra Bal and Punyadhara Pani

CRC Press
Taylor & Francis Group
Boca Raton London New York

CRC Press is an imprint of the
Taylor & Francis Group, an **informa** business

Designed cover image: Naresh Chandra Bal and Punyadhara Pani

First edition published 2025
by CRC Press
2385 NW Executive Center Drive, Suite 320, Boca Raton FL 33431

and by CRC Press
4 Park Square, Milton Park, Abingdon, Oxon, OX14 4RN

CRC Press is an imprint of Taylor & Francis Group, LLC

ISBN: 978-1-032-45634-8 (hbk)
ISBN: 978-1-032-48931-5 (pbk)
ISBN: 978-1-003-39147-0 (ebk)

DOI: 10.1201/9781003391470

Typeset in Times
by SPi Technologies India Pvt Ltd (Straive)

Contents

Preface

In recent years biochemical aspects of thermogenesis, and more specifically non-shivering thermogenesis (NST), have become an important area of research. NST is now considered part of the whole-body metabolic rate. Many laboratories around the globe project someday to target these mechanisms to counter several chronic metabolic diseases including obesity and type 2 diabetes. A simple search in Pubmed shows that the research on NST, both on skeletal muscle and brown adipose tissue (BAT), has increased exponentially in the past 15 years. However, when we tried to search for a book that gives a comprehensive overview of all aspects of NST, we could not find a suitable one. I myself faced a lot of challenges to comprehend the biochemical concepts of NST when I started my journey as a new researcher to study mechanisms of muscle NST in 2007. Since then I have been keen on putting together a book that deals with various aspects of NST in a lucid manner for beginning researchers in the field. Over the past decade, I have completed several projects to elucidate biochemical mechanisms of NST in both skeletal muscle and BAT. I am happy that my student, Ms. Punyadhara Pani, also became interested to join this project as a coeditor. We hope this—is a reference volume that has long been needed in the field.

This book is mainly aimed at undergraduates, postgraduates, and PhD research students who are trying to enter the field of physiology. It is also helpful for instructors looking for one-stop-shop updates in the field of biochemical aspects of thermogenesis. We have attempted to provide both basic information and a more straightforward explanation of advanced concepts in order to cater to the needs of a variety of readers. The subjects are covered in detail, from the fundamentals to in-depth understanding of biochemistry.

To enable students to grasp the important points of the chapters, keynotes have been included and a brief summary provided at the beginning of each chapter. The content is categorized into three units: (1) Non-Shivering Thermogenesis through Brown Fat and Allied Mechanisms; (2) Skeletal Muscle Non-Shivering Thermogenesis: Evolutionary, Mechanistic and Ecological Aspects; and (3) Biomedical Relevance of Non-Shivering Thermogenesis. Each unit is accompanied by an overview for easy comprehension.

In Unit 1, we discuss various aspects of NST in BAT and its role during the evolution of endothermy in mammals. One of our chapters provides information about the biochemical mechanisms, including uncoupling protein-1 (UCP1), that operate inside the brown adipocytes and are responsible for heat production. Others deal with how neurohormonal mechanisms have developed to regulate BAT-based NST in a precise manner and their intricate functional interplay with the skeletal muscle in hibernating mammals. The chapters are authored by scientists who have conducted their research on the topic and have gained significant knowledge on the molecular aspects of adipose tissue function. An objective of Unit 1 is to provide information about the biochemical basis of the BAT function; more recent ideas on utilizing these mechanisms to directly or indirectly counter obesity are not discussed in detail.

Besides classical BAT-based NST, alternative NST mechanisms, namely those originating in skeletal tissue, have attracted attention of the scientific community. However, how muscle NST is regulated and whether it can be adaptive remains a matter of long-lasting controversy. Unit 2 of the book deals with this subject in more detail. We have taken considerable care in preparing illustrations for each chapter to excite the curiosity of readers. These illustrations should make it easier for students to develop an intuitive feel for the understanding of how critical metabolic thermogenesis is for the survival of endotherms. Each of the chapters is authored by researchers who are internationally acclaimed for their contributions to the field. We feel privileged to have the opportunity to edit their chapters.

This edition of *Biochemistry of Non-Shivering Thermogenesis in Vertebrates: Evolutionary Perspectives* offers a wide selection of high-quality supplements to assist students and instructors.

We are optimistic to see the satisfaction of students and teachers aspiring to have clear concepts and knowledge in the area of biochemistry. The authors of this project are some of the most interesting and thought-provoking individuals in the field. We look forward to getting some positive feedback from our readers.

The editors and authors confirm that this book content has been created with no conflict of interest.

Dr. Naresh Chandra Bal
Ms. Punyadhara Pani
School of Biotechnology
KIIT University, Odisha, India

About the Editors

Dr. Naresh Chandra Bal has more than 15 years of research experience in muscle and brown adipose tissue (BAT) thermogenesis. He also possesses more than 20 years of teaching experience in physiology- and biochemistry-related subjects at different levels. He has designed and performed studies on several transgenic mouse models using manipulation of thermogenic capacities of both BAT and muscle. In addition to these transgenic models he has employed other strategies such as pharmacological, surgical, and diet intervention to address questions related to thermogenesis, especially non-shivering thermogenesis. Dr. Bal has authored more than 30 articles on this topic published in top-tier journals such as *Nature Medicine*, *Trends in Endocrinology and Metabolism*, *Biological Reviews*, *Journal of Biological Chemistry*, and *Biochemical Journal*. He has published 60 journal articles in total, and his papers have been cited by top researchers in the field from all over the world. Currently, his H-index is 30 with total citations of more than 3,000. For his pioneering research Dr. Bal has been awarded several honors including Physiological Genomics Fellowship, American Physiological Society (APS), USA; K01 Award from National Institutes of Health (NIH), USA; Ramalingaswami Re-entry Fellowship, Government of India; and Early Career Research Award, Government of India. He serves as associate editor of *Frontiers in Physiology* and editor of *Metabolites & Molecular Medicine Reports*. Due to his scientific contributions he has been accepted as Fellow of the Royal Society of Biology, UK.

Ms. Punyadhara Pani is currently working in the laboratory of Dr. Bal in delineating molecular mechanisms of muscle non-shivering thermogenesis (NST) in birds. She completed her Bachelor's degree at Utkal University and her Master's degree (*summa cum laude*) at Ravenshaw University. She taught undergraduate students for about five years. Due to her excellent academic distinctions, she has been awarded one of the most competitive predoctoral research funding fellowships, called DST-Inspire fellowship, from the Government of India. Her research has been published in several renowned journals such as, among others, *Biological Reviews* and *Biochemical Journal*. She also has successfully presented her work as poster and oral presentations at both national and international conferences, for which she has received recognition and awards. She has done extensive research on skeletal muscle adaptation to environmental factors, pharmacological agents, and cytokines, leading to several recent publications in internationally reputed journals. For these studies she has used different animal models, including mammalian and avian, providing biochemical/genetic insight for NST in the muscle.

Contributors

Naresh Chandra Bal
School of Biotechnology
KIIT University, Bhubaneswar
Odisha, India

Kênia Cardoso Bicego
Department of Animal Morphology and Physiology
School of Agriculture and Veterinary Sciences
São Paulo State University
Jaboticabal, Brazil

José Eduardo Pereira Wilken Bicudo
Department of Physiology
Bioscience Institute
University of São Paulo
São Paulo, Brazil

Jessica L. Braun
Centre for Bone and Muscle Health
Department of Kinesiology
Brock University
St. Catharines, Ontario, Canada

Sun-ying Cheng
Institute of Pharmacology
Department of Pharmaceutical Sciences
Zhejiang University of Technology
Hangzhou, China

Val A. Fajardo
Centre for Bone and Muscle Health
Department of Kinesiology
Brock University
St. Catharines, Ontario, Canada

Jens P.C. Franck
Department of Biology
University of Winnipeg
Winnipeg, Manitoba, Canada

Mia S. Geromella
Centre for Bone and Muscle Health
Department of Kinesiology
Brock University
St. Catharines, Ontario, Canada

Gordon Grigg
School of Environment (Biological Sciences)
University of Queensland
Brisbane, Queensland, Australia

Sam D. Houman
Touro University
California College of Osteopathic Medicine
Vallejo, California

Moriah Hunstiger
University of Alaska Fairbanks
Fairbanks, Alaska

Abraham Karimi
Touro University
California College of Osteopathic Medicine
Vallejo, California

Julius L. Larion
Touro University
California College of Osteopathic Medicine
Vallejo, California

Bradley S. Launikonis
School of Biomedical Sciences
The University of Queensland
Brisbane, Queensland, Australia

Han-bing Li
Institute of Pharmacology
Department of Pharmaceutical Sciences
Zhejiang University of Technology
Hangzhou, China

Bianca M. Marcella
Centre for Bone and Muscle Health
Department of Kinesiology
Brock University
St. Catharines, Ontario, Canada

Aldo Meizoso-Huesca
School of Biomedical Sciences
The University of Queensland
Brisbane, Queensland, Australia

Shona A. Mookerjee
Touro University
California College of Osteopathic Medicine
Vallejo, California
and
Buck Institute for Research on Aging
Novato, California

Colleen M. Novak
Department of Biological Sciences
Kent State University
Kent, Ohio

S. Ryan Oliver
United States Naval Academy
Annapolis, Maryland

Anuttama Pal
School of Biotechnology
KIIT University, Bhubaneswar
Odisha, India

Punyadhara Pani
School of Biotechnology
KIIT University, Bhubaneswar
Odisha, India

Sunil Pani
School of Biotechnology
KIIT University, Bhubaneswar
Odisha, India

Muthu Periasamy
Medical Center
University of Central Florida
Orlando, Florida

Sean W.M. Robinson
Department of Biology
University of Winnipeg
Winnipeg, Manitoba, Canada

Leslie A. Rowland
Program in Molecular Medicine
University of Massachusetts
Chan Medical School
Worcester, Massachusetts

Bijayashree Sahu
School of Biotechnology
KIIT University, Bhubaneswar
Odisha, India

Yuan-hai Sun
Institute of Pharmacology
Department of Pharmaceutical Sciences
Zhejiang University of Technology
Hangzhou, China

Christina A. Watts
School of Biomedical Sciences
Kent State University
Kent, Ohio

Teng-jiao Zhao
Institute of Pharmacology
Department of Pharmaceutical Sciences
Zhejiang University of Technology
Hangzhou, China

Unit 1

Non-Shivering Thermogenesis through Brown Fat and Allied Mechanisms

OVERVIEW

Mammals are the most evolved vertebrates, yet their neonates are the most vulnerable and may even not reach adulthood if they are not able to overcome these vulnerability factors. One of these factors is massive heat loss and the need for high rate of heat production within the body, i.e., thermogenesis. To overcome this limitation, mammals have evolved several molecular, structural, and behavioral innovations that gave them the advantage in their struggle for existence. One such advancement is the genesis of a unique organ, the brown adipose tissue (BAT), which is present among many of the taxa belonging to eutherian mammals. The evolution of BAT has not only reduced the cost of maintaining a constant body temperature but also provided a major evolutionary advantage to survive the cold stress of birth, reducing the requirement of parental care. The appearance of BAT during the Cenozoic era facilitated diversification of several eutherian groups into different ecological niches.

BAT was first formally described in 1551 by the Swiss naturalist Conrad Gessner. It was originally termed the "hibernating gland," and over the following 400 years several different functions were attributed to the tissue. It was only in the 1950s that BAT was firmly identified as a thermogenic organ—the key site of heat production without the involvement of shivering, a process termed a *non-shivering thermogenesis* (NST). A decade later there was considerable interest, particularly centered on the mechanisms by which heat is generated in this tissue. Later studies showed that BAT serves as the effector organ of heat production, which channels chemical energy from food materials to be converted into heat energy by modifying the mitochondria, the powerhouse of the cell. The mitochondria are abundant in BAT and express a specialized protein celled uncoupling protein-1 (UCP1), initially termed *thermogenin*.

This unit deals with several aspects of BAT and its physiological relevance. Chapter 1 deals with the evolutionary and developmental origin of BAT during the early origins of eutherian mammals. The anatomical, morphological, and molecular features that make the adult BAT an easily

DOI: 10.1201/9781003391470-1

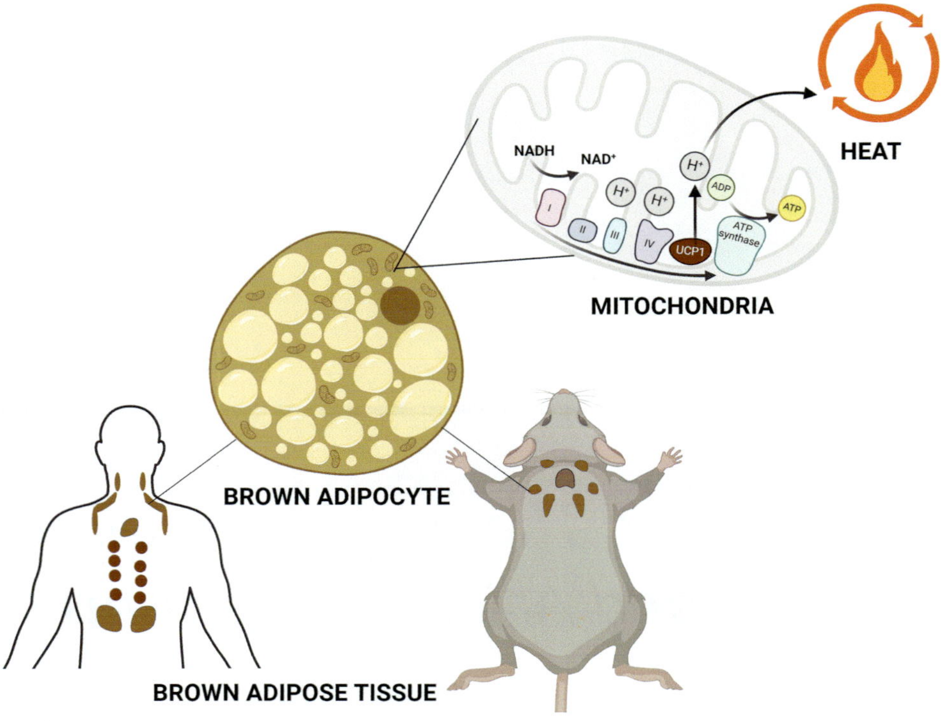

activatable organ in adulthood are also discussed in this chapter. In Chapter 2 we focus on the underlying mechanisms by which heat is produced in the brown fat, concentrating on intracellular regulation of UCP1-mediated heat production and the role of fatty acids in activating it. In addition, BAT is highly integrated with the central nervous system via the sympathetic nervous system. Expanding on this idea, Chapter 3 focuses on central neural thermoregulation, particularly addressing our current understanding of the importance of hypothalamic neural signaling in the regulation of BAT thermogenesis. Last but not least, Chapter 4 aims to expand the readers' understanding of the mechanisms regulating the cyclic metabolic demands during the hibernation in Arctic ground squirrels, an ideal model for hibernation.

Since this unit deals with one of the most rapidly progressing areas of applied science, it remains critical for students not only to have the most current information available but also to understand the experimental nature of contemporary research in NST through BAT and allied mechanisms. With this goal in mind, representative experiments are discussed throughout the text to illustrate the kinds of approaches on which our current understanding has been built. In addition, each chapter contains a Key Experiment essay that describes a seminal paper and its background in detail, with the intent of giving the reader a sense of "doing science." Finally, we have tried to point out the limits of our knowledge and to identify outstanding unanswered questions throughout the text. We hope that, together, these approaches give the student an appreciation of the ways in which progress in understanding the unit has been made, and convey the excitement and challenges of participating in this frontier of scientific exploration.

1 Evolutionary and Developmental Origin of Brown Adipose Tissue

Sunil Pani and Naresh Chandra Bal
School of Biotechnology, KIIT School of Biotechnology, India

1.1 INTRODUCTION

Brown adipose tissue (BAT) was described for the first time by a Swiss physician Conrad Gesner in 1551. He observed a unique tissue in the interscapular region of hibernating marmots and explained the architecture of the tissue as neither fat nor muscle (1, 2). Initially, the tissue was termed "hibernation gland," suggesting an endocrine role of the organ in the regulation of hypometabolism and arousal of hibernators. After 400 years, the importance of this tissue was explained, and it was named *brown adipose tissue* (BAT), as it is rich in mitochondria imparting it the brown color and contains fat droplets like white adipose tissue, which are, however, multilocular and much smaller. The multilocular lipid droplets of BAT provide an increased lipid surface area, which allows easy lipolysis for energy generation and lipid oxidation, in contrast to conventional white adipose tissue (WAT), which primarily stores nutrients in the form of huge unilocular fat droplets (3, 4). Several studies during the 1960s showed that the thermoregulatory mechanism was found to be a precisely regulated process in BAT involving mitochondria, termed *non-shivering thermogenesis* (NST). During the 1970s, a unique protein responsible for BAT-based NST was identified, which was originally named thermogenin and later renamed *uncoupling protein 1* (UCP1) (5). UCP1 is an inner mitochondrial membrane protein that uncouples the oxidative phosphorylation pathway from ATP synthesis and generates heat via the proton channeling from intermembrane space to matrix (5, 6). Knockout studies of UCP1in a mouse model further confirmed its role in NST (7); the mechanism of heat production by UCP1 is discussed in Chapter 3. However, interest in researching UCP1 waned after human studies showed BAT to express only during the neonatal period and not during adulthood. Surprisingly, the conclusive crystallographic structure of UCP1 is lacking to date. During 2007–2009, several studies showed the presence of BAT in adult humans by PET scanning studies, which reignited the research on BAT worldwide (8). The idea that BAT is a major calorie sink and can be targeted to treat metabolic diseases, including obesity, drove the research further. Later on, another crucial function of BAT came to light, specifically endocrine regulation of other organs including muscle, liver, and WAT.

To target BAT-mediated NST pathways for treating metabolic diseases, a better understanding of its comparative expression, evolution, early origin, development, and differentiation will be very useful. In this chapter we will discuss evolutionary and developmental origins in details. The major focus will be on Prototheria and Metatheria class, as they are suggested to be at the boundary between the animals possessing BAT and the ones not having it. Further we will discuss the effect of mammalian radiation on reliance on BAT-mediated NST, specifically how proto-endothermic metabolism fueled the BAT thermogenesis in eutherians. Covering the developmental aspect, we will also discuss the origin of BAT during early life, what distinguishes BAT from its counterparts, transcriptional regulation of its development, and its importance during adulthood.

DOI: 10.1201/9781003391470-2

1.2 ORIGIN OF BROWN ADIPOCYTE IN EARLY MAMMALS

The evolution of endothermy in birds and mammals is one of the crucial evolutionary milestones. Even after more than a century of research, the debate on its ancestry is still a hot topic. While classical research suggested endothermy evolved independently in birds and mammals, recent findings on muscle-mediated thermogenesis indicate sharing of common molecular mechanisms. Grigg and colleagues have suggested that evolution of endothermy may pre-date even the divergence of synapsids from amniotes (elaborated in Unit 2)(9). However, BAT-mediated NST has not been observed in any other clades and is restricted to only some groups of eutherian mammals. Contrastingly, UCP1 gene has been present since 400 million years in the genome of common ancestors of ray-finned and lobe-finned vertebrates. However, UCP1 in these vertebrates does not seem to play a role in NST, as the cold challenge did not upregulate its expression. Moreover, they lack BAT, express UCP1 in non-adipose cells, and are ectothermic in nature. Therefore, thermogenic function of UCP1 must have been infused once BAT appeared and got incorporated into the BAT during mammalian evolution. The specialized features of BAT such as extensive mitochondrial density, sympathetic innervation, vascularization, and lipolytic activity complemented the role of UCP1 in NST (10). The role of BAT and UCP1 in prototherians and marsupials is still under debate. Most researchers consider monotremes to lack BAT-like tissue, and thus to lack BAT-based NST and also maintain lesser BMR as compared to the placentals. Many monotremes such as echidnas display fluctuation in the Tc depending on the environmental temperature, and hence are regarded as proto-endotherms by many scientists (11). In contrast, platypus (also a monotreme), which neither hibernates nor torpors, is able to maintain its body temperature in a narrow range even in fluctuations of environmental temperature as large as 10°C. This reflects the thermogenic strategies can also be diverse within the monotreme clade and that those mammals possibly rely other NST mechanisms like the ones originating in the skeletal muscle. Recently UCP1 has been reported to be present (both in genomic data and mRNA) in platypus (12), but its thermogenic ability has not been studied.

On the other hand, a few marsupial species have been reported to bear a BAT-like tissue and may hold answer into integration of UCP1 function into BAT during evolution. Once UCP1 was functionally established, its expression in brown adipocytes and neuro-hormonal regulation might have intricately evolved with the BAT, and that novel tissue became associated with adaptive radiation in eutherian mammals. The UCP1 expression in BAT has taken several ups and downs during radiation of mammals. While UCP1 plays a key role in NST of most of the smaller-sized and hibernating mammals, it is nonfunctional/truncated in several clades of larger mammals (pig, elephant, and sea cows), aquatic mammals (whales and dolphins), and mammals with lower metabolic rates (pangolins, sloths) (1, 13). In some other mammals, such as humans, cows, goat, and sheep, UCP1 is highly expressed during neonatal life having a well-defined BAT, which, however, gets depleted with increasing age and becomes negligible during adult life. Although some studies in the last 15 years have shown active BAT in adult human, its functionality in NST is yet to be substantiated. The addition of NST functionality to UCP1 only after the amalgamation of UCP1 and BAT is supported by two arguments. (1) Interestingly, UCP-like sequences are present in several plant and invertebrate genomes. Upon comparison, these sequences appear closer to UCP1 (of mammals) than other UCP isoforms. Given that a thermogenic function cannot be conceived for such sequences in plants and invertebrates, it can be considered these may have some unresolved properties. If such a primitive function existed for UCP-like sequences, it may be retained in case of UCP1 of ectothermic vertebrates. (2) UCP1 in non-mammalian vertebrates is expressed mostly in tissues other than adipose. For example, the liver of cyprinids (freshwater fish) expresses UCP1 at the highest level while the kidneys and brain also express UCP1in significant amounts. In contrast to mammals, cold water causes liver UCP1 mRNA levels to drop but brain UCP1 mRNA levels to rise. This suggests that transcriptional regulation of UCP1 gene among ectothermic vertebrates might have happened quite early during evolution but was not integrated into adipose tissue until evolution of a brown adipocyte

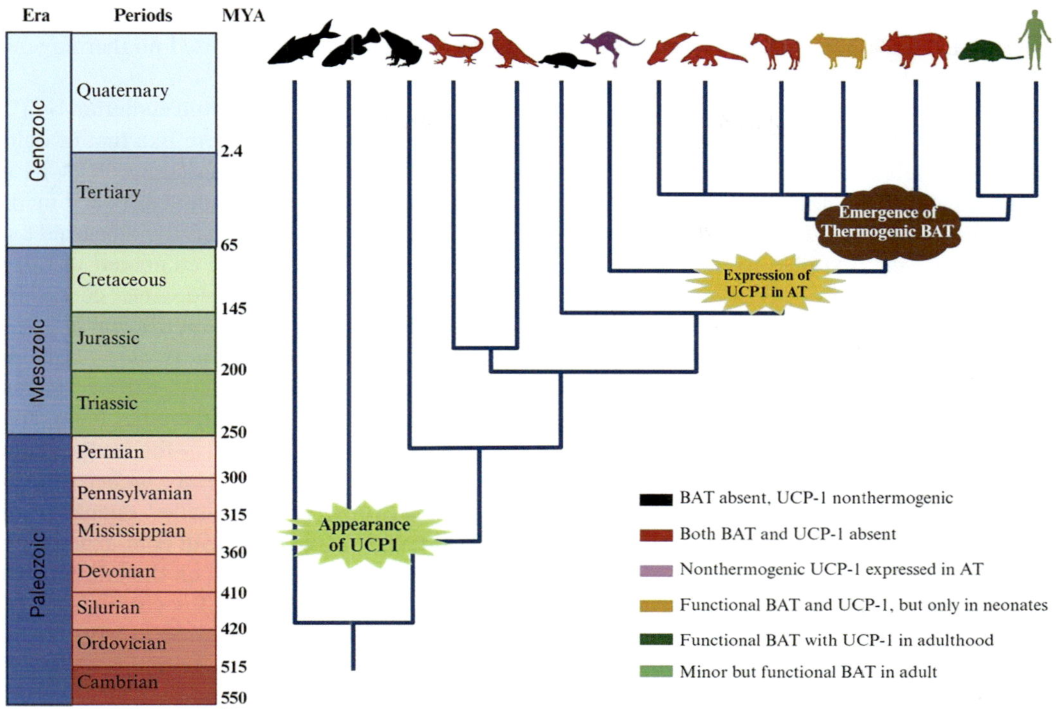

FIGURE 1.1 The evolution of BAT-mediated NST during the chordate evolution. The UCP1 gene has been estimated to have emerged approximately 400 million years ago in the genomes of vertebrates, presenting an intriguing evolutionary tale. Initially, it appeared to have no role in regulating core body temperature, with its expression detected in sites other than adipose tissue (AT). As evolution unfolded, nonthermogenic UCP1 was also present in amphibian genomes but conspicuously absent in reptiles and birds. During early mammalian evolution, UCP1 made its first appearance in AT, yet its precise role in thermogenesis remains a topic of debate. Across mammalian clades, UCP1 played a pivotal role in thermogenesis in some instances while in others it was either absent or truncated. The differential patterns of UCP1 expression are visually represented by color-coded vertebrates, each hue signifying a specific expression pattern as detailed in the legends.

in eutherians (14–16). Thus, several decades of research on expression of UCP1 in a brown adipocyte has been somewhat of a roller-coaster ride, and a conclusive origin of BAT is yet to be identified (see Figure 1.1).

1.2.1 Marsupials and the Development of Archetypal BAT

Monotremes and marsupials have been among the most crucial clades during the evolution of endothermy in mammals. While the appearance of a first BAT-prototype can be regarded to have happened after the divergence of monotremes, the search for brown adipocytes in marsupials can provide a key insight on this matter. The views on presence of BAT in marsupials have remained quite skeptical; while some species have been shown to possess both UCP1 and BAT, others do not. Surprisingly, some of the marsupials express UCP1 in other tissues even though they lack BAT (8). Rowlatt and colleagues in 1971 performed a descriptive study on several mammals and suggested BAT to be absent in marsupials (17). Later studies by a few other groups reported the existence of BAT-like tissue in some marsupials (14, 18). However, a consensus is lacking on whether this BAT-like tissue in marsupials is functionally active and can produce heat. Researchers searching for evidence of NST in marsupials have observed a thermogenic response to cold in eastern barred bandicoot, eastern bettong, long-nosed potoroo, and Red-necked wallaby (6, 18–20), although the role of UCP1 in these marsupials has not been studied. Intriguingly, other marsupials such as gray short-tailed opossum, common brushtail possum, and brown antechinus exhibited no thermogenic response upon adrenergic stimulation (21, 22).

Another theory about the marsupial UCP1 is that it may operate differently from eutherian UCP1. According to molecular phylogenetic analyses, marsupial UCP1 functions more like that of cold-blooded vertebrates than of eutherian mammals (8). It is possible that looser restrictions on UCP1 in common ancestors of marsupials and eutherians caused higher mutational accumulation and incorporation of novel activities in eutherians, such as thermogenesis. Several scientists hypothesized that marsupials lack BAT due to their restricted habitat with moderate winter as compared to many eutherians, which rely heavily on BAT-mediated thermogenesis to cope with cold winter conditions (8). However, the presence of persistent classical BAT in rodents residing in hot tropical climates undermines this argument. Interestingly, the striped-faced dunnart (*Sminthopsis macroura*), an Australian marsupial mouse, showed significantly elevated heat production when exposed to low temperature, suggesting the existence of adaptive NST mechanisms (23). On the other hand, anatomical observations by Rowlett and colleagues showed the absence of BAT in the interscapular region of several marsupials (17). A study by Hope et al. showed comparatively higher level of brownness of the interscapular adipose tissue as compared to tail fat in fat-tailed dunnart (24). Jastroch et al. also noticed an increase in brownness of the interscapular fat with cold exposure in fat-tailed dunnart (*S. crassicaudata*), another Australian marsupial (14). Based on the current state of knowledge, it can be proposed that although the appearance of archetypal BAT occurs in marsupials, more studies on its metabolic networks and signaling pathways are required to gain further understanding about UCP1 integration into BAT-based NST function (25).

1.2.2 Proto-eutherian Mammals Bear a Rudimentary Type of BAT

Thermogenically competent BAT has been described in the primitive eutherians such as afrotherians (26). The discovery of thermogenic BAT expressing UCP1 in the afrotherian species implies that this type of BAT existed in common placental ancestors. Whether BAT in these organisms possesses a cold-inducible UCP1 has not been carefully tested. Answering this question will clarify whether the transcriptional metabolic regulation of UCP1 was already integrated in the functional BAT in these primitive taxa that must have originated quite close to the marsupial–eutherian divergence. Such a viewpoint is also supported by molecular evolution studies that indicate thermogenic UCP1 may only be present in eutherians and did not evolve in marsupials. After the marsupial–eutherian separation, between 70 million and 150 million years ago, most of earth's landmasses were in a warm-to-hot greenhouse climate without polar glaciations. This might have provided the conditions for the evolution of UCP1 thermogenesis and transcriptional integration with BAT cells. Long branch lengths and unusual molecular phylogenetic traits in UCP1 point to a high substitution rate that may be responsible for the protein's thermogenic action. BAT might have first developed to help with thermoincubation or to keep babies' bodies warm. With the continental drift and creation of cooler ecological niches, eutherian mammals had the evolutionary benefits to migrate and establish themselves in these climates. The Cretaceous–Paleogene (K-Pg) mass extinction, which occurred about 70 million years ago, is thought to have provided the eutherian progenitor a considerable advantage, and the development of thermogenic BAT may have played a role in this (27).

Thermogenic BAT may have been used by the underground-dwelling eutherian ancestors to outcompete other species in the short-term cooler climate, assuring quick rewarming from hypometabolic conditions. It is suggested BAT might have improved fitness in small-sized eutherians assisting in incubation of young and in hibernation. So, the evolution of eutherian UCP1 may have been impacted by the selection forces at the K-Pg boundary. However, since thermogenic trait is often mass dependent and ineffective in larger animals, it is still unknown why eutherian mammals maintain BAT even when they are enormous. This begs the question of whether thermogenic BAT is being lost in big-sized eutherians. It has been proposed that the evolutionary constraints on eutherian

UCP1 and the abundance of adult BAT may have loosened in the post–ice age era, with the exception of species whose UCP1 confers a key benefit, such as those that inhabit cold climates. Both the Laurasiatheria and Euarchontoglires superorders, which include humans that express BAT from infancy to adulthood, have well-established BAT-mediated NST. More studies on transcriptional and neurohormonal control of BAT thermogenesis in more recently evolved eutherian species will throw light on this debatable area.

1.3 DEVELOPMENTAL ORIGIN OF BAT IN EUTHERIANS

The formation of BAT during gestational and postnatal stages is another important factor controlling its thermogenic action that is associated with survival of the individual. Human neonates have substantial BAT, mostly found in the interscapular, supraclavicular, and internal organ regions. Even though BAT makes up only 2–4% of birth weight, it is essential for newborns because of its remarkable ability to generate heat during birth. The development process of BAT differs considerably between smaller rodents and larger mammals. The precocial behavior of the majority of larger mammals leads to early development of the hypothalamic-pituitary-adrenal (HPA) axis during the prenatal period. In contrast, altricial species like mice and rats rely on huddling for heat regulation due to an immature HPA axis (28, 29). Because of the advanced development of BAT in altricial animals, they can maintain body temperature by quickly initiating NST in response to exposure to cold environments. Notably, hypothermia can result from preterm delivery or from failing to activate BAT, highlighting the vital function BAT plays in thermoregulation for these animals. Thus, species differing in terms of maturity, duration of gestation, and body composition at birth exhibit differences in the timing of BAT maturation during their perinatal phase. Therefore, in this section, we will discuss the tissue lineage of BAT, the establishment of neural connections during this period, and the molecular mechanism guiding this process of development.

1.3.1 Tissue That Gives Rise to BAT During Early Development

During prenatal development, the blastula is organized into a three-layered structure during gastrulation, consisting of endoderm (inner lining), mesoderm (middle layer), and ectoderm (outside surface). In the mesoderm layer, mesenchymal progenitor cells differentiate into specialized cell types like chondrocytes, myocytes, and adipocytes. It was believed in the early 20th century that the adipocytes, BAT and WAT, have similar developmental origins (30, 31). In the early 21st century, however, several studies demonstrated that brown adipocytes have a developmental origin closer to myocytes than to white adipocytes (32–35). This suggests that progenitor cells of BAT and SkM were delineated from white adipocyte progenitor cells earlier than expected. A unique protein named myogenic factor 5 (Myf5) acts as the checkpoint during this differentiation between white adipocytes and progenitor cells of BAT and SkM. White adipocytes originate from $Myf5^-$ mesenchymal progenitor cells, whereas brown adipocytes and myocytes come from $Myf5^+$ cells (33, 34). Later PRDM16 acts as a switch for the divergence between BAT and SkM. The expression of PRDM16 leads to the formation of brown adipocytes, whereas cells lacking it originate into myocytes. The transformation of cultured myoblasts into brown adipocytes upon forced expression of PRDM16 provided more evidence that both BAT and SkM share a common progenitor cell. Furthermore, myocyte formation was stimulated by PRDM16 knockdown in primary brown adipocytes (33). It is important to note that heterogeneity in BAT depots is observed at sites other than the traditional interscapular site. While all the brown adipocytes of the interscapular depot possessed Myf5, its expression is inconsistent in other BAT depots. Certain adipocytes in the visceral BAT depots express Myf5, whereas others lack it, even though they bear multilocular fat droplets and a higher mitochondrial density. These adipocyte types are primarily produced by adrenergic stimulation and are mostly seen scattered among white adipocytes (36). This progenitor brown adipocyte inside white fat depot has become a major area of metabolic research.

1.3.2 Establishment of Neural Connections on Developing BAT

The development of BAT relies heavily on proper innervation so that the central nervous system can effectively control the tissue function. Before the appearance of brown adipocytes, the neural crest cells migrate toward the developing fat pad that later paves the way for developing nerve fibers by secreting several crucial factors such as the nerve growth factor, brain-derived neurotrophic factor (BDNF), ciliary neurotrophic factor (CNTF), and glial cell line–derived neurotrophic factor (GDNF), among others (37, 38). Differentiated BAT is then innervated by both the sympathetic nervous system (SNS) and the sensory nervous system (discussed further in Chapter 3). While SNS is the primary mediator of UCP1-mediated NST in BAT, sensory neurons have also been reported to monitor lipolysis thereby modulating UCP1 activity (39). Upon arrival of the neural crest, brown adipocytes start to express β-adrenergic receptors that communicate with the neuron terminals and set the stage for the innervation. Out of all the β-adrenergic receptors, mice and rats rely predominantly on β3 type whereas humans rely equally on β1 and β2 types (40). Following the signals laid down before, as BAT matures, SNS fibers branch like small tree roots, spreading out at the adipocyte level. Once the nerves reach BAT depots, they establish synapses with BAT cells. This synaptic connection allows for communication between the nervous system and BAT. The neurotransmitter norepinephrine, released by sympathetic nerves, plays a crucial role in activating BAT cells.

Neurotrophin-3 (NT-3) secreted by BAT acts as a crucial mediator for the establishment and survival of SNS (41). Interestingly, adequate blood supply during this period is essential for establishing proper development and innervation of BAT. So, the innervation of BAT during development is a complex process that involves the coordinated growth and differentiation of neurons, Schwann cells, and blood vessels. Although the timeline varies between precocial and altricial mammals, the mechanistic basis is similar. Over time, the innervation of BAT matures, and the tissue becomes more efficient in responding to signals from the SNS. This maturation process ensures that the body can effectively regulate heat production in response to changes in temperature and energy demands. The neuronal circuits and subcellular pathways by which cold sensation triggers BAT thermogenesis have been intensively studied in the past. Cold-sensitive thermoreceptors on the skin as well as in the body core, spine, and brain transmit thermal afferent signals to the preoptic area in the hypothalamus (42) and to control regions in the brain stem (43, 44), leading to the activation of postganglionic sympathetic fibers that release noradrenaline to activate BAT (45). Noradrenaline binds to specific β3-adrenergic receptors on the surface of brown adipocytes and activates Gs-protein coupled signaling pathways to stimulate lipolysis for release of free fatty acids. These free fatty acids serve as substrates for lipid oxidation while simultaneously acting as direct activators of UCP1 and heat production at the mitochondrial level (45).

1.4 TRANSCRIPTIONAL REGULATION OF BROWN ADIPOCYTE DEVELOPMENT

BAT shares a major part of the transcriptional cascade with WAT despite of their functional and origin differences. The development of adipose tissue undergoes a cascade of events starting from preadipocyte hyperplasia followed by lipid assimilation, hypertrophy, and angiogenesis. The master regulators of the development and differentiation of brown adipocytes are peroxisome proliferator-activated receptor-γ (PPARγ) and CCAAT/enhancer binding proteins (C/EBP) family proteins. The delineation of BAT from myogenic lineage starts with the expression of PRDM16 (Figure 1.2). In the absence of PRDM16, other PRDM family proteins can compensate for its function (45). During the initial phases of development, the proliferation and expansion of white preadipocytes start early during the gestational period with the expression of the dividing cell marker: Ki-67 (46). Next, Homeobox (HOX) and Bone morphogenetic protein (BMP) genes are

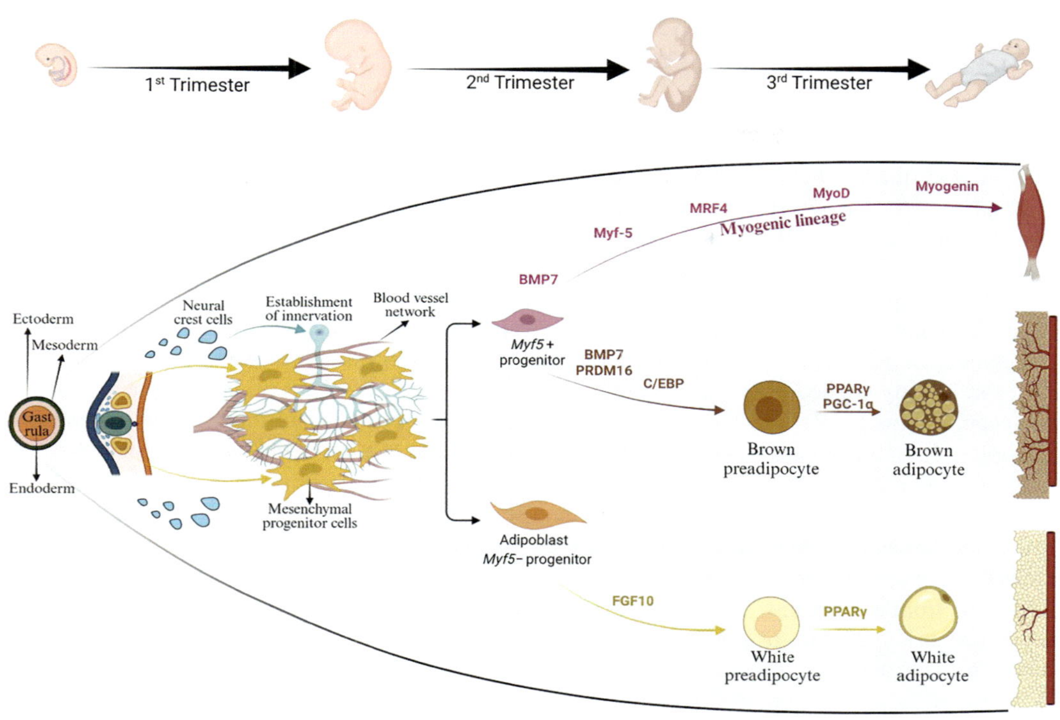

FIGURE 1.2 Developmental origin of BAT. Brown adipose tissue shares its developmental origin with other metabolic sites such as white fat and skeletal muscle. Mesenchymal progenitor cells, originating from paraxial and lateral mesoderm, undergo differentiation upon innervation by neural crest cells. Concurrently, the induction of blood vascularization aids in this differentiation process. These mesenchymal progenitor cells differentiate into two distinct cell types, with one subset expressing Myf5 while the other does not. Cells expressing Myf5 can further differentiate into either brown fat or myogenic lineage. The expression of PRDM16 serves as a key determinant, guiding progenitor cells toward differentiation into brown adipocytes. Additionally, the process of brown adipogenesis is intricately regulated by the expression of various factors including PPARγ and BMP7. This coordinated interplay of molecular signals orchestrates the development and specialization of BAT, contributing to its metabolic role within the body.

expressed at the mid-gestation period promoting the generation of adipocytes from preadipocytes (47). The PRDM16 and C/EBP family proteins play critical roles in adipogenesis and maturation of preadipocytes into brown and white adipocytes, respectively (32, 48). However, these two factors do not promote the expression of UCP1; rather, a concomitant rise in two major mitochondrial regulators, PPAR-γ and PPAR-gamma coactivator (PGC)-1α, induces UCP1 expression during this stage (46). It is also proposed that the expression of adrenergic receptors in adipocytes reaches their peak during late gestation, thereby promoting the generation of brown adipocytes over white ones (49).

BAT markers like Cell Death Inducing DFFA Like Effector A (CIDEA), PRDM16, PGC-1α and prolactin receptor (PRLR), and UCP1 are expressed at maximal levels up to birth and decline immediately thereafter (48, 50). Simultaneously, an elevation in the expression level of the factors promoting white adiposity, like C/EBP and BMP, and white adipocyte markers like leptin, adiponectin, and PPARα is observed, which drives the expansion of WAT while BAT is reduced (46, 51, 52). BAT adipocytes are uniquely different from white adipocytes in being transcriptionally responsive to thyroid hormones. T3 regulate several of the mitochondrial protein expression in brown adipocytes as well as the epinephrine release in the SNS terminals (53, 54).

1.5 MORPHOLOGICAL AND MOLECULAR FEATURES OF BAT DURING ADULTHOOD

The most widely studied BAT depot in rodents is the interscapular one, although in humans the supraclavicular region (especially the neck) is the primary (55). A comparable depot has now been identified in mice, which shows a number of similarities to interscapular fat from 18.5 days post-conception (56). This includes some molecular and morphological (e.g., mitochondrial) characteristics, although its postnatal growth is constrained. Furthermore, gene expression of specific putative beige or white markers are very different in the supraclavicular region compared with interscapular, inguinal, and epididymal (i.e., white fat) depots. For example, HoxC8 and Zic1 were absent from the supraclavicular depot, which also showed a separate principal component analysis (56). In sheep, the equivalent depot is found in the sternal region and has a very different postnatal ontogeny from the more widely studied perirenal-abdominal depot (57). The sternal depot, unlike other fat depots in sheep, retains UCP1 into adulthood, where it exhibits a pronounced thermogenic response to feeding (57). In this regard it is comparable to the supraclavicular depot in humans, which increases in temperature after a meal (58). The unique gene profile of supraclavicular fat shown when candidate genes were measured in both mice (56) and sheep (57) has recently been extended to the five major fat depots in young sheep when analyzed using microarrays (59). Undertaking microarray analysis at two important time points in the early postnatal life of sheep, coincident with the transition of fat from a depot in which brown characteristics dominate (i.e., 7 days of age) to when brown fat is scarce (i.e., 28 days of age) (60), has illustrated the unique nature of each depot (59). Machine learning algorithms, in conjunction with weighted gene co-expression network analysis, demonstrated that the five depots examined (i.e., sternal, perirenal, pericardial, subcutaneous, omental) could be segregated into defined sets of modules containing co-expressed genes, indicative of separate functions. The developmental changes markedly differed between depots despite showing a similar macroscopic morphology (59). This means that at 28 days, when, in sheep, fat is considered primarily white (60), each adipose depot kept a distinct gene expression profile (59). Consequently, although adipose tissue has been considered a metabolic organ with important functions beyond lipid storage, this varies between depots, especially during development. Adipose tissue, therefore, has a range of functions depending on location (50). This concept was formulated by using a computer-assisted supervised learning algorithm to demonstrate that, during postnatal development, each fat depot contains a transcriptome that forms dynamic networks with unique sets of genes (61). Over the first month of life, in sheep, these gene networks are reorganized by accommodating novel members and/or losing some of their original components (59). Dynamic changes in gene regulation with age are rarely examined but do enable the identification of important regulatory relationships. These will have crucial depot-specific roles to enable the differentiation and adaptation necessary to modulate metabolic homeostasis (62). Despite recent efforts to elucidate the cellular and transcriptome composition of different fat depots (63, 64), the influence of genetic, endocrine or environmental factors on fat development remains largely unknown. It is further likely that these differences are mediated in part by sympathetic innervation that has recently been shown to include dense arborizations within adipose tissue that regulate the beiging process.

1.6 FUTURE DIRECTIONS

Decades of intensive research on BAT have yielded invaluable insights into its multifaceted roles related thermogenesis and metabolism. Moreover, some insight has been gained about the developmental regulation and evolutionary relevance of BAT. However, there are still a number of important issues that need to be addressed in order to fully comprehend the dynamic role of BAT. The most important of them is the need to map out the evolutionary history of UCP1's coupling with eutherian NST. Further, the mysterious loss of UCP1 in several eutherian clades raises intriguing—and still unanswered—questions. Determining the timescale of UCP loss in different eutherian lineages should be useful in explaining possible relationships between climatic conditions and the UCP1 truncation or deletion. Another matter that needs clarification is whether the loss of UCP1 function spurred eutherians to adopt alternative NST mechanisms, such as skeletal muscle–mediated thermogenesis and UCP1–independent mechanisms within adipose tissue. Furthermore, it is still to be tested whether the archetypal BAT in marsupials is functional, neuronally controlled, and exhibits cold-induced UCP1 expression. The existence of UCP1 in exotherms, suggesting that it may play roles other than NST, is another significant factor that future research needs to resolve. Addressing these outstanding questions will enrich our understanding of BAT physiology and offer fertile ground for future research endeavors.

REFERENCES

1. Speakman JR. Fifty shades of brown: The functions, diverse regulation and evolution of brown adipose tissue, 1–5, 2019. 10.1016/j.mam.2019.07.006
2. Trayhurn P. Origins and early development of the concept that brown adipose tissue thermogenesis is linked to energy balance and obesity. *Biochimie* 134: 62–70, 2017.
3. Keipert S, and Jastroch M. Brite/beige fat and UCP1—is it thermogenesis? *Biochimica et Biophysica Acta (BBA)-Bioenergetics* 1837: 1075–1082, 2014.
4. Cannon B, and Nedergaard J. Brown adipose tissue: function and physiological significance. *Physiological Reviews* 2004. https://doi.org/10.1152/physrev.00015.2003
5. Ricquier D, and Kader J-C. Mitochondrial protein alteration in active brown fat: a sodium dodecyl sulfate-polyacrylamide gel electrophoretic study. *Biochemical and Biophysical Research Communications* 73: 577–583, 1976.
6. Nicholls DG, Bernson VS, and Heaton GM. The identification of the component in the inner membrane of brown adipose tissue mitochondria responsible for regulating energy dissipation. In: *Effectors of Thermogenesis: Proceedings of a Symposium held at Geneva (Switzerland) on 14 to 16 July 1977*, Springer, 1978, p. 89–93.
7. Enerbäck S, Jacobsson A, Simpson EM, Guerra C, Yamashita H, Harper M-E, and Kozak LP. Mice lacking mitochondrial uncoupling protein are cold-sensitive but not obese. *Nature* 387: 90–94, 1997.
8. Jastroch M, Oelkrug R, and Keipert S. Insights into brown adipose tissue evolution and function from non-model organisms. *Journal of Experimental Biology* 221: jeb169425, 2018.
9. Grigg G, Nowack J, Bicudo JEPW, Bal NC, Woodward HN, and Seymour RS. Whole-body endothermy: ancient, homologous and widespread among the ancestors of mammals, birds and crocodylians. *Biological Reviews* 97: 766–801, 2022.
10. Klingenspor M, Fromme T, Hughes Jr DA, Manzke L, Polymeropoulos E, Riemann T, Trzcionka M, Hirschberg V, and Jastroch M. An ancient look at UCP1. *Biochimica et Biophysica Acta (BBA)-Bioenergetics* 1777: 637–641, 2008.
11. Nicol SC. Energy Homeostasis in Monotremes. *Front Neurosci* 11: 195, 2017.
12. Gaudry MJ, and Campbell KL. Evolution of UCP1 transcriptional regulatory elements across the mammalian phylogeny. *Frontiers in Physiology* 8: 670, 2017.
13. Gaudry MJ, Jastroch M, Treberg JR, Hofreiter M, Paijmans JL, Starrett J, Wales N, Signore AV, Springer MS, and Campbell KL. Inactivation of thermogenic UCP1 as a historical contingency in multiple placental mammal clades. *Science Advances* 3: e1602878, 2017.
14. Jastroch M, Withers KW, Taudien S, Frappell PB, Helwig M, Fromme T, Hirschberg V, Heldmaier G, McAllan BM, and Firth BT. Marsupial uncoupling protein 1 sheds light on the evolution of mammalian nonshivering thermogenesis. *Physiological Genomics* 32: 161–169, 2008.
15. Jastroch M, Wuertz S, Kloas W, and Klingenspor M. Uncoupling protein 1 in fish uncovers an ancient evolutionary history of mammalian nonshivering thermogenesis. *Physiological Genomics* 22: 150–156, 2005.
16. Trzcionka M, Withers KW, Klingenspor M, and Jastroch M. The effects of fasting and cold exposure on metabolic rate and mitochondrial proton leak in liver and skeletal muscle of an amphibian, the cane toad Bufo marinus. *Journal of Experimental Biology* 211: 1911–1918, 2008.
17. Rowlatt U, Mrosovsky N, and English A. A comparative survey of brown fat in the neck and axilla of mammals at birth. *Neonatology* 17: 53–83, 1971.

18. Loudon A, Rothwell N, and Stock M. Brown fat, thermogenesis and physiological birth in a marsupial. *Comparative Biochemistry and physiology A, Comparative Physiology* 81: 815–819, 1985.
19. Ikonomopoulou MP, and Rose RW. The development of endothermy during pouch life in the eastern barred bandicoot (Perameles gunnii), a marsupial. *Physiol Biochem Zool* 79: 468–473, 2006.
20. Rose RW, West AK, Ye J-M, McCormack GH, and Colquhoun EQ. Nonshivering thermogenesis in a marsupial (the Tasmanian bettong Bettongia gaimardi) is not attributable to brown adipose tissue. *Physiological and Biochemical Zoology* 72: 699–704, 1999.
21. Nicol S, Pavlides D, and Andersen N. Nonshivering thermogenesis in marsupials: absence of thermogenic response to β3-adrenergic agonists. *Comparative Biochemistry and Physiology Part A: Physiology* 117: 399–405, 1997.
22. Dawson TJ, and Dawson WR. Metabolic scope and conductance in response to cold of some dasyurid marsupials and Australian rodents. *Comparative Biochemistry and Physiology Part A: Physiology* 71: 59–64, 1982.
23. Clements F, Hope P, Daniels C, Chapman I, and Wittert G. Thermogenesis in the marsupial Sminthopsis crassicaudata: effect of catecholamines and diet. *Australian Journal of Zoology* 46: 381–390, 1999.
24. Hope PJ, Pyle D, Daniels CB, Chapman I, Horowitz M, Morley JE, Trayhurn P, Kumaratilake J, and Wittert G. Identification of brown fat and mechanisms for energy balance in the marsupial, Sminthopsis crassicaudata. *American Journal of Physiology-Regulatory, Integrative and Comparative Physiology* 273: R161–R167, 1997.
25. Jastroch M, Polymeropoulos ET, and Gaudry MJ. Pros and cons for the evidence of adaptive non-shivering thermogenesis in marsupials. *Journal of Comparative Physiology B*: 2021: 1–11, 2021.
26. Mzilikazi N, Jastroch M, Meyer CW, and Klingenspor M. The molecular and biochemical basis of non-shivering thermogenesis in an African endemic mammal, Elephantulus myurus. *American Journal of Physiology-Regulatory, Integrative and Comparative Physiology* 293: R2120–R2127, 2007.
27. O'Leary MA, Bloch JI, Flynn JJ, Gaudin TJ, Giallombardo A, Giannini NP, Goldberg SL, Kraatz BP, Luo Z-X, and Meng J. The placental mammal ancestor and the post–K-Pg radiation of placentals. *Science* 339: 662–667, 2013.
28. Bienboire-Frosini C, Wang D, Marcet-Rius M, Villanueva-García D, Gazzano A, Domínguez-Oliva A, Olmos-Hernández A, Hernández-Ávalos I, Lezama-García K, Verduzco-Mendoza A, Gómez-Prado J, and Mota-Rojas D. The Role of Brown Adipose Tissue and Energy Metabolism in Mammalian Thermoregulation during the Perinatal Period. *Animals (Basel)* 13: 2173, 2023.
29. Lezama-García K, Mota-Rojas D, Martínez-Burnes J, Villanueva-García D, Domínguez-Oliva A, Gómez-Prado J, Mora-Medina P, Casas-Alvarado A, Olmos-Hernández A, and Soto P. Strategies for hypothermia compensation in altricial and precocial newborn mammals and their monitoring by infrared thermography. *Veterinary Sciences* 9: 246, 2022.
30. Ailhaud G, Grimaldi P, and Negrel R. Cellular and molecular aspects of adipose tissue development. *Annual Review of Nutrition* 12: 207–233, 1992.
31. Hahn P, and Novak M. Development of brown and white adipose tissue. *Journal of Lipid Research* 16: 79–91, 1975.
32. Kajimura S, Seale P, Kubota K, Lunsford E, Frangioni JV, Gygi SP, and Spiegelman BM. Initiation of myoblast to brown fat switch by a PRDM16–C/EBP-β transcriptional complex. *Nature* 460: 1154–1158, 2009.
33. Seale P, Bjork B, Yang W, Kajimura S, Chin S, Kuang S, Scimè A, Devarakonda S, Conroe HM, Erdjument-Bromage H, Tempst P, Rudnicki MA, Beier DR, and Spiegelman BM. PRDM16 controls a brown fat/skeletal muscle switch. *Nature* 454: 961–967, 2008.
34. Pani S, Dey S, Pati B, Senapati U, and Bal NC. Brown to white fat transition overlap with skeletal muscle during development of larger mammals: Is it a coincidence? *Journal of the Endocrine Society* 6: bvac151, 2022.
35. Cannon B, and Nedergaard J. Neither fat nor flesh. *Nature* 454: 947–948, 2008.
36. Jung SM, Sanchez-Gurmaches J, and Guertin DA. Brown adipose tissue development and metabolism. *Brown Adipose Tissue* 251:3–36, 2019.
37. Song HD, Kim SN, Saha A, Ahn SY, Akindehin S, Son Y, Cho YK, Kim M, Park JH, Jung YS, and Lee YH. Aging-Induced Brain-Derived Neurotrophic Factor in Adipocyte Progenitors Contributes to Adipose Tissue Dysfunction. *Aging Dis* 11: 575–587, 2020.
38. Billon N, Iannarelli P, Monteiro MC, Glavieux-Pardanaud C, Richardson WD, Kessaris N, Dani C, and Dupin E. The generation of adipocytes by the neural crest. *Development* 134: 2283–2292, 2007.

39. Bartness TJ, Vaughan CH, and Song CK. Sympathetic and sensory innervation of brown adipose tissue. *International Journal of Obesity* 34: S36–S42, 2010.
40. Chusyd DE, Wang D, Huffman DM, and Nagy TR. Relationships between Rodent White Adipose Fat Pads and Human White Adipose Fat Depots. *Front Nutr* 3: 10, 2016.
41. Cui X, Jing J, Wu R, Cao Q, Li F, Li K, Wang S, Yu L, Schwartz G, and Shi H. Adipose tissue-derived neurotrophic factor 3 regulates sympathetic innervation and thermogenesis in adipose tissue. *Nature Communications* 12: 5362, 2021.
42. Morrison SF, Madden CJ, and Tupone D. Central neural regulation of brown adipose tissue thermogenesis and energy expenditure. *Cell Metabolism* 19: 741–756, 2014.
43. Nautiyal KM, Dailey M, Brito N, Brito M NDA, Harris RB, Bartness TJ, and Grill HJ. Energetic responses to cold temperatures in rats lacking forebrain-caudal brain stem connections. *American Journal of Physiology-Regulatory, Integrative and Comparative Physiology* 295: R789–R798, 2008.
44. Bartness T, Vaughan C, and Song C. Sympathetic and sensory innervation of brown adipose tissue. *International Journal of Obesity* 34: S36–S42, 2010.
45. Lowell BB, and Spiegelman BM. Towards a molecular understanding of adaptive thermogenesis. *Nature* 404: 652–660, 2000.
46. Symonds ME. Brown adipose tissue growth and development. *Scientifica* 2013(1): 305763, 2013.
47. Tseng Y-H, Kokkotou E, Schulz TJ, Huang TL, Winnay JN, Taniguchi CM, Tran TT, Suzuki R, Espinoza DO, and Yamamoto Y. New role of bone morphogenetic protein 7 in brown adipogenesis and energy expenditure. *Nature* 454: 1000–1004, 2008.
48. Seale P, Kajimura S, Yang W, Chin S, Rohas LM, Uldry M, Tavernier G, Langin D, and Spiegelman BM. Transcriptional control of brown fat determination by PRDM16. *Cell Metabolism* 6: 38–54, 2007.
49. Casteilla L, Muzzin P, Revelli J, Ricquier D, and Giacobino J. Expression of β 1-and β 3- adrenergic-receptor messages and adenylate cyclase β-adrenergic response in bovine perirenal adipose tissue during its transformation from brown into white fat. *Biochemical Journal* 297: 93–97, 1994.
50. Symonds M, Pope M, Sharkey D, and Budge H. Adipose tissue and fetal programming. *Diabetologia* 55: 1597–1606, 2012.
51. Gilsanz V, Hu HH, and Kajimura S. Relevance of brown adipose tissue in infancy and adolescence. *Pediatric Research* 73: 3–9, 2013.
52. Ponrartana S, Aggabao PC, Chavez TA, Dharmavaram NL, and Gilsanz V. Changes in brown adipose tissue and muscle development during infancy. *The Journal of Pediatrics* 173: 116–121, 2016.
53. Sentis SC, Oelkrug R, and Mittag J. Thyroid hormones in the regulation of brown adipose tissue thermogenesis. *Endocrine Connections* 10: R106–R115, 2021.
54. Hu J, and Christian M. Hormonal factors in the control of the browning of white adipose tissue. *Hormone Molecular Biology and Clinical Investigation* 31: 20170017, 2017.
55. Cypess AM, White AP, Vernochet C, Schulz TJ, Xue R, Sass CA, Huang TL, Roberts-Toler C, Weiner LS, and Sze C. Anatomical localization, gene expression profiling and functional characterization of adult human neck brown fat. *Nature Medicine* 19: 635–639, 2013.
56. Mo Q, Salley J, Roshan T, Baer LA, May FJ, Jaehnig EJ, Lehnig AC, Guo X, Tong Q, and Nuotio-Antar AM. Identification and characterization of a supraclavicular brown adipose tissue in mice. *JCI Insight* 2: 11, 2017.
57. Henry BA, Pope M, Birtwistle M, Loughnan R, Alagal R, Fuller-Jackson J-P, Perry V, Budge H, Clarke IJ, and Symonds ME. Ontogeny and thermogenic role for sternal fat in female sheep. *Endocrinology* 158: 2212–2225, 2017.
58. Scotney H, Symonds ME, Law J, Budge H, Sharkey D, and Manolopoulos KN. Glucocorticoids modulate human brown adipose tissue thermogenesis in vivo. *Metabolism* 70: 125–132, 2017.
59. Fainberg HP, Birtwistle M, Alagal R, Alhaddad A, Pope M, Davies G, Woods R, Castellanos M, May ST, and Ortori CA. Transcriptional analysis of adipose tissue during development reveals depot- specific responsiveness to maternal dietary supplementation. *Scientific Reports* 8: 9628, 2018.
60. Symonds ME, Pope M, and Budge H. The ontogeny of brown adipose tissue. *Annual Review of Nutrition* 35: 295–320, 2015.
61. Barabasi A-L, and Oltvai ZN. Network biology: understanding the cell's functional organization. *Nature Reviews Genetics* 5: 101–113, 2004.
62. Macotela Y, Emanuelli B, Mori MA, Gesta S, Schulz TJ, Tseng Y-H, and Kahn CR. Intrinsic differences in adipocyte precursor cells from different white fat depots. *Diabetes* 61: 1691–1699, 2012.

63. Lidell ME, Betz MJ, Leinhard OD, Heglind M, Elander L, Slawik M, Mussack T, Nilsson D, Romu T, and Nuutila P. Evidence for two types of brown adipose tissue in humans. *Nature Medicine* 19: 631–634, 2013.
64. Rockstroh D, Landgraf K, Wagner IV, Gesing J, Tauscher R, Lakowa N, Kiess W, Bühligen U, Wojan M, and Till H. Direct evidence of brown adipocytes in different fat depots in children. *PloS One* 10: e0117841, 2015.

2 Molecular Mechanisms of Heat Production in Brown Adipose Tissue

Bijayashree Sahu and Naresh Chandra Bal
School of Biotechnology, KIIT School of Biotechnology, India

2.1 INTRODUCTION

BAT in eutherians is so named due to its peculiar characteristic brown appearance primarily caused by mitochondrial abundance. Unlike lipid-laden white adipose tissue (WAT), it consists of multilocular adipocytes with centrally situated nuclei. In addition, BAT is highly vascularized for substrate supply to support its non-shivering thermogenesis (NST) and quick dispersion of heat throughout the body, thereby supporting an increase in core body temperature (Tc). Brown adipocytes originate from myogenic precursor cells expressing $MYF5^+$ during the embryonic stage by the expression of an adipogenic molecular switch PRDM 16 (Sahu et al. 2022). The origin of BAT as well as its physiological function are tightly regulated by many transcriptional regulators and neurohormonal mediators. It is believed that BAT has evolved to maximize the combustion of fatty acid (FA) in a short period and probably depends on the surface-to-body ratio of animals. In its mitochondria, the brown adipocyte expresses a unique protein, termed *uncoupling protein 1* (UCP1, previously known as thermogenin), in its inner membrane. UCP1 dissipates the proton gradient across the inner mitochondrial membrane generated through oxidative phosphorylation (OXPHOS), there by dampening the production of ATP and simultaneously leading to heat production (thermogenesis) (Cannon and Nedergaard 2004). The rodents and newborn babies have an abundance of BAT, mostly in the interscapular region. Interestingly, as discussed in Chapters 1 and 3, interscapular BAT persists in adult rodents but diminishes gradually in adult humans. In the last decade, studies showed that adult humans also have dispersed clusters of brown adipocytes expressing UCP1 in subcutaneous, pericardiac, and periaortic adipose tissue depots (Carlos Gerngroß et al. 2017). Further studies have shown that brown adipocytes precursor cells expressing UCP1 sporadically exist in WAT depots that give rise to brown-like characteristics upon activation, and such cells are termed *beige adipocytes*. Its activation is regulated not only by intrinsic factors (catecholamines, thyroid) but also by many extrinsic factors such as cold stimulation, exercise, and diet (Machado et al. 2022). Since the identification of BAT, the delineation of the molecular basis of heat production in brown adipocytes has attracted serious research efforts. The discovery of beige cells having BAT-like features has led to the discovery of several UCP1-independent mechanisms of heat production. Currently there is an active debate in the field to define the relative contribution of these mechanisms. However, a potential consensus is building that proposes UCP1 as the primary molecular mechanism of heat production in the BAT. In this chapter we will discuss the molecular mechanisms proposed for BAT and beige cells in the production of heat and the regulation of this mechanism in different metabolic adaptations.

2.2 MOLECULAR BASIS OF HEAT PRODUCTION IN BAT

As stated earlier, the mechanism of heat production in brown adipocytes is principally associated with the structure and functionality of UCP1. Therefore, we first delineate the structural features of

DOI: 10.1201/9781003391470-3

the UCP1 complex and investigate the proposed mechanism of action for each part of the substructures. UCP1 functions as a transmembrane protein, forming a channel for protons to leak across the membrane. This action disrupts the electrochemical H^+ gradient, effectively uncoupling the proton-motive force from ATP synthesis. As a result of this uncoupling, the mitochondrial respiration rate increases and heat generates (Ghesmati et al. 2023) (Figure 2.1).

2.2.1 UCP1 Structure

UCP1 is a small 32 kDa membrane protein belonging to mitochondrial carrier protein and consists of 307 amino acids. Basic amino acid residues cluster in the cavity toward the matrix. In contrast, acidic amino acid residues are distributed uniformly throughout the protein. The polypeptide chain clusters as six hydrophobic membranes spanning α-helices and forms the core structure of the protein. On the other hand, the hydrophilic the C- terminus and N-terminus residues protrude into intermembrane space (Cannon and Nedergaard 2004; Jacobsen, Lydersen, and Khandelia 2023). Each pair of transmembrane helices constitute a repeat with having two hydrophobic areas in each repeat and constitutes the tripartite structure of UCP1 (see Figure 2.2) (Krauss, Zhang, and Lowell 2005). The tripartite structure of UCP1 also consists of two conserved salt bridge networks facing the matrix forming the central cavity and nucleotide-binding pockets, which is essential for alternating regulation of access to the substrate-binding site.

- **Matrix-exposed loops**: The long hydrophilic loop connecting the two helices in each repeat is located on the matrix side. These loops likely play a role in transporting protons across the inner mitochondrial membrane.
- **Active site**: The proton-conducting activity of UCP1 is attributed to a specific region within its transmembrane domains, termed *core proton-conducting site*. This region forms a channel through which protons can pass, allowing for the uncoupling of oxidative phosphorylation. Recent advances in molecular and computational biology suggested monomoric UCP1 function in H^+ transporter and ATP/ADP uncoupling (Crichton, Lee, and Kunji 2017; Lee et al. 2015).
- **Regulating site**: On UCP1, cysteine-253 acts as a regulatory site for increasing protein activity by thermogenic ROS. Structural modeling indicates that C253 is situated at the hydrophobic interface in the inactive protein. The binding of ROS to C253 disrupts the inactive state of UCP1 and activates the thermogenic mechanism. A deficiency of C253 leads to altered oxidative homeostasis of BAT and beige fat in adipose tissue, which leads to significant immune cell infiltration into the tissues (Mills et al. 2022).
- **Amino acid sequences and homology**: UCP1 shares homology with other mitochondrial carrier proteins (specifically UCP2 and UCP3) in its tripartite structure and amino acid sequences conserved between the 100–amino acid sequences. However, two highly conserved amino acid sequences are exclusive to UCP1, found in the middle region of the central hydrophilic loop and the terminal segment of the COOH terminus. The COOH-terminal region is immunogenic and a prime target for the development of specific antibodies.

2.2.2 Intracellular Events: The Production of Intracellularly Derived Substrate and UCP1 Activation for Thermogenesis

UCP1 activation is a complex process, influenced by several factors, and involves intricate intracellular signaling pathways. A major factor that activates BAT is the level of catecholamines like NE, either as a neurotransmitter in the tissue or at the systemic level. Various other elements, such as FAs

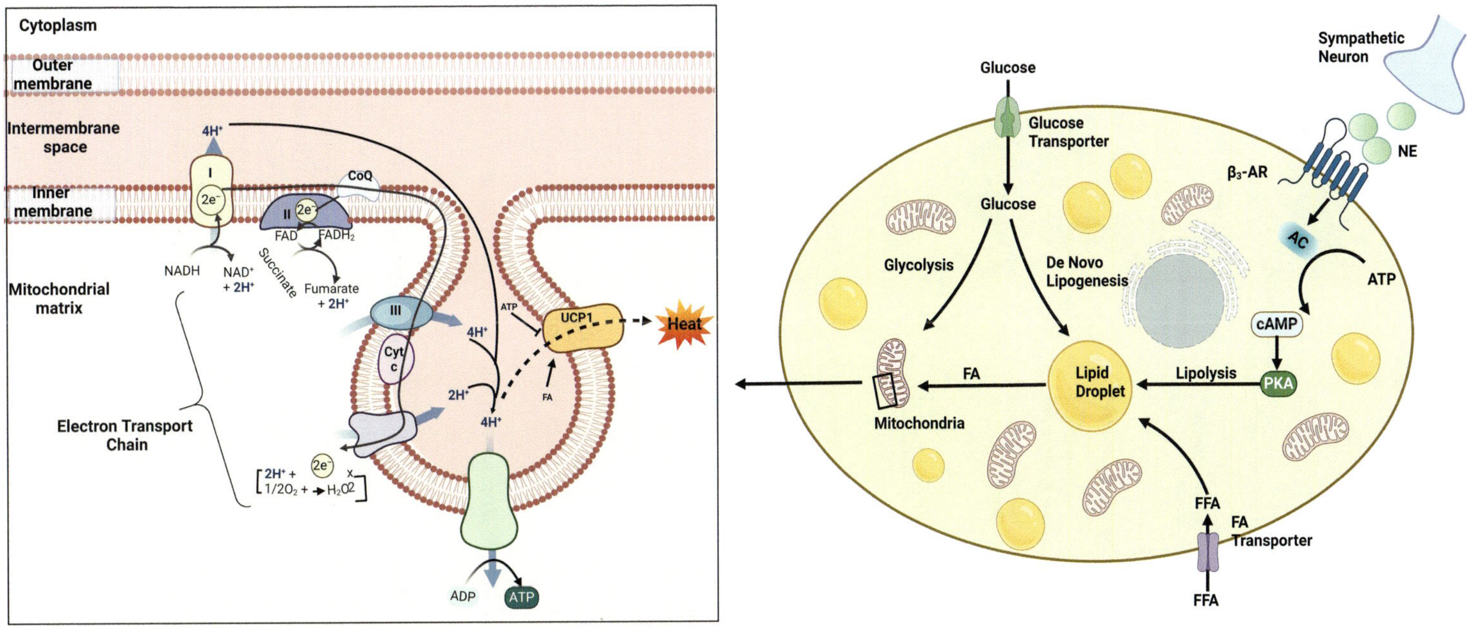

FIGURE 2.1 **Oxidative phosphorylation and thermogenic role of uncoupling protein 1 in brown adipocyte.** Similar to other mammalian cells, electrons are generated in the form of reduced hydrogen carriers (NADH and FADH2) and donated to different ETC protein complexes within the mitochondrial inner membrane. These electrons are subsequently transported through complexes and eventually reduce molecular oxygen to water. Along with electron progression, a specific number of protons (H^+) are actively driven from the mitochondrial matrix into the intermembrane space, creating a proton gradient across the mitochondrial inner membrane. The energy stored in this proton gradient serves as the driving force for ATP synthesis from ADP and inorganic phosphate (Pi) by ATP synthase as protons go back to the mitochondrial matrix. The unique UCP1 in BAT and beige fat, which uncouples the generation of the proton gradient from ATP synthesis, leads to dissipation of energy as thermogenesis (heat production) instead of ATP. The activation of UCP1 regulated by ATP and FAs. This process is activated by the release of norepinephrine (NE) from sympathetic neurons from the synapse. NE binds to β_3-AR, leading to activation of adenylyl cyclase (AC). AC convert ATP to cyclic AMP (cAMP), in turn activating protein kinase A (PKA). PKA facilitates the release of FAs from lipid stores. FA transported to mitochondria attach to UCP1 and activate the UCP1 protein. Further circulatory FFAs and glucose are transported into the brown adipocyte via FA transporters and glucose transporters, respectively, contributing to cellular respiration and/or *de novo* lipogenesis. *CI–IV*, complex I–IV; *CoQ*, co-enzyme Q; *Cyto C*, cytochrome C; e^-, electron; H^+, hydrogen ion.

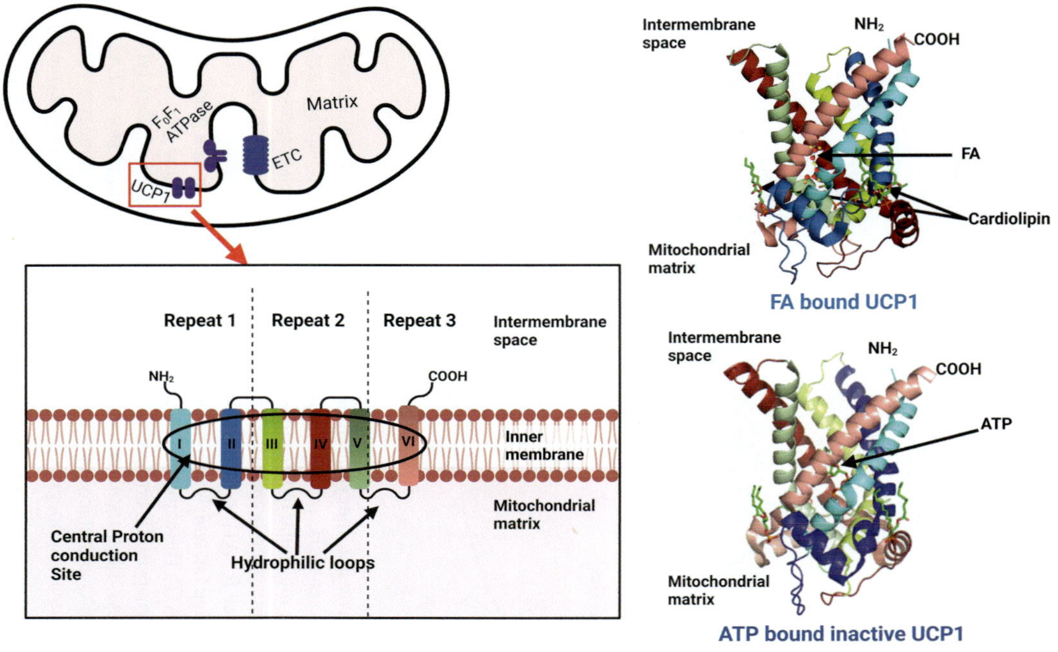

FIGURE 2.2 UCP1 structure and its conformational change with ATP binding inhibitory state. (a) UCP1 has a tripartite structure where each repeat contains two transmembrane α-helices. Therefore, a total six α-helical regions (color coded) span the mitochondrial inner membrane and form the proton-conducting channels. A long hydrophilic loop connects the two attached helices of each repeats, orientated toward the matrix. These loops act as gates for the proton-conducting channel. (b) Binding of three cardiolipin lipids to UCP1 stabilizes the UCP1 protein and increases its ability to transport H^+. The UCP1 protein remains in an inactive state in cellular milieu due to abundancy of purine nucleotide. The binding of purine nucleotide (mostly ATP) increases inward bending of H1, H4, H5, and H6 and results in a more compact inactive protein. Upon stimulation, the FA sited at the H1–H6 interface (closer to the matrix side) activates the UCP1 protein with displacement of ATP from UCP1. Part (a) of the figure is adapted from Krauss, Zhang, and Lowell (2005). Part (b) of the figure is reproduced based on the finding of Jacobsen, Lydersen, and Khandelia (2023).

and their metabolites, thyroxine, and oxidative stress, contribute to the activation of BAT. These factors are as follows:

- **Sympathetic nervous system (SNS)-mediated activation pathway and cyclic AMP (cAMP) signaling:** A pivotal role in BAT activation is played by the SNS. When the body is exposed to cold, the SNS releases NE, which interacts with β3-adrenergic receptors (β3-ARs) on brown adipocytes and initiates a cascade of events. It starts with the conversion of ATP to cAMP by adenylate cyclase, leading to increased cAMP levels (see Figure 2.1) (Ghesmati et al. 2023).
- **Protein kinase A (PKA) activation:** Elevated cAMP levels activate PKA, leading to the phosphorylation of various proteins, including hormone-sensitive lipase (HSL). HSL's role is to break down stored triglycerides into free fatty acids (FFAs) and glycerol.
- **FA activation:** These liberated FFAs are taken up by mitochondria, where they are activated and broken down through β-oxidation to generate acetyl-CoA. Additionally, the elevated local concentrations of long-chain FAs overcome purine nucleotide inhibition and induce H^+ leak via a FA/H^+ symport mechanism, thereby helping thermogenesis (Zhao et al. 2017).
- **UCP1 activation and thermogenesis**: Elevated levels of FFAs and increased cAMP signaling trigger the activation of UCP1, which forms a proton channel in mitochondria,

enabling the influx of protons to the mitochondrial matrix. This process uncouples oxidative phosphorylation, disrupting ATP production and generating heat instead. This heat is then distributed throughout the body ensuring warmth and thermal equilibrium. It is paramount to emphasize that UCP1 is inactive in the cellular milieu, as it is maintained in a purine nucleotides-bound state. Activators like FAs and dinitro 2,4-dinitrophenol (DNP) attach within the central cavity, establishing contact with transmembrane helix (H) 1/2 and 6 (Jacobsen, Lydersen, and Khandelia 2023; Zhao et al. 2017). By contrast, ATP competitively overlaps with the binding site within the same cavity, promoting conformational changes in H2 and causing an inward bending of H1, H4, H5, and H6 and resulting in a more compact UCP1 (Jacobsen, Lydersen, and Khandelia 2023).

2.2.3 FAs and Purine Nucleotides as the Regulator of UCP1-mediated Thermogenesis

The conventional perspective in the context of adrenergically driven thermogenesis is that lipid-laden brown adipocytes serve as proximal repositories for fuel and play a crucial role in the activation of the UCP1 proton gradient dissipation activity. Most probably high cardiolipin content in the brown adipocyte mitochondrial inner membrane facilitates the dissociation of purine nucleotides (ADP, GDP) and UCP1 while FA fuels the activation of UCP1 (Crichton, Lee, and Kunji 2017). Various models are proposed to explain the FA-mediated activation of BAT; some of the common ones are:

- **Model 1:** In the allosteric interaction model, the FAs, or mostly their derivative acyl-CoA, compete with bound purine nucleotides and engage in allosteric interactions that activate the protein. This activation may be analogous to "lowering the threshold for a Zener diode" in bioenergetic terms, but convincing evidence for a physiological relevance for such an effect is lacking (Shabalina et al. 2004).
- **Model 2:** According to the cofactor theory, FAs are localized near the binding sites within the proton-conducting channel of UCP1. In this model, the acid moieties of FAs serve as effective "stepping stones" for protons during their passage through the membrane. However, no interaction with the purine nucleotide-binding inhibitory site is formulated, and specific intra-channel FA binding sites for this model have yet to be identified (Winkler and Klingenberg 1994).
- **Model 3:** According to the shuttling theory, rather than freely transporting through UCP1, protons (re)enter the mitochondria in the form of undissociated FAs. Subsequently, FAs in their anionic form (re)exit the mitochondria with the assistance of UCP1. However, questions arise regarding the attainability of sufficiently high free levels of FAs for this process and an abundance of FA-binding proteins in the cytosol of brown adipocytes. Furthermore, the role of the inhibitory GDP-binding site on UCP1 is unsolved in this model (Fedorenko, Lishko, and Kirichok 2012).

None of the three models fully elucidates the regulation and protonophoric properties of UCP1. The lack of direct competition between FAs and the GDP-binding site indicates that the competition might be functional in nature. Rial, Poustie, and Nicholls (1983) suggest that UCP1 transforms by both FFAs and GDP, resulting in two distinct states. Further research is needed to validate and refine these proposed models for a more comprehensive understanding of UCP1 function and regulation.

2.3 TRANSCRIPTIONAL REGULATION OF UCP1 UPON EXPOSURE TO COLD

As discussed earlier, BAT is essential and well recruited during cold acclimation via the NE-stimulated mechanism and augments the NE-induced UCP1-dependent thermogenesis. Notably, cold acclimation and cold-recruited NE-induced thermogenesis occur concurrently and contribute to up to 300% increase in energy expenditure. For this NE must reach the adrenergic receptors (ARs) at sympathetic nerve synapses. BAT thermogenesis is finely regulated through dual mechanisms: first, at the transcriptional level involving the recruitment of transcription factors and epigenetic modifications to drive gene expression; and second, at the protein activity level within the mitochondrion to ensure effective thermogenic function. Upon exposure to cold, PKA activation in brown adipocytes triggers the phosphorylation of CREB and activates transcription factor 2 (ATF2), thereby initiating the transcription of UCP1 and peroxisome proliferator-activated receptor (PPAR) co-activator 1α (ppargc1a; encoding PGC1α). Moreover, PGC1α collaborates with transcriptional factors such as CEBPβ and NRF1, synergistically inducing the transcription of thermogenic genes, including UCP1 (Cao et al. 2004; Wicksteed and Dickson 2017). Furthermore, the expression of Adenylate cyclase3 and downregulation of its inhibitor RGS2 genes enhance adrenergic signal transduction. Additionally, damage-specific DNA binding protein 1 (DDB1) binds UCP1 and Ppargc1a promoters, recruiting positive transcriptional elongation factor b (P-TEFb). P-TEFb. phosphorylates Pol II at Ser2 of its C-terminal repeat and releases promoter-proximally paused RNA polymerase II (Pol II), thereby enabling rapid and synchronized transcription of thermogenic genes upon acute exposure to cold (Wang et al. 2022). Further zinc finger protein (Zfp)516 induced by SNS through cAMP-CREB/ATF2 pathway directly interacts with PRDM16 and the proximal region of the UCP1 promoter (Dempersmier et al. 2015). Zfp 516 interacts with its partner lysine-specific demethylase 1 (LSD1) and recruits it to UCP1 and PGC1a to function as a co-activator by demethylating H3K9 (Sambeat et al. 2016). Beyond DDB1 and P-TEFb, various epigenetic and transcriptional factors play crucial roles in thermogenic gene transcription (Inagaki, Sakai, and Kajimura 2016). For instance, The 5’ flanking regions of rat, mouse, and human UCP1 genes contain cis-acting elements, including a potent 200-bp enhancer for tissue-specific expression and hormonal regulation. UCP1 mRNA levels rise in BAT a few minutes after exposure to cold. This exposure concurrently increases the expression of genes involved in FA synthesis and oxidation in BAT.

2.4 ALTERNATE MECHANISMS OF HEAT PRODUCTION IN BAT

Besides the activation of UCP1-mediated thermogenesis, other mechanisms were discovered to facilitate thermogenesis in brown and beige adipocytes. Figure 2.3 explains the existing thermogenic mechanisms.

2.4.1 Futile Ca^{2+} Cycling and SERCA2b

In beige fat, a UCP1-independent thermogenic mechanism involves an enhanced ATP- dependent Ca^{2+} cycling orchestrated by sarco/endoplasmic reticulum Ca^{2+}-ATPase2b (SERCA2b), ryanodine receptor 2 (RyR2), and inositol 1,4,5-trisphosphate receptors (IP_3Rs) such as IP_3R1 and IP_3R2 (Ikeda et al. 2017). Upon exposure to cold, NE binds to α1-AR and β3-AR, triggering an increased intracellular Ca^{2+} flux in beige adipocytes by activating SERCA2b and RyR2. SERCA2b efficiently transports Ca^{2+} back to the ER and likely to the mitochondria via the voltage-dependent anion channel and mitochondrial calcium uniporter. The heightened mitochondrial Ca^{2+} levels activate pyruvate dehydrogenase phosphatase PDP1c38, leading to the activation of the PDH enzyme and subsequent ATP synthesis. Additionally, PKA signaling phosphorylates RyR2, causing the dissociation of Calstabin2 from RyR2 and facilitating enhanced Ca^{2+} release from the ER. This intricate interplay highlights the multifaceted pathways involved in the UCP1-independent thermogenic process in beige adipocytes (Brownstein et al. 2022; Ikeda et al. 2017).

2.4.2 Creatine Phosphate Futile Cycling

Wallimann and colleagues elucidated the utilization of the phosphocreatine (PCr) / creatine kinase (CK) cycle by thermogenic adipocytes to drive respiration (Wallimann, Tokarska-Schlattner, and Schlattner 2011). Within this process, mitochondrial CK produces PCr, facilitating its diffusion

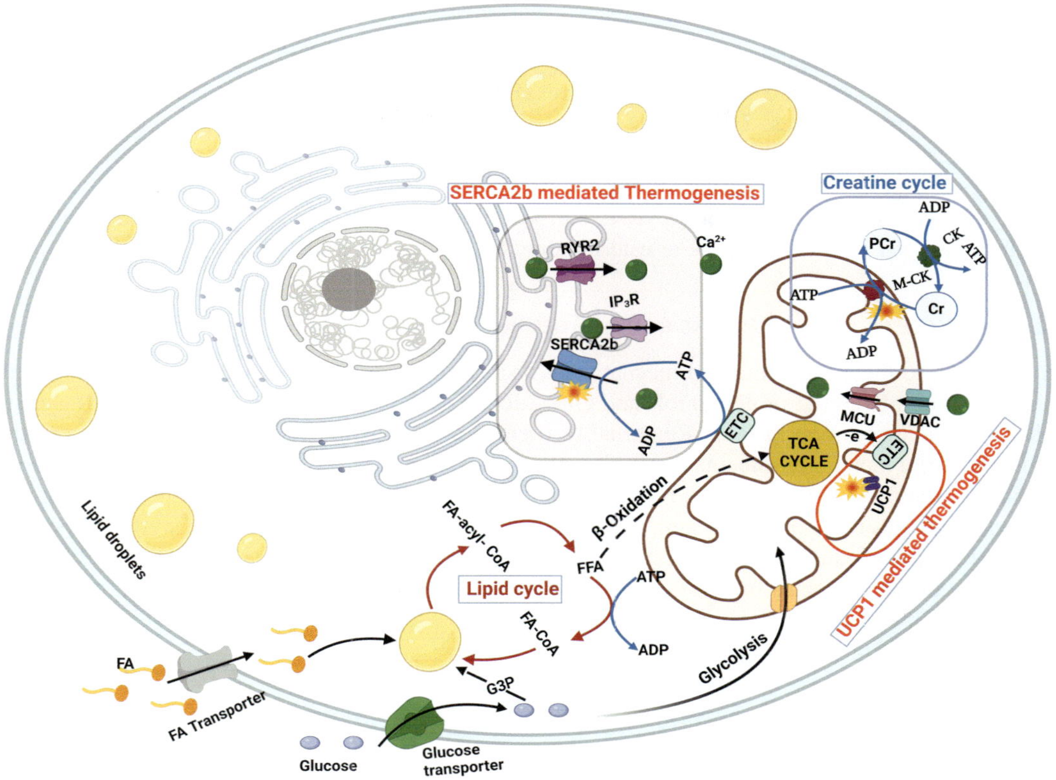

FIGURE 2.3 Digramatic representation of canonical and non-canonical mechanisms of thermogenesis in brown adipocytes. As explained earlier, UCP1 is the principal molecule for thermogenesis by acting as ADP/ATP exchanger and H^+ transporter in the BAT mitochondria. In reduced activity of UCP1, other thermogenic mechanisms are recruited to maintain Tc. In beige adipocytes norepinephrine (NE)– mediated stimulation of adrenergic receptors stimulates futile Ca^{2+} cycling through activation of SERCA2b, RyR2, and IP_3R. The Ca^{2+} efflux from the endoplasmic reticulum (ER) begins with higher demand of Ca^{2+} in mitochondria. Calcium signaling is essential for the respiration and Ca^{2+} entry facilitated by VDAC (outer membrane) and MCU (inner membrane). In resting state, when cytosolic Ca^{2+} level increases, the reuptake of Ca^{2+} to ER via SERCA2b requires hydrolysis of ATP. Additionally, creatine (Cr) and phosphocreatine (PCr) cycling is coupled with creatine kinase (CK)-mediated hydrolysis of ATP to provide temporal and spatial high-energy buffer to maintain cellular energy homeostasis. The mitochondrial CK (in inner-membrane space) functions as an ADP/ATP exchanger to control the cellular ATP pool. This cycling dissipates energy in the form of heat with simultaneous stimulation of mitochondrial ATP turnover. In the absence of mitochondrial pyruvate import, brown adipocytes activate lipolysis, which induces β-oxidation and futile lipid cycling. Further, glucose is converted to G3P and contributes to lipogenesis. This lipid cycling also dissipates energy with hydrolysis of ATP for the generation of FA-CoA. *G3P*, Glycerol-3-phosphate.

throughout the cell to sustain high ATP/ADP ratios near ATP consumption sites and the SERCA pump for calcium cycling support. PCr undergoes direct hydrolysis, or indirectly controls ATP turnover through creatine (Cr). Notably, the thermogenic effects of Cr manifest primarily under conditions of ADP limitation. The findings suggest a potential link between Cr-dependent thermogenesis in human BAT and the biomarker creatinine, which is a direct product of PCr metabolism (Rahbani et al. 2021). Additionally, thermogenic fat cells exhibit robust PCr phosphatase activity attributed to mitochondria-localized tissue-nonspecific alkaline phosphatase (TNAP). TNAP expression is significantly induced in response to cold conditions, where it hydrolyzes PCr, initiating a futile cycle of Cr dephosphorylation and phosphorylation (Sun et al. 2021).

2.4.3 Futile Lipid Cycling

Brown adipocytes sustain the thermogenic process in the absence of UCP1 through a futile substrate cycle of lipolysis and lipogenesis (Brownstein et al. 2022). In the cyclic process, ATGL mediates the hydrolysis of triglycerides (TGs), while long-chain acyl-CoA synthetase catalyzes the activation of ATP-consuming Fas, and diacylglycerol O-acyltransferase 1 (DGAT1) facilitates the esterification of FAs into diglycerides. A portion of glucose is converted to glycerol-3-phosphate (G3P) via dihydroxyacetone phosphate instead of pyruvate production. So, in response to adrenergic stimulation, G3P can act as a backbone for the re-esterification of FAs (Oeckl et al. 2022).

2.4.4 Role of Other Organs in BAT Thermogenesis

In addition to the aforementioned mechanisms, increased levels of ROS in adipocytes, along with the oxidation of cellular protein thiols and glutathione pools, play a significant role in driving acute thermogenic respiration in BAT. Remarkably, there is a notable and selective accumulation of the mitochondrial metabolite succinate, which acts as a potent source for ROS production, further contributing to thermogenesis in brown adipocytes (Mills et al. 2018). Furthermore, the gut hormone secretin emerges as a novel non-sympathetic activator of BAT in mediating prandial thermogenesis and unveils a non-canonical gut-secretin-BAT-brain axis in regulation of satiation (Li et al. 2018). In another scenario, exercise increases the production of lactate from anaerobic oxidation of glucose and secretion of myokines (Irisin, BAIBA). These molecules are transported to BAT and augment the thermogenic effect (Song et al. 2017).

2.5 ADAPTIVE THERMOGENESIS IN BAT: MECHANISM OF RESPONSE TO DIFFERENT STRESSOR

In homeothermic endotherms, thermogenesis is closely associated with ambient temperature. BMR (that depends on lean body mass, age, sex, and thyroid function) is the chief source for thermogenesis to maintain Tc. In addition to BMR, adaptive thermogenesis is dynamically recruited to maintain Tc under metabolically changing situations like cold exposure, exercise, and diet. Cold is the primary stimulator of BAT-based thermogenesis, which depends on the activation of UCP1, SERCA2b, and creatine by different mechanisms. When the demand for heat generation arises, creatine acts as a store of energy. It is fascinating that cold exposure response occurs via β3-AR but the degree of low temperature and duration dictate the mechanisms that will be recruited for adaptive thermogenesis. The acute cold exposure acts on protein levels that activate UCP1 via a cAMP-PKA-mediated FA release. On the other hand, chronic cold mostly works at a transcriptional level via cAMP-PKA-p38MAPK-mediated phosphorylation of PGC1α and ATF-2 to increase UCP1 transcription. Concurrently, FAs also exert adaptive changes by activation of PPARγ and promote transcription of PGC1α, PRDM16, and ATF-2 to increase UCP1 transcription (Evans et al. 2019). BAT and UCP1 are proposed to be the major mediators of diet-induced thermogenesis (DIT), occurring as a result of chronic calorie overload. DIT is thought to be due to the activation of UCP1 in brown and beige adipocytes by gut hormones and circulating FAs along with NE secretion from the SNS in response to a meal (Saito et al. 2020). Gut hormone secretin is released from the small intestine in response to a meal while bile acids are released from the liver in response to fat digestion, and they bind to secretin receptors and Takeda G-protein receptor 5 (TGR5) on brown adipocytes, respectively (Vítek and Haluzík 2016). All these biomolecules increase UCP1 expression and activity in brown adipocytes by similar cAMP-dependent pathways as exposure to cold. By contrast, circulating FAs can also activate BAT thermogenesis similar to chronic cold stress through binding to PPARα on brown adipocytes. Interestingly, exercise induces thermogenesis in brown adipocytes by augmenting mitochondrial biogenesis along with NE induction and increased FA mobilization from white fat to BAT and beige adipocytes (Sepa-Kishi and Ceddia 2016; Stanford and Goodyear

2016). Further exercised muscle secrete cytokines (termed *myokines*) such as irisin, IL-6, BAIBA, and IL-15, which are supposed to regulate thermogenesis and mitochondrial metabolism in BAT and beige adipocytes (Sahu et al. 2022).

2.6 CONCLUSIONS

Although UCP1 has traditionally been considered the primary mediator of NST in mammals, recent findings challenge this belief and provide evidence in favor of activation of additional operations insides BAT. The degree of BAT UCP1 activation is affected by the physiological state of other organs. Emerging findings suggest that BAT may play a more multifaceted role in thermoregulation and metabolic control than previously thought. BAT not only functions as a heat generator but also appears to interact and synchronize with various other physiological processes within the body. Importantly, the degree of UCP1 activation in BAT seems to be influenced by the physiological state of other organs.

This evolving perspective on BAT and UCP1 highlights the dynamic nature of thermoregulation and energy metabolism in mammals. It underscores the interconnectedness of various physiological systems, demonstrating that BAT's functions extend beyond simply generating heat. As our understanding of BAT's role continues to evolve, it opens up new avenues of research and potential therapeutic applications for managing metabolic health and thermoregulation.

REFERENCES

Brownstein, Alexandra J, Michaela Veliova, Rebeca Acin-Perez, Marc Liesa, and Orian S Shirihai. 2022. "ATP-Consuming Futile Cycles as Energy Dissipating Mechanisms to Counteract Obesity." *Reviews in Endocrine and Metabolic Disorders* 23 (1): 121–31.

Cannon, Barbara, and Jan Nedergaard. 2004. "Brown Adipose Tissue: Function and Physiological Significance." *Physiological Reviews* 84 (1): 277–359. https://doi.org/10.1152/physrev.00015.2003.

Cao, Wenhong, Kiefer W Daniel, Jacques Robidoux, Pere Puigserver, Alexander V Medvedev, Xu Bai, Lisa M Floering, Bruce M Spiegelman, and Sheila Collins. 2004. "P38 Mitogen-Activated Protein Kinase Is the Central Regulator of Cyclic AMP- Dependent Transcription of the Brown Fat Uncoupling Protein 1 Gene." *Molecular and Cellular Biology* 24 (7): 3057–67.

Carlos, Gerngroß, Johanna Schretter, Martin Klingenspor, Markus Schwaiger, and Tobias Fromme. 2017. "Active Brown Fat During ^{18}F-FDG PET/CT Imaging Defines a Patient Group with Characteristic Traits and an Increased Probability of Brown Fat Redetection." *Journal of Nuclear Medicine* 58 (7): 1104. https://doi.org/10.2967/jnumed.116.183988.

Crichton, Paul, Yang Lee, and Edmund Kunji. 2017. "The Molecular Features of Uncoupling Protein 1 Support a Conventional Mitochondrial Carrier-like Mechanism," March. https://doi.org/10.17863/CAM.10904.

Dempersmier, Jon, Audrey Sambeat, Olga Gulyaeva, Sarah M Paul, Carolyn SS Hudak, Helena F Raposo, Hiu-Yee Kwan, Chulho Kang, Roger HF Wong, and Hei Sook Sul. 2015. "Cold-Inducible Zfp516 Activates UCP1 Transcription to Promote Browning of White Fat and Development of Brown Fat." *Molecular Cell* 57 (2): 235–46.

Evans, Bronwyn A., Jon Merlin, Tore Bengtsson, and Dana S. Hutchinson. 2019. "Adrenoceptors in White, Brown, and Brite Adipocytes." *British Journal of Pharmacology* 176 (14): 2416–32. https://doi.org/10.1111/bph.14631.

Fedorenko, Andriy, Polina V Lishko, and Yuriy Kirichok. 2012. "Mechanism of Fatty-Acid- Dependent UCP1 Uncoupling in Brown Fat Mitochondria." *Cell* 151 (2): 400–413.

Ghesmati, Zeinab, Mohsen Rashid, Shabnam Fayezi, Frank Gieseler, Effat Alizadeh, and Masoud Darabi. 2023. "An Update on the Secretory Functions of Brown, White, and Beige Adipose Tissue: Towards Therapeutic Applications." *Reviews in Endocrine and Metabolic Disorders*, December. https://doi.org/10.1007/s11154-023-09850-0.

Ikeda, Kenji, Qianqian Kang, Takeshi Yoneshiro, Joao Paulo Camporez, Hiroko Maki, Mayu Homma, Kosaku Shinoda, Yong Chen, Xiaodan Lu, and Pema Maretich. 2017. "UCP1-Independent Signaling Involving SERCA2b-Mediated Calcium Cycling Regulates Beige Fat Thermogenesis and Systemic Glucose Homeostasis." *Nature Medicine* 23 (12): 1454–65.

Inagaki, Takeshi, Juro Sakai, and Shingo Kajimura. 2016. "Transcriptional and Epigenetic Control of Brown and Beige Adipose Cell Fate and Function." *Nature Reviews Molecular Cell Biology* 17 (8): 480–95.

Jacobsen, Luise, Laura Lydersen, and Himanshu Khandelia. 2023. "ATP-Bound State of the Uncoupling Protein 1 (UCP1) from Molecular Simulations." *The Journal of Physical Chemistry B* 127 (45): 9685–96. https://doi.org/10.1021/acs.jpcb.3c03473.

Krauss, Stefan, Chen-Yu Zhang, and Bradford B Lowell. 2005. "The Mitochondrial Uncoupling-Protein Homologues." *Nature Reviews Molecular Cell Biology* 6 (3): 248–61.

Lee, Yang, Chrissie Willers, Edmund RS Kunji, and Paul G Crichton. 2015. "Uncoupling Protein 1 Binds One Nucleotide per Monomer and Is Stabilized by Tightly Bound Cardiolipin." *Proceedings of the National Academy of Sciences* 112 (22): 6973–78.

Li, Yongguo, Katharina Schnabl, Sarah-Madeleine Gabler, Monja Willershäuser, Josefine Reber, Angelos Karlas, Sanna Laurila, et al. 2018. "Secretin-Activated Brown Fat Mediates Prandial Thermogenesis to Induce Satiation." *Cell* 175 (6): 1561–1574. https://doi.org/10.1016/j.cell.2018.10.016.

Machado, Sabrina Azevedo, Gabriel Pasquarelli-do-Nascimento, Debora Santos da Silva, Gabriel Ribeiro Farias, Igor de Oliveira Santos, Luana Borges Baptista, and Kelly Grace Magalhães. 2022. "Browning of the White Adipose Tissue Regulation: New Insights into Nutritional and Metabolic Relevance in Health and Diseases." *Nutrition & Metabolism* 19 (1): 61. https://doi.org/10.1186/s12986-022-00694-0.

Mills, Evanna L., Cathal Harmon, Mark P. Jedrychowski, Haopeng Xiao, Anja V. Gruszczyk, Gary A. Bradshaw, Nhien Tran, et al. 2022. "Cysteine 253 of UCP1 Regulates Energy Expenditure and Sex-Dependent Adipose Tissue Inflammation." *Cell Metabolism* 34 (1): 140–157. https://doi.org/10.1016/j.cmet.2021.11.003.

Mills, Evanna L, Kerry A Pierce, Mark P Jedrychowski, Ryan Garrity, Sally Winther, Sara Vidoni, Takeshi Yoneshiro, Jessica B Spinelli, Gina Z Lu, and Lawrence Kazak. 2018. "Accumulation of Succinate Controls Activation of Adipose Tissue Thermogenesis." *Nature* 560 (7716): 102–6.

Oeckl, Josef, Petra Janovska, Katerina Adamcova, Kristina Bardova, Sarah Brunner, Sebastian Dieckmann, Josef Ecker, et al. 2022. "Loss of UCP1 Function Augments Recruitment of Futile Lipid Cycling for Thermogenesis in Murine Brown Fat." *Molecular Metabolism* 61 (July): 101499. https://doi.org/10.1016/j.molmet.2022.101499.

Rahbani, Janane F, Anna Roesler, Mohammed F Hussain, Bozena Samborska, Christien B Dykstra, Linus Tsai, Mark P Jedrychowski, Laurent Vergnes, Karen Reue, and Bruce M Spiegelman. 2021. "Creatine Kinase B Controls Futile Creatine Cycling in Thermogenic Fat." *Nature* 590 (7846): 480–85.

Rial, Eduardo, Alison Poustie, and David G Nicholls. 1983. "Brown-adipose-tissue Mitochondria: The Regulation of the 32 000-Mr Uncoupling Protein by Fatty Acids and Purine Nucleotides." *European Journal of Biochemistry* 137 (1–2): 197–203.

Sahu, Bijayashree, Ojas Tikoo, Benudhara Pati, Unmod Senapati, and Naresh C Bal. 2022. "Role of Distinct Fat Depots in Metabolic Regulation and Pathological Implications." *Reviews of Physiology, Biochemistry and Pharmacology*, 135–76. https://doi.org/10.1007/112_2022_73

Saito, Masayuki, Mami Matsushita, Takeshi Yoneshiro, and Yuko Okamatsu-Ogura. 2020. "Brown Adipose Tissue, Diet-Induced Thermogenesis, and Thermogenic Food Ingredients: From Mice to Men." *Frontiers in Endocrinology* 11. https://www.frontiersin.org/journals/endocrinology/articles/10.3389/fendo.2020.00222

Sambeat, Audrey, Olga Gulyaeva, Jon Dempersmier, Kevin M Tharp, Andreas Stahl, Sarah M Paul, and Hei Sook Sul. 2016. "LSD1 Interacts with Zfp516 to Promote UCP1 Transcription and Brown Fat Program." *Cell Reports* 15 (11): 2536–49.

Sepa-Kishi, Diane M., and Rolando B. Ceddia. 2016. "Exercise-Mediated Effects on White and Brown Adipose Tissue Plasticity and Metabolism." *Exercise and Sport Sciences Reviews* 44 (1). https://journals.lww.com/acsm-essr/fulltext/2016/01000/exercise_mediated_effects_on_white_and_brown.8.aspx

Shabalina, Irina G, Anders Jacobsson, Barbara Cannon, and Jan Nedergaard. 2004. "Native UCP1 Displays Simple Competitive Kinetics between the Regulators Purine Nucleotides and Fatty Acids." *Journal of Biological Chemistry* 279 (37): 38236–48.

Song, No-Joon, Seo-Hyuk Chang, Dean Y Li, Claudio J Villanueva, and Kye Won Park. 2017. "Induction of Thermogenic Adipocytes: Molecular Targets and Thermogenic Small Molecules." *Experimental & Molecular Medicine* 49 (7): e353–e353. https://doi.org/10.1038/emm.2017.70.

Stanford, Kristin I., and Laurie J. Goodyear. 2016. "Exercise Regulation of Adipose Tissue."*Adipocyte* 5 (2): 153–62. https://doi.org/10.1080/21623945.2016.1191307.

Sun, Yizhi, Janane F Rahbani, Mark P Jedrychowski, Christopher L Riley, Sara Vidoni, Dina Bogoslavski, Bo Hu, Phillip A Dumesic, Xing Zeng, and Alex B Wang. 2021. "Mitochondrial TNAP Controls Thermogenesis by Hydrolysis of Phosphocreatine." *Nature* 593 (7860): 580–85.

Vítek, Libor, and Martin Haluzík. 2016. "The Role of Bile Acids in Metabolic Regulation." *Journal of Endocrinology* 228 (3): R85–96.

Wallimann, Theo, Malgorzata Tokarska-Schlattner, and Uwe Schlattner. 2011. “The Creatine Kinase System and Pleiotropic Effects of Creatine.” *Amino Acids* 40: 1271–96.

Wang, Xu, Shen-Ying Liu, Guo-Sheng Hu, Hao-Yan Wang, Guo-Liang Zhang, Xiang Cen, Si-Ting Xiang, Wen Liu, Peng Li, and Haobin Ye. 2022. “DDB1 Prepares Brown Adipocytes for Cold-Induced Thermogenesis.” *Life Metabolism* 1 (1): 39–53.

Wicksteed, Barton, and Lorna M Dickson. 2017. “PKA Differentially Regulates Adipose Depots to Control Energy Expenditure.” *Endocrinology* 158 (3): 464–66.

Winkler, Edith, and Martin Klingenberg. 1994. “Effect of Fatty Acids on H+ Transport Activity of the Reconstituted Uncoupling Protein.” *Journal of Biological Chemistry* 269 (4): 2508–15.

Zhao, Linlin, Shuqing Wang, Qianli Zhu, Bin Wu, Zhijun Liu, Bo OuYang, and James J. Chou. 2017. “Specific Interaction of the Human Mitochondrial Uncoupling Protein 1 with Free Long-Chain Fatty Acid.” *Structure* 25 (9): 1371–1379. https://doi.org/10.1016/j.str.2017.07.005.

3 Neurohormonal Regulation of Heat Production in Brown Adipose Tissue

Anuttama Pal and Naresh Chandra Bal
School of Biotechnology, KIIT School of Biotechnology, India

3.1 INTRODUCTION

Brown adipose tissue (BAT) is a highly specialized and well-documented organ found exclusively in eutherian mammals. Its function is to help the body in the process of biochemical method of thermogenesis, i.e., generation of heat inside the body. It is proposed that BAT thermogenesis is responsible for endothermy in mammals and maintenance of core body temperature (Tc). Biochemically, it is completely independent of the skeletal muscle and not associated with shivering. Therefore, heat production by BAT, as discussed in Chapters 1 and 2 of this volume, is called non-shivering thermogenesis (NST). The BAT-based NST plays a vital role in cold-induced thermogenesis (CIT), and is also possibly required for maintaining the energy balance in the organism and is activated upon overeating [1]. This phenomenon is known as diet-induced thermogenesis (DIT). BAT can be identified by its distinct "brown" color, which is reflected from the large number of mitochondria containing high amounts of iron in the brown adipocytes compared to white adipocytes. It is characterized by multilocular intercellular lipid droplets, the large number of mitochondria, regulated expression, and activity of a specialized thermogenic proteinuncoupling protein 1 (UCP1) [2–4]. UCP1, previously termed thermogenin, is located in the inner mitochondrial membrane of BAT, and its mechanism of action has been discussed in detail in Chapter 2. UCP1 expression is tightly regulated by transcriptional machinery, and this protein is expressed in the mitochondria of BAT and not in any other organ. The greater number of small lipid droplets provides an increase in lipid droplet surface area promoting metabolite exchange with mitochondria [5]. When less active or not engaged in thermogenesis, BAT morphology somewhat resembles white adipocyte by fusion of lipid droplets becoming unilocular-like. These less used BAT cells do not lose the epigenetic cellular identity of a brown adipocyte, but upon exposure to cold, BAT quickly upregulates UCP1 expression and undergoes structural and functional changes contributing to NST. These structural changes include mitochondrial biogenesis and lipid droplet utilization leading to multilocular appearance [6, 7].

To study BAT function, physiology, and biochemistry, rodents (mainly mice) have been used widely as a model organism. Newborn humans have active brown adipocytes present at birth localized in the interscapular and perirenal regions conspicuously aggregated into BAT, which possibly helps in maintaining Tc and may also have other functions that are not yet known. In adult humans, these become segregated into tiny pads present especially in the neck and shoulder regions. In contrast to rodents, humans have a distinct distribution of BAT as well as specific biochemical properties of it, as discussed in Chapter 1. In rodents, BAT is located in the dorsal anterior scapular region (interscapular [iBAT], subscapular [sBAT], cervical [cBAT]) and near the large arteries (periaortic BAT depot [paBAT], perirenal BAT depot [prBAT]) [8–11]. These locations are advantageous, as they help in heat production essential for the survival of small mammals in cold environments; they are also needed for arousal from hibernation. It must be noted that, depending on the age, gender,

 DOI: 10.1201/9781003391470-4

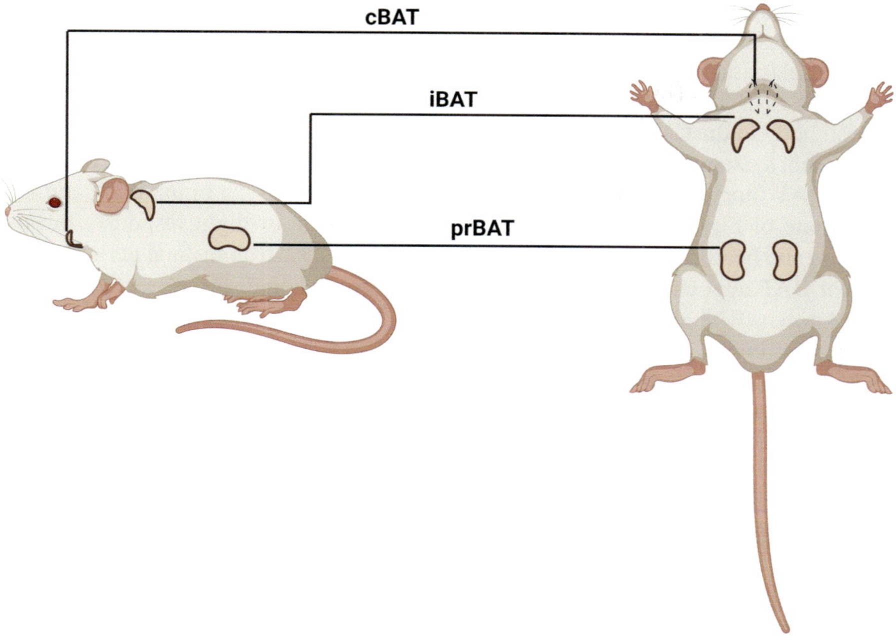

FIGURE 3.1 Locations of BAT in rodents. BAT is present both at the neonatal stage and in adulthood. It is suggested to help in thermoregulation, as these creatures are of small size and lose body heat rapidly and require a robust thermogenic organ. The more conspicuous BAT depots are: cervical brown adipose tissue (cBAT), interscapular brown adipose tissue (iBAT), and perirenal brown adipose tissue (prBAT). In addition, few dispersed brown adipocytic aggregates may be found in some white adipose tissue (WAT) depots.

and mouse strain background, the size and composition of the individual BAT depot differ. Interestingly, the iBAT is usually similar in size (as a ratio to body size) in most of the rodent models and is present throughout life even in adulthood. In these mammals, the iBAT constitutes more than 60% of total BAT and contains two distinct lobes placed between the two scapulae [12]. It is highly vascularized to facilitate the dispersal of heat produced throughout the body. In addition, it is richly innervated for quick turning on and off of the thermogenic process (Figure 3.1).

Thermogenesis by BAT is an essential part of the homeostatic property of maintaining body temperature at low or critically low environment temperature. It also plays an important role in increasing body temperature (fever) as a response to infection. BAT has a critical role in the processes of CIT and DIT that is mainly due to the dense innervation by nerves of the sympathetic nervous system (SNS). The iBAT is most prominent in this function due to its size, accessibility, and innervation, so that it can be swiftly turned on and off by neurotransmitters. This role of the SNS is regulated by neural networks in the CNS in response to cutaneous and deep body thermoreceptor signals. Following the elevation of Tc, multiple functions such as wakefulness, immune responses, and stress are also boosted up, which gives survival advantages. Once Tc reaches the upper limit of the optimal zone, the NST in the BAT needs to be shut off. BAT consumes a large amount of substrate for NST and after initial intracellular lipolysis starts taking up glucose rapidly from the circulation that contributes to the whole-body energy balance and is associated with metabolic diseases.

3.2 AUTONOMIC NERVOUS SYSTEM AND BAT

Research in the last several decades has established that BAT function is under the direct control of neural inputs [13, 14]. This neural regulation of BAT is by the autonomic nervous system (ANS), which is a part of the peripheral nervous system (PNS). ANS is well known for regulating multiple involuntary processes such as heart rate, blood pressure, respiration, digestion, and sexual arousal. ANS is further classified into sympathetic (SNS), parasympathetic (PNS), and enteric nervous system (ENS). The SNS and PNS contain both afferent and efferent fibers containing a preganglionic neuron, a cell body in the CNS, and a postganglionic neuron innervating the target tissues. The ENS is an independent system responsible for regulation of the digestive processes and does not have any influence on BAT function. SNS innervates nearly every living tissue in the body and is known to activate the state of high alert of the body known as the fight-or-flight response. In this process, blood pressure and heart rate increase, glycogenolysis occurs, gastrointestinal peristalsis ceases, etc.

All rodent BAT depots are innervated by SNS fibers, although the role of PNS fibers is still not clearly known. BAT thermogenesis is activated by SNS upon norepinephrine (NE) release followed by β-adrenergic receptor stimulation [15]. The β-adrenergic receptors are a class of G protein–coupled receptors that respond to the binding of catecholamines, such as adrenaline and NE. There are three main isoforms or subtypes of β-adrenergic receptors, commonly referred to as β1, β2, and β3 receptors. These isoforms have distinct tissue distributions and physiological functions. The β1 receptors mediate cardiac activity (heart rate, force of contraction, rate of relaxation) and are located primarily in the heart [16]. The β2 receptors control different aspects of metabolic activity and induce smooth muscle relaxation, with a key role in vasodilation, and these are located diversely in many organ systems. β3 receptors are the most abundant of the three and are involved in inducing the breakdown of fat cells. Additionally, in the urinary bladder, β3 receptors are involved in relaxation of the detrusor muscle [17]. High densities of β receptors are found in the caudate, putamen, different cortical areas, and the hippocampal formation. Low densities are present in other areas such as the thalamus, hypothalamus, midbrain, and cerebellar cortex. It must also be noted that the distribution of beta-adrenergic receptor subtypes in the brain has certain species-specific differences. In the human hippocampus, β2 receptors predominate, while in the rat hippocampus β1 receptors are more prevalent [18]. Additionally, in the rat olfactory bulb, both β1 and β2 receptors are present, with distinct spatial distributions [19]. In the human brain, the putamen is enriched in β1 receptors, while the cerebellum contains predominantly β2 receptors [20]. Researchers have used blockade or deficiency strategies to demonstrate the role of β receptor isoform in regulation of BAT function [21]. It has been shown that β1 and β2 receptors are indirect and mild regulators of BAT NST while β3 receptors play a direct pivotal function [22].

Mechanism: Decades of studies have shown that many factors, such as cold, diet, and infection, recruit BAT-based NST. Among these factors that activate BAT, the mechanism for cold is probably the most detailed. Once cold is perceived by skin surface receptors, it relies on the CNS and hypothalamus, which finally activate BAT via SNS nerves. Upon increased SNS activity, there is a surge of NE and ATP, which are released near the nerve endings near adipocytes in BAT that bind to the β3 adrenergic receptors found more exclusively on brown adipocytes. Following this, ATP is converted to adenosine (co-transmitter), which in turn activates the adenosine receptor, A2A [23]. The β3 adrenergic receptor and A2A receptor collectively stimulate cAMP production, thus activating protein kinase A (PKA). PKA drives lipolysis of the intracellular lipid droplets to generate fatty acids via phosphorylation of hormone-sensitive lipase (HSL). The fatty acids then enter mitochondria and simultaneously activate two processes. First, they enter into beta oxidation and electron transport chain (ETC) boosting proton gradient. Second, they bind to UCP1 allosterically, leading to its activation for proton gradient dissipation. Consequently, there is a phosphorylation of cAMP response element binding protein (CREB) and p-38–mediated phosphorylation of activating transcription factor 2 (ATF2). This in turn induces the transcription of proteins associated with mitochondrial biosynthesis and thermogenesis, including UCP1. The process of gene expression in brown adipocytes is also regulated by signaling of hormones such as thyroid (T3/T4) glucocorticoids as discussed in previous sections.

3.2.1 Innervations of BAT

Sensing and controlling the energy status, including thermogenesis, of an organism is carried out primarily by the CNS. The primary integrating, superior master regulator of the body's overall energy expenditure and balance is the hypothalamus [24]. The innervation of BAT by the SNS is undoubted, and the activation of the SNS is a stimulator of BAT thermogenesis. Although neurons originating in the hypothalamus (or brain) and terminating in each individual BAT depots have not completely been traced yet, there are several lines of evidence to suggest the innervation of BAT sites. The innervation of SNS into iBAT has been extensively researched and established over the past few years [25]. The SNS innervation into iBAT has been fully established. Each individual lobe of iBAThas unilateral postganglionic SNS innervation. Using transneuronal retrograde tracing by pseudorabies virus, a group of scientists demonstrated that the brain areas where iBAT's SNS innervation begins are those that substantially express central melanocortin 4 receptor (MC4-R), which has a role in the central control of a pituitary function [26]. It has also been observed in this study that the injection of an MC4-R agonist directly into the brain results in iBAT thermogenesis activation.

It has been suggested that the vagus nerve mediates peripheral signals to the central nervous system and has a major role in innervation. Vagus nerve stimulation (VNS) has also been reported to generate weight loss, and hence it has been hypothesized that VNS increases energy expenditure by activating BAT [27]. The vagus nerve is the longest and probably most important nerve of the parasympathetic system and controls various vital organ functions such as heart activity and digestion. Additionally, tyrosine hydroxylase (TH) has been studied for its role in NST. TH, which has been demonstrated to be generated in BAT during cold adaptation, rate-limits the release of the neurotransmitter epinephrine in the sympathetic neuron terminals that innervate BAT [13].

3.2.2 Experiments Confirming Innervation of BAT

In rodents, especially mice, iBAT is a conspicuous organ and has two distinct lobes dorsally outside the body cavity having separate nerves for each. This has been utilized to gather evidence for SNS innervation of BAT. While one of the iBAT pads is surgically denervated, the other one is left unaltered and utilized as control called sham-operated. Denervation can happen in a tissue due to injury, disorders, or by surgical procedures. If the denervated tissue functionality is dependent on the nerve or neuronal communication, then the physiological properties of the tissue will be altered, and that can be characterized. This unilateral surgical denervation model has been exploited by several researchers to study the role of innervation. As tissue from the same animal is used as experimental and control, this ensures that genetics, behavior, nutritional status, and circulating factors remain identical [28]. Using the unilateral denervation model, the following characteristics were shown to be blocked or greatly diminished in the cold-induced denervated lobe compared to the innervated sham-operated pad (where they increase by cold):

1. UCP1 concentration
2. Mitochondriogenesis and cristae density
3. Guanosinediphosphate binding
4. Cytochrome oxidase activity (equivalent to mitochondrial mass)
5. Increase in blood flow
6. Glucose uptake
7. Abundance of transporter proteins

In denervated BAT, the above (expected) cold acclimation responses can be replicated by external application of norepinephrine, the primary neurotransmitter that increases via SNS activation. Similar to cold exposure, overfeeding or fever also shows some of the analogous adaptive changes in the denervated BAT pad [29]. These experiments confirm that BAT is indeed deeply innervated by SNS nerves, and this has a major role in thermogenic responses in cold, overfeeding, or fever. However, one drawback is that along with hampering the SNS innervation to BAT, it also affects the connection from iBAT to the brain.

3.3 HORMONES REGULATING BAT FUNCTION

As discussed earlier, nervous input via SNS forms the first major way that BAT is recruited. However, chronic regulation and fine-tuning are primarily driven by the hormonal axis. Therefore, the action of hormones is intricately intertwined with various BAT functions. Hormones modulate several aspects of the brown adipocyte properties including protein expression and NST capacity that affect not just tissue level function but also whole body energy balance having implication toward metabolic diseases [21]. Moreover, hormonal factors can influence SNS amplitude affecting functional state of BAT and can be of medicinal values against metabolic diseases. Hence, this area has become an active topic of current research.

a. **Systemic (nor)epinephrine:** Adrenal medulla is a primary source of epinephrine and norepinephrine, belonging to the catecholamine family, found in the circulatory fluid in the body. These hormones are secreted upon activation by the hypothalamus, and they act as hormones of distress during the fight-or-flight response. The systemic level of epinephrine is elevated following acute cold exposure that binds to the receptors located on the brown adipocyte cell surface activating intracellular signal cascade. This stimulation triggers intracellular lipolysis and UCP1-mediated thermogenesis to maintain Tc. The role of systemic epinephrine was more profoundly established when its level was found upregulated in urine of wild-type mice acclimated to mild cold of 18°C and continued to rise with a drop in ambient temperature [30]. Thus, systemic epinephrine is important for optimal BAT activity and thermogenesis, especially during prolonged cold exposure.
b. **Thyroid hormones (thyroxine [T4] and triiodothyronine[T3])**: Thyroid hormones are the other type known to be essential for the maintenance of metabolic rate and greatly impact BAT activity. They upregulate the expression of UCP1 and other genes in brown fat cells and promote the expression of enzymes involved in fatty acid oxidation increasing lipolysis. In addition, thyroid hormones stimulate mitochondrial biogenesis that facilitates BAT-based NST, so called as a synergetic partner of SNS-mediated recruitment of BAT function. Hence, during hypothyroidism, a condition characterized by low thyroid hormone levels and lowered metabolic rates, BAT activity is also found to be compromised. BAT abundantly expresses type2 deiodinase (Dio2), a protein that converts T4 into the active thyroid hormone T3. Inside the brown adipocyte, T3 is known to promote the expression of peroxisome proliferator-activated receptor gamma co-activator 1-alpha (PGC-1α), the master regulator of mitochondrial biogenesis. Interestingly, direct T4 treatment increases body temperature in cold without affecting the UCP1 activity, suggesting that the thyroid hormone additionally mediates UCP1-independent thermogenesis [31].
c. **Glucocorticoids:** Another type of hormone whose role has been implicated in BAT function is glucocorticoids. These include corticosterone and cortisol, which are secreted by the adrenal gland upon stimulation by adrenocorticotropic hormone (ACTH) from the pituitary gland. Classically, glucocorticoids have been suggested to inhibit and/or downregulate the BAT function based on studies in mouse models. Latter studies proposed a modified hypothesis that distinct BAT depots may respond differently to glucocorticoids based on local pathophysiological conditions [32]. However, a recent study shows that glucocorticoids-mediated upregulation of metabolism is not due to BAT activation but may be due to skeletal muscle NST. Therefore, the functional relevance of glucocorticoids in BAT activity still needs to be fully investigated.

d. **GIP-gut hormones**: Some hormones released by the gut, especially the intestine, in response to food intake can indirectly affect BAT activity by modulating appetite and overall energy expenditure. Used in clinical settings to treat type 2 diabetes mellitus, medications based on the gut hormone glucagon-like peptide (GLP)-1 have also been found to cause weight loss. GLP-1 from the gut binds to its specific receptor (called GLP-1R) found in neurons of thalamic and hypothalamic regions that also modulate their SNS input into BAT. Due to this, current research is being done on additional gut hormone–based therapeutics for treatment of obesity by activating energy utilization in BAT [33].

3.4 NOVEL CYTOKINES AND CHEMOKINES IN BAT FUNCTION MODULATION

Research in the last few decades has uncovered that the BAT function can be further modulated by circulating factors called cytokines and chemokines [34]. Cytokines are signaling molecules that play diverse roles in the immune system, inflammation, and metabolic regulation. Some cytokines that have been investigated for their potential effects on brown adipose tissue are:

a. **Interleukin-6 (IL-6)**: IL-6 is a cytokine with pleiotropic effects, including roles in inflammation and metabolic regulation. Studies have suggested that IL-6 may influence BAT function. For example, IL-6 can promote thermogenesis in brown adipocytes and may be involved in browning of WAT[35], a process termed *browning*. The process of browning can occur by converting progenitor white adipocyte into a more metabolically active, brown-like state or amplifying the precursor brown adipocytes inside WAT.
b. **Tumor necrosis factor-alpha (TNF-α)**: TNF-α is a pro-inflammatory cytokine that can influence BAT function. It has been shown to affect BAT thermogenesis negatively in some studies by inhibiting the expression of UCP1 [36]. This primarily happens due to the fact that the downstream transcription factor of TNF-α, nuclear factor kappa beta (NfkB), usually downregulate PGC1α that directly regulate UCP1 expression in brown adipocyte. However, the relationship between TNF-α and BAT function is complex and may vary depending on the context.
c. **Interleukin-15 (IL-15)**: Another cytokine that has been studied recently in the context of BAT is IL-15. It appears to have a role in promoting BAT thermogenesis and enhancing the metabolic activity of brown adipocytes. The pro-inflammatory cytokine IL-15 may increase energy expenditure to shield the body from obesity and type 2 diabetes according to data from both human and animal studies [37]. IL-15 treatment in rats has been observed to cause an important increase in UCP1 expression. Certain effects of IL-15 on BAT may include upregulating PPARδ and PPARα, which in turn are suggested to activate the production of UCPs and proteins involved in the metabolism of fatty acids regulating mitochondrial metabolism [38].
d. **Fibroblast growth factor 21 (FGF21)**: FGF21 is a hormone-like cytokine secreted by various tissues, including the liver, adipose tissue, and skeletal muscle. It has been shown to increase thermogenesis in brown adipocytes through UCP1-dependent and -independent mechanisms, and is considered a potential target for obesity and metabolic disorders [39]. However, ambiguity was introduced when it was observed that FGF21-knockout mice did not exhibit alteration in BAT function. Another unknown aspect is whether FGF21-mediated recruitment of BAT works differently in rodents than in humans. Whether FGF21 secreted from different tissues affects BAT differently is also unclear. Therefore, FGF21 action on BAT needs further investigation to define its role.
e. **Irisin**: While not a traditional cytokine, irisin is a myokine (a cytokine-like signaling molecule secreted by muscle tissue) that has received attention for its potential role in browning WAT and activating BAT. Irisin is proposed to be secreted in response to exercise and increases whole body energy expenditure by promoting the browning of WAT [40] (Figure 3.2).

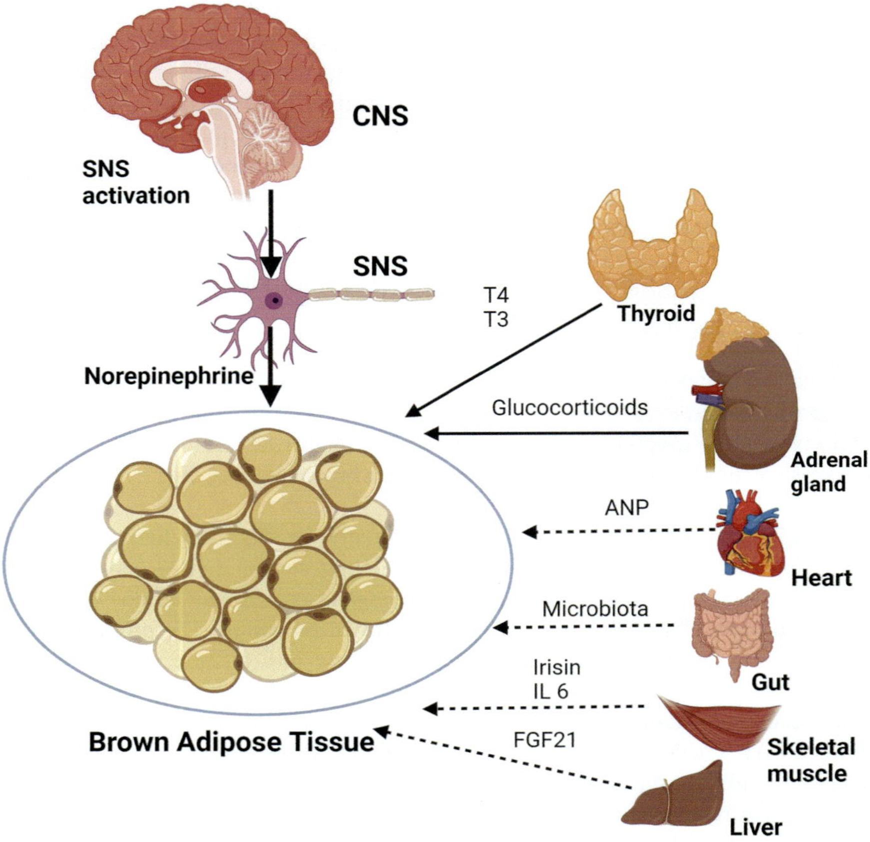

FIGURE 3.2 Multifaceted regulation of BAT activity. Thermogenesis in BAT is primarily regulated by CNS through neurotransmitters like NE from the SNS coming mainly from the thalamic/hypothalamic region of the brain. However, recent research has demonstrated that there are several secondary players that modulate the NST inside BAT by influencing its biochemical properties. Classically, it has been shown that catecholamines from adrenal and thyroid hormones activate the BAT function. Cytokines from liver, skeletal muscle, heart, and gut have also been shown to modify the BAT function.

3.5 FUTURE FIELDS OF APPLICATION

Advances in the understanding of the neurohormonal control of BAT show promising outcomes in multiple avenues beneficial to the fields of medicine and research. Some of the primary ones are:

Identification of novel neurohormones: It is possible for researchers to identify novel neurohormones or signaling pathways that affect BAT activation and function. Gaining insight into these pathways may help identify new therapeutic intervention targets.

Precision medicine approaches: A shift toward individualized therapies based on a person's neurohormonal profile would occur as research advances. Customizing therapy to an individual's unique neuronal and hormonal balance may improve the efficacy of BAT-related treatments.

Therapeutic targets for metabolic disorders: FGF-21 and IL-6 contribute to cardiovascular protection and metabolic regulation, thus a better understanding of their functionality in BAT could be a valuable focus area to develop treatments for metabolic disorders and obesity.

Integration with other therapies: Future research may also examine the synergistic effects of combining other therapy modalities, including exercise, dietary changes, or other metabolic interventions, with neurohormonal interventions.

Noninvasive monitoring techniques: Technological developments in imaging and monitoring may make noninvasive evaluation of BAT activity possible. This could allow scientists and medical experts to monitor the efficacy of neurohormonal therapies in real time.

Neurostimulation techniques: Potential approaches to modify the neurohormonal pathways influencing BAT include neurostimulation and neuromodulation. To improve BAT activation could entail electrically or medicinally modifying neuronal pathways.

Understanding crosstalk between tissues: It would be crucial to look at how BAT communicates with other tissues, including muscle, liver, and WAT. Comprehensive therapeutic solutions may result from an understanding of the systemic effects of neurohormonal control on overall metabolism.

Exploring environmental influences: The impact of temperature and light exposure on the neurohormonal control of BAT may be the subject of future research. This could possibly shed light on lifestyle modifications that would consequentially encourage BAT activity.

3.6 CONCLUSION

Despite its relatively tiny size, BAT stands out as an exceptional performer in the body's thermogenic system and through that majorly impacts the metabolic rate. This chapter explored the complex network of neurohormonal regulation controlling BAT's role in thermogenesis. The basal autonomic neural system plays a crucial role in regulating Tc in demanding situations due to its rapid reaction to neurotransmitters such as norepinephrine and epinephrine, which activate the SNS. In addition, the endocrine system—which includes hormones like glucocorticoids and thyroid hormones—is a key player in controlling the BAT function. Contemporary research has uncovered the role of cytokines such as TNF-α, IL-15, and IL-6 in regulating BAT activity, demonstrating the intricacy of its control. Overall, BAT's significance in preserving metabolic homeostasis is demonstrated by its capacity to assist with energy balance, adipose depot regulation, and glucose utilization. BAT presents promising opportunities for therapeutic interventions in obesity and metabolic disorders as our understanding of it develops.

REFERENCES

1. Saito, M., et al., *Brown Adipose Tissue, Diet-Induced Thermogenesis, and Thermogenic Food Ingredients: From Mice to Men*. Front Endocrinol (Lausanne), 2020. **11**: p. 222.
2. Aquila, H., T.A. Link, and M. Klingenberg, *The uncoupling protein from brown fat mitochondria is related to the mitochondrial ADP/ATP carrier. Analysis of sequence homologies and of folding of the protein in the membrane*. EMBOJ, 1985. **4**(9): p. 2369–76.
3. Bouillaud, F., J. Weissenbach, and D. Ricquier, *Complete cDNA-derived amino acid sequence of rat brown fat uncoupling protein*. J Biol Chem, 1986. **261**(4): p. 1487–90.
4. Heaton, G.M., et al., *Brown-adipose-tissue mitochondria: photoaffinity labelling of the regulatory site of energy dissipation*. Eur J Biochem, 1978. **82**(2): p. 515–21.
5. Puri, V., et al., *Cidea is associated with lipid droplets and insulin sensitivity in humans*. Proc Natl Acad Sci U S A, 2008. **105**(22): p. 7833–8.
6. Bartness, T.J., et al., *Sensory and sympathetic nervous system control of white adipose tissue lipolysis*. Mol Cell Endocrinol, 2010. **318**(1–2): p. 34–43.
7. Rothwell, N.J. and M.J. Stock, *Effects of denervating brown adipose tissue on the responses to cold, hyperphagia and noradrenaline treatment in the rat*. J Physiol, 1984. **355**: p. 457–63.
8. Frontini, A. and S. Cinti, *Distribution and development of brown adipocytes in the murine and human adipose organ*. Cell Metab, 2010. **11**(4): p. 253–6.

9. de Jong, J.M., et al., *A stringent validation of mouse adipose tissue identity markers*. Am J Physiol Endocrinol Metab, 2015. **308**(12): p. E1085–105.
10. Waldén, T.B., et al., *Recruited vs. nonrecruited molecular signatures of brown, "brite," and white adipose tissues*. Am J Physiol Endocrinol Metab, 2012. **302**(1): p. E19–31.
11. Cinti, S., *The adipose organ*. Prostaglandins Leukot. Essent. Fat. Acids, 2005. **73**(1): p. 9–15.
12. Grimpo, K., et al., *Brown adipose tissue dynamics in wild-type and UCP1-knockout mice: in vivo insights with magnetic resonance*. J Lipid Res, 2014. **55**(3): p. 398–409.
13. Morrison, S.F. and C.J. Madden, *Central nervous system regulation of brown adipose tissue*. Compr Physiol, 2014. **4**(4): p. 1677–713.
14. Morrison, S.F., C.J. Madden, and D. Tupone, *Central neural regulation of brown adipose tissue thermogenesis and energy expenditure*. Cell Metab, 2014. **19**(5): p. 741–756.
15. Ueta, C.B., et al., *β(1) Adrenergic receptor is key to cold- and diet-induced thermogenesis in mice*. J Endocrinol, 2012. **214**(3): p. 359–65.
16. Brodde, O.-E., *Beta-adrenoceptors in cardiac disease*. Pharmacology &Therapeutics, 1993. **60**(3): p. 405–430.
17. do Vale, G.T., et al., *Three Generations of β-blockers: History, Class Differences and Clinical Applicability*. Curr Hypertens Rev, 2019. **15**(1): p. 22–31.
18. Duncan, G.E., et al., *Beta-adrenergic receptor distribution in human and rat hippocampal formation: marked species differences*. Brain Res, 1991. **561**(1): p. 84–92.
19. Woo, C.C. and M. Leon, *Distribution and development of beta-adrenergic receptors in the rat olfactory bulb*. J Comp Neurol, 1995. **352**(1): p. 1–10.
20. Pazos, A., A. Probst, and J.M. Palacios, *Beta-adrenoceptor subtypes in the human brain: autoradiographic localization*. Brain Res, 1985. **358**(1–2): p. 324–8.
21. Reddy, N.L., et al., *Brown adipose tissue: endocrine determinants of function and therapeutic manipulation as a novel treatment strategy for obesity*. BMC Obes, 2014. **1**: p. 13.
22. Cero, C., et al., *β3-Adrenergic receptors regulate human brown/beige adipocyte lipolysis and thermogenesis*. JCI Insight, 2021. **6**(11).
23. Bartness, T.J., C.H. Vaughan, and C.K. Song, *Sympathetic and sensory innervation of brown adipose tissue*. Int J Obes (Lond), 2010. **34**(1): p. S36–S42.
24. Timper, K. and J.C. Brüning, *Hypothalamic circuits regulating appetite and energy homeostasis: pathways to obesity*. Dis Model Mech, 2017. **10**(6): p. 679–689.
25. François, M., et al., *Sympathetic innervation of the interscapular brown adipose tissue in mouse*. 2019. **1454**(1): p. 3–13.
26. Song, C.K., G.J. Schwartz, and T.J. Bartness, *Anterograde transneuronal viral tract tracing reveals central sensory circuits from white adipose tissue*. Am J Physiol Regul Integr Comp Physiol, 2009. **296**(3): p. R501–11.
27. Vijgen, G.H., et al., *Vagus nerve stimulation increases energy expenditure: relation to brown adipose tissue activity*. PLoS One, 2013. **8**(10): p. e77221.
28. Carlson, B.M., *The Biology of Long-Term Denervated Skeletal Muscle*. Eur J Transl Myol, 2014. **24**(1): p. 3293.
29. Young, J.B., et al., *Effect of diet and cold exposure on norepinephrine turnover in brown adipose tissue of the rat*. J Clin Invest, 1982. **69**(5): p. 1061–71.
30. Sharara-Chami, R.I., et al., *Effect of epinephrine deficiency on cold tolerance and on brown adipose tissue*. Mol Cell Endocrinol, 2010. **328**(1–2): p. 34–9.
31. Sentis, S.C., R. Oelkrug, and J. Mittag, *Thyroid hormones in the regulation of brown adipose tissue thermogenesis*. Endocr Connect, 2021. **10**(2): p. R106–r115.
32. Swarbrick, M., H. Zhou, and M. Seibel, *Mechanisms in endocrinology: Local and systemic effects of glucocorticoids on metabolism: new lessons from animal models*. Eur J Endocrinol, 2021. **185**(5): p. R113–R129.
33. Scott, R., T. Tan, and S. Bloom, *Gut hormones and obesity: physiology and therapies*. Vitam Horm, 2013. **91**: p. 143–94.
34. Omran, F. and M. Christian, *Inflammatory Signaling and Brown Fat Activity*. Front Endocrinol (Lausanne), 2020. **11**: p. 156.
35. Kristóf, E., et al., *Interleukin-6 released from differentiating human beige adipocytes improves browning*. Exp Cell Res, 2019. **377**(1–2): p. 47–55.

36. Nisoli, E., et al., *Tumor necrosis factor alpha mediates apoptosis of brown adipocytes and defective brown adipocyte function in obesity*. Proc Natl Acad Sci U S A, 2000. **97**(14): p. 8033–8.
37. Ye, J., *Beneficial metabolic activities of inflammatory cytokine interleukin 15 in obesity and type 2 diabetes*. Front Med, 2015. **9**(2): p. 139–45.
38. Almendro, V., et al., *Effects of IL-15 on rat brown adipose tissue: uncoupling proteins and PPARs*. Obesity (Silver Spring), 2008. **16**(2): p. 285–9.
39. Cuevas-Ramos, D., R. Mehta, and C.A. Aguilar-Salinas, *Fibroblast Growth Factor 21 and Browning of White Adipose Tissue*. Front Physiol, 2019. **10**: p. 37.
40. Yano, N., Y.T. Zhao, and T.C. Zhao, *The Physiological Role of Irisin in the Regulation of Muscle Glucose Homeostasis*. Endocrines, 2021. **2**(3): p. 266–83.

4 Thermogenic Determinants in Hibernation Models

Moriah Hunstiger
University of Alaska Fairbanks, Fairbanks, Alaska, USA

S. Ryan Oliver
United States Naval Academy, Annapolis, Maryland, USA

4.1 ENDOTHERMY: HOMEOTHERMY VS. HETEROTHERMY

Endothermy is a physiologic state primarily seen in birds and mammals, where heat is produced through cellular respiration, movement (muscle contraction), and thermodynamic processes (Boyles et al., 2013; Grigg et al., 2022; 2004). This allows body temperature to exceed ambient temperatures, increasing geographical niches and available foraging times. Maintaining a state of high constant body temperature and continual metabolic work is known as homeothermy (Grigg et al., 2022; 2004). This is fueled by a high metabolic rate and requires consistent food sources. Smaller animals need more food relative to their size due to heat loss from their relatively high surface-to-volume ratio (Grigg et al., 2022). However, in some animals, an alternative phenotype known as heterothermy can be utilized (Boyles et al., 2013; Grigg et al., 2022; 2004). Heterothermy occurs when an animal reduces the demand of maintaining a high body temperature and allows the environment to dictate body temperature (Boyles et al., 2013; Grigg et al., 2022). It involves a depression in metabolism to conserve fuel sources and mitigate heat loss to the environment, resulting in a reduction in body temperature known as torpor (Heldmaier et al., 2004).

Torpor occurs when adverse environmental conditions and/or decreased food availability result in the energetic costs of thermoregulation and metabolism to exceed energy intake via foraging (Grigg et al., 2004; Ruf & Geiser, 2015; Staples, 2016; Wang & Wolowyk, 1988). Physical indications of torpor include a depression in metabolism and, in some cases, a concomitant decrease in body temperature (Ruf & Geiser, 2015; Staples, 2016). Torpor has been recorded in a minimum of seven orders of mammals and falls into two distinct patterns of expression: seasonal (hibernation) or non-seasonal (daily torpor) (Ferris & Gregg, 2019; Wang & Wolowyk, 1988; Table 4.1). Most endotherms that experience daily torpor do so during perceived cold environmental conditions, but there is a specific subset that undergoes metabolic suppression during hot and dry conditions (Geiser, 2007; Staples, 2016). This type of daily torpor is classified as estivation and seems to be predominantly driven by water availability (Staples, 2016). Depending on the length of suppression, type of torpor, body size, and environmental conditions, energy savings can range from 10% to 98% (Buck & Barnes, 2000; Cranford, 1978; Geiser, 2007; Jonasson & Willis, 2012; Wang & Wolowyk, 1988). Hibernators will typically experience more extreme environmental conditions, temperature variance, and increased metabolic suppression during torpor than their daily torpor counterparts (Ruf & Geiser, 2015). Hibernators are heavier than daily torpor animals and tend to be found at higher latitudes (Porter & Kearney, 2009; Ruf & Geiser, 2015). Compared to daily heterotherms, average maximum torpor bout length in hibernators is more than 30-fold longer (Ruf & Geiser, 2015). Mean torpor bout duration is also more than 25-fold longer in hibernators vs. daily heterotherms (Ruf & Geiser, 2015). During daily torpor, basal metabolic rate (BMR) can fluctuate, falling as low as 35% on average, but during hibernation, BMR can drop to as low as 1–6% (Buck et al., 2008; Ruf & Geiser, 2015; Tøien et al., 2001). Mean body temperature falls by an average of 13°C during daily torpor but as much as 40–44°C during hibernation (Barnes &

DOI: 10.1201/9781003391470-5

TABLE 4.1
Distribution of Types of Torpor in Mammals

Group	Common Name	Hibernation	Daily	Estivation
Marsupials	Pygmy Possum & Dunnart	x	x	
Afrotheria	Tenrec	x	x	
Insectivora	Hedgehog	x	x	
Carnivora	Grizzly and Black Bear	x	x	
Chiroptera	Fruit Bat and Little Brown Bat	x	x	
Primates	Fat-tailed Dwarf Lemur	x	x	
Rodentia	Marmot, Arctic Ground Squirrel, 13-lined Ground Squirrel, Dormice	x	x	x

Hibernation and daily torpor are seen in seven clades of mammals. Estivation is only seen in Rodentia. Adapted from Letunic & Bork (2007).

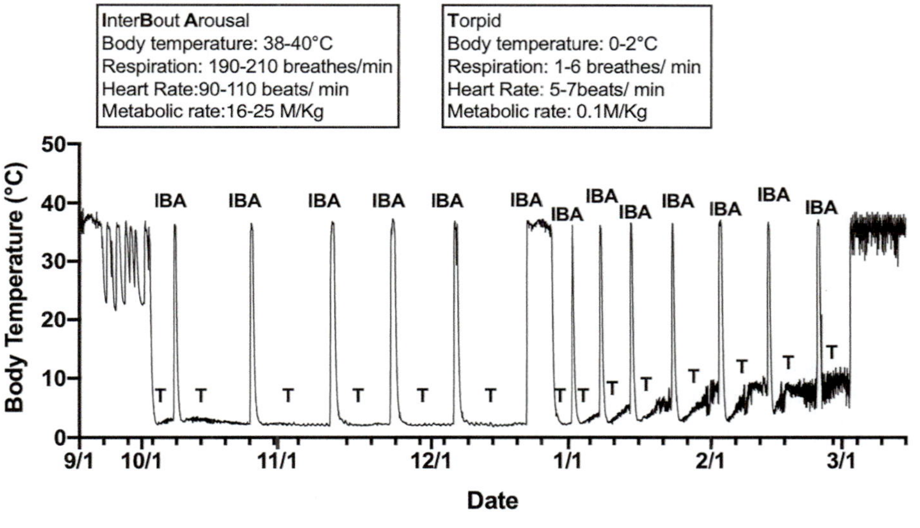

FIGURE 4.1 **Body temperature during repeated cycles of torpor in an Arctic ground squirrel.** Body temperature over the hibernation season is shown as a black line. Body temperature, respiration rate, heart rate, and metabolism during torpor and interbout arousal are in the boxes, and torpor and interbout arousals are marked with T and IBA, respectively.

Buck, 2000; Barnes & Ritter, 2019; Long et al., n.d.; Ruf & Geiser, 2015). Arctic ground squirrels are an extreme case, whereby they reach a minimum body temperature of –2.9°C from an aroused body temperature of approximately 38–40°C (Barnes, 1989; Barnes & Ritter, 2019; Buck & Barnes, 1999b; Williams, Barnes, & Buck, 2012; Figure 4.1).

4.2 HIBERNATION: BODY COMPOSITION, METABOLISM, AND ENERGY EXPENDITURE

Hibernation is energetically costly and occurs during a time when food sources are limited or non-existent and typically coincides with extreme cold temperature conditions. To ensure adequate fuel sources throughout the winter season, animals will prepare for hibernation during the spring and summer with pre-hibernation fattening, hyperphagia, and caching of food sources depending on the species. One of the smallest known hibernators, the little brown bat (*Myotis lucifugus*), increases total body mass by 38%, from 6.8 to 9 g, during mid and late August and increases total adipose tissue from 10% to more than 25% of total body mass (Kronfeld-Schor et al., 2000). A slightly larger mammal, the Belding ground squirrel (*Urocitellus beldingi*), has juveniles that increase from 100 g to just over 200 g prior to the hibernation season (Morton et al., 1974). Similarly, AGSs increase total body mass from 639 g to 904 g, or 42%, for males and 492 g to 815 g, or 62%, for females during the active season (Sheriff et al., 2013). AGS adipose mass increases seven- to eight-fold or between 20% and 30% by the beginning of the hibernation season and sometimes reaches more than 50% of adipose tissue by total body mass (Sheriff et al., 2013).

The largest known hibernators are ursids, or bears. Grizzly bears, *Ursus arctos horribilis*, can increase their total body mass by up to 50% during the active season with up to 81% of gain being adipose tissue (Rigano et al., 2017). Pre-hibernation fattening is ubiquitous across many clades of hibernators, and these endogenous sources of fuel, in the form of fat, are essential for survival. The fat stores can be used as fuel during times of food scarcity and be metabolized to defend body temperature both during aroused cold exposure and hibernation. For example, polar bears, *Ursus maritimus*, expend energy at roughly the same rate compared to other ursid counterparts due to similar size constraints (Robbins et al., 2012). However, during the summer and fall, polar bears experience walking hibernation, a period of reduced food availability, fasting, and reduced activity due to receding sea ice (Robbins et al., 2012). A 500-kg polar bear will lose approximately 1,300 g of mass per day during walking hibernation compared to brown bears of a similar size transitioning between active state and hibernation that lose only 800 g/day and those that are hibernating that only lose 500 g/day (Robbins et al., 2012). Additionally, during walking hibernation, lean body mass loss constitutes more of the total mass loss compared to hibernation, where adipose tissue loss constitutes more of the total mass loss (Robbins et al., 2012). Thus, hibernation is highly efficient in conserving energy and mass during periods of food scarcity.

When comparing hibernating and active-state metabolism based on total body mass, active-state metabolic rate scales allometrically whereas hibernating metabolism scales isometrically (Heldmaier et al., 2004; Kleiber, 1932; Figure 4.2). This results in hibernation being of greater benefit to smaller mammals and leads to generally longer bouts of torpor, larger depressions in body temperature, and overall larger energy savings in smaller mammals (Figure 4.2). Hibernating bears can reduce metabolism by up to 75% of BMR during torpor compared to AGS, which can depress metabolism by up to 98% (Buck & Barnes, 2000; Nespolo et al., 2022). Other small mammals such as the little brown bat, pigmy possum (*Burramyidae*), and dormice (*Gliridae*) experience similar reductions of 95–98% in metabolism during hibernation (Cranford, 1978; Geiser, 2007; Jonasson & Willis, 2012). For example, in AGS, this results in a decrease in body temperature from 38–40°C to between –2.9°C and 2°C (Barnes, 1989; Figure 4.1), while in hibernating black bears (*Ursus americanus*), active core body temperature drops from between 37–38°C to an average of 33°C during torpor (Tøien et al., 2015). In AGS, depressed metabolism and temperature last 1–4 weeks while in black bears, torpor lasts 1.6–7.3 days (Frank et al., 2008; Tøien et al., 2015). In many smaller species, periods of torpor are interrupted by brief periods of increased metabolic activity and body temperature, known as interbout arousals (IBA), before rapidly returning to a metabolically depressed torpor state (Reinhardt, 2022; Ruf et al., 2022, Figure 4.1). In smaller hibernators, the majority of endogenous energy storage is utilized during periods of IBA, accounting for a minimum of 70% of energy expenditure, to increase body temperature and metabolism quickly (Reinhardt, 2022; Ruf et al., 2022). In AGS, increasing the body temperature by 38°C during an IBA is estimated to also increase metabolic rate sixfold and constitutes up to 86% of total energy expended over the hibernation season (Karpovich et al., 2009). In the 13-lined ground squirrel (*Ictidomys tridecemlineatus*), this corresponds to a 400-fold increase in metabolic rate during the initial phase of arousal (Regan et al., 2019). Thus, IBAs tend to be short and, in AGS, lasts up to 24 hours before a complete return to torpor, with repeated cycles continuing until the spring terminal arousal (Frank et al., 2008; Karpovich et al., 2009, Figure 4.1). In AGS, males will usually arouse first around March and

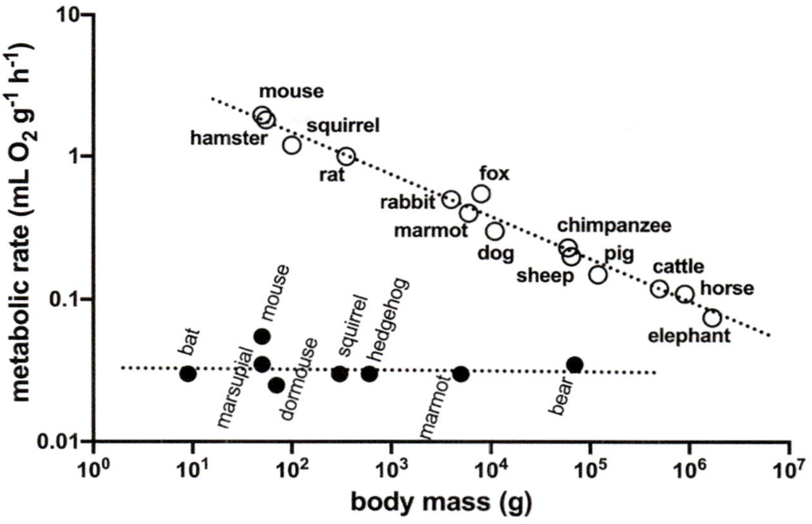

FIGURE 4.2 Body mass and metabolic rate relationship in heterothermic (closed circles, depression during hibernation) and euthermic species (open circles). Hibernation shown was at 2–7°C ambient. Adapted from Heldmaier et al. (2004).

females will emerge 3–4 weeks later (Buck & Barnes, 2000). Mating will occur within 24–48 hours, and then animals will begin to pre-fatten again (Buck & Barnes, 1999a). Due to the large changes in body composition and body temperature during hibernation, many hibernators utilize multiple mechanisms to contribute to overall body metabolism and maintain thermoregulation.

4.3 MECHANISMS OF THERMOGENESIS

4.3.1 Shivering

Shivering is a mechanism employed by many mammals in response to acute or chronic cold stress (Hohtola, 2004). Shivering is an isometric muscular movement, or repetitive contraction action of muscle fibers, that produces heat by myosin cross-bridge cycling, calcium cycling, and sodium/potassium transport (Herpin et al., 2002). Actin-myosin interactions occur when Ca^{2+} is released down its concentration gradient from the sarcoendoplasmic reticulum (SR) via the ryanodine receptor resulting in actin binding sites being exposed. The myosin head then binds via myosin ATPase, allowing for recruitment of motor units within the muscle (Hohtola, n.d.; Hohtola & González-Alonso, 2020; Muthu et al., 2008). This results in muscle tremors or micro-vibrations that will vary in amplitude depending on the synchronization and number of motor units recruited (Hohtola, n.d.; Hohtola & González-Alonso, 2020). Shivering can be categorized by (1) low-intensity prolonged shivering that occurs in type I muscle fibers and utilizes predominantly lipids and (2) high-intensity bursts in type II muscle groups that rely on carbohydrate and lipid fuel sources (Haman, 2006). Type I and IIa muscle fibers are preferentially recruited for shivering, as they are more fatigue resistant, more heat efficient, and rely on lipid fuel sources (Bell et al., 1992). Vigorous shivering can increase heat production 600% above basal for short periods of time, but this can result in myoelectric fatigue, muscle atrophy, and is very energetically expensive (Giesbrecht, 1994).

During longer periods of cold stress, shivering can push metabolic expenses 200% above basal for prolonged periods of time (Horvath et al., 1956; Periasamy et al., 2017a). As a result, shivering is recruited within very specific short time periods during hibernation, resulting in rapid whole-body temperature fluctuations. For instance, in American black bears, a larger hibernator, shivering was observed to occur three to four times a day during torpor with a duration of 20–30 minutes (Harlow et al., 2004). Conversely, in AGS, a smaller hibernator, maximal shivering occurs during IBAs after body temperature has reached approximately 5°C and lasts until the animal has reached full arousal, ~35°C (Karpovich et al., 2009). In the bat species *Myotis myotis*, shivering heat production also occurs during arousal but does so at body temperatures of 10–17°C (Mejsnar & Janský, 1970). When rewarming at higher hibernaculum conditions of 25°C, shivering intensity increases and lasts for longer periods of time (Mejsnar & Janský, 1970). Thus, both shivering recruitment and overall body temperature vary across species and size. Shivering alone is not sufficient for rewarming at lower body temperatures, especially in smaller mammals. Other sources of heat production are known, such as non-shivering thermogenesis (NST), that support proper thermoregulation and muscle integrity during periods of increased thermogenic demands (Horvath et al., 1956; Periasamy et al., 2017b).

4.3.2 Adipose Tissue

4.3.2.1 Mechanism

Brown adipose tissue (BAT) is postulated to have developed very early in mammalian evolution (Cannon & Nedergaard, 2004). Its first documentation on record is in an observational anatomical inspection of a marmot almost 500 years ago in 1551 and was described as "neither fat nor meat but somewhere in between" (Montani et al., 2018). BAT is suggested to have greatly diversified mammalian evolution by providing the necessary energy and heat production capacity to expand active foraging hours into the night or during periods of cold stress (Cannon & Nedergaard, 2004; Grigg et al., 2004). Involvement of BAT is highlighted in the maintenance of body temperature and survival of mammalian neonates (Cannon et al., 1988; Symonds, 2013). Although comprising at most ~4% total body weight, unique anatomical placement and specific evolutionary development makes it capable of producing large amounts of heat (Cannon & Nedergaard, 2004). Locations of BAT will vary depending on species, but some of the most common areas include the thoracic cavity, intrascapular and subscapular regions, and the kidneys (Cannon & Nedergaard, 2004; Symonds et al., 2015). BAT has a characteristic brown color that occurs because of high iron, increased vascularization, and increased mitochondrial density. High mitochondrial density allows for large increases in heat production via UCP1, and the increased vascularization allows for greater efficiency in heat transfer and shuttling of fuel sources to BAT (Cannon & Nedergaard, 2004; Symonds et al., 2015). Thermogenesis occurs by beta-adrenergic activation of UCP1, located in the inner mitochondrial membrane (Cannon & Nedergaard, 2004; Nicholls, 2021). UCP1 functions by allowing hydrogen ions to leak back into the matrix, effectively uncoupling the electrochemical gradient created by the electron transport chain from ATP production and resulting in the production of heat (Cannon & Nedergaard, 2004; Nicholls, 2021). The presence of UCP1 in BAT is estimated to increase the thermogenic capacity by up to 20% (Cannon & Nedergaard, 2012).

In comparison, white adipose tissue (WAT) on its own is not naturally associated with thermogenic capacity but rather as a lipid fuel source used by muscles, BAT, and other thermogenic tissues to produce heat (Wu et al., 2013). However, recent data have shown that in certain cases subcutaneous WAT can be an inducible thermogenic tissue that expresses UCP1 and increases its mitochondrial density, resulting in a beiging or browning effect. The beige adipose tissue produces energy in a manner similar to that seen in BAT (Symonds, 2013; Vargas-Castillo et al., 2021; Wu et al., 2013). Although beige adipocytes use UCP1 to produce energy, they only contain ~10% of the UCP1 abundance seen in BAT and are thus more limited in thermogenic capacity (Wu et al., 2013). There are two prevailing theories of how induction of thermogenesis in WAT occurs. The first, *de novo* differentiation, postulates that beige adipocytes originate from smooth muscle progenitor cells that are committed to being a beige adipocyte, and that stimuli of inducibility will only affect cells from this progenitor line to adopt beige characteristics (Vargas-Castillo et al., 2021). The second, transdifferentiation, theorizes that a mature adipocyte experiencing sufficient stimuli can activate a "dormant" beige phenotype; when the stimuli is removed, such as adaptation to a warm environment, the phenotype reverts to a traditional white adipocyte (Vargas-Castillo et al., 2021). Although beige adipocytes are distinct from brown adipocytes in origin, anatomical location, phenotypic plasticity, and thermogenic capacity, they may share similar thermogenic mechanisms, dependence on UCP1 uncoupling, and fatty acid utilization for thermogenesis.

4.3.2.2 Expression Neonates

At birth, mammals experience a period of cold stress as they transition from the mother's body temperature to ambient temperature (Bienboire-Frosini et al., 2023; Symonds, 2013). In order to survive, rapid heat generation is necessary, and heat produced via UCP1 uncoupling in BAT is highly relied upon (Bienboire-Frosini et al., 2023; Symonds, 2013). BAT is highly efficient at heat production and can produce up to 300 watts/kg of heat, via UCP1 uncoupling, compared to 1 watt/kg in other tissues, and is thus an optimal thermogenic effector (Symonds, 2013). Mammals vary widely in their development at birth, and thus the recruitment and timing of BAT usage vary (Bienboire-Frosini et al., 2023; Symonds, 2013). Altricial species, or species that are born helpless and dependent on the parent, such as rats, dogs, and mice, have young that are born without or with underdeveloped BAT tissues (Bienboire-Frosini et al., 2023; Symonds, 2013; Trevathan, 2017). BAT is activated shortly after birth, and in rat pups UCP1 RNA increases fivefold between 2 and 20 hours post birth (Obregón et al., 1989). In contrast, precocial species, or those that birth relatively mature young with most senses fully developed, like calves and guinea pigs, are born with BAT depots that are fully intact and can begin to produce heat emergently (Bienboire-Frosini et al., 2023; Symonds, 2013; Trevathan, 2017). However, some precocial species, in particular those that are large and have reduced metabolic rates, like elephants, horses, and pigs, no longer express UCP1 (Bienboire-Frosini et al., 2023; Symonds, 2013). Humans are known as secondarily altricial, as they are born sharing the characteristics of both altricial and precocial mammals (Symonds, 2013; Trevathan, 2017). However, BAT tissues in humans can be seen to develop as early as 29 weeks of gestational age, and infants are born with BAT depots in interscapular, neck, axillae, and intra-abdominal regions that are fully developed (Lidell, 2018). However, as these infants age, the BAT depots shrink and reliance on them as thermogenic effectors decreases (Lidell, 2018).

4.3.2.3 Adults

In many adults of eutherian lineages, BAT depots decrease in size and functionality during the postnatal stage (Bienboire-Frosini et al., 2023; Symonds, 2013), although in many small rodents, bats, and others, BAT remains present in the adult stage and is essential for maintaining metabolism, body temperature, and body composition (Nedergaard & Cannon, 2018). In humans, it was thought for many years that they did not express UCP1 after the ages of early childhood (8–10 years) (Lee et al., 2013; Nedergaard et al., 2010; Porter, 2017). However, recent studies have shown that adults *do* express BAT but to a lesser extent, and that not only is it expressed in human adults, but gram for gram, human UCP1 function in BAT is comparable to UCP1 function in rodents (Lee et al., 2013; Nedergaard et al., 2010; Porter, 2017). In mice and the great roundleaf bat, UCP1 shows similar levels of expression, increasing in BAT as temperatures decrease and increasing in abundance when the duration of lower body temperatures increased (Wang et al., 2014). However, when UCP1 is removed, as in the case of UCP1-knockout mice, they become obese at thermoneutrality and when fed high-fat diets (Lodhi & Semenkovich, 2009). Similarly, in humans, defects in UCP1 and beta-3 adrenergic receptors have been linked to induced additive effects on weight gain and obesity (Cinti, 2006; Nedergaard et al., 2010). Additionally, obese individuals show lower levels of UCP1 mRNA than their lean counterparts, which may indicate that lower levels of functional UCP1 are linked to disease progression (Cinti, 2006).

4.3.2.4 Hibernators

Unlike humans or some other mammalian lineages, adult hibernating mammals maintain large depots of BAT with functional UCP1 even while maintaining an "obese" phenotype (Fröbert et al., 2020; Wu et al., 2013; Yan et al., 2021). Successful torpor and arousal are dependent on both expression of UCP1 and controlled activation of UCP1 by beta-adrenergic receptors (Cannon & Nedergaard, 2004; Kitao & Hashimoto, 2012; Nicholls, 2021; Swoap et al., 2006). In *Dbh–/–* mice, those that do not secrete some neurotransmitters, daily torpor could not be induced under fasted conditions, but when fasted mice were given beta-adrenergic agonists, torpor was expressed (Swoap et al., 2006). In the same study when control mice (*Dbh+/–* mice), those that could experience normal torpor, were given beta3-specific antagonist SR 59230A, torpor depth and duration were blunted (Swoap et al., 2006). In Syrian hamsters (*Mesocricetus auratus*), a hibernator, administration of beta-adrenergic agonist, CL316,243, increased BAT thermogenesis in animals at both 36°C and 12°C but resulted in larger metabolic ratios above basal in the 12°C group (Kitao & Hashimoto, 2012). This agonist was also able to shorten arousal time and induce arousal in animals in deep hibernation (Kitao & Hashimoto, 2012). Syrian hamsters that received the antagonist SR 59230A took longer to arouse or did not arouse from hibernation at all, highlighting the importance of UCP1 activation for successful hibernation (Kitao & Hashimoto, 2012). While the activation of UCP1 is essential for successful arousal, animals can experience a wide range of ambient temperatures and fuel availability over the hibernation season requiring differential expression of UCP1 to successfully preserve body temperature and determine hibernation outcomes.

Expression of UCP1 is increased in many hibernators during the hibernation season versus the active season (Ballinger & Andrews, 2018; Hindle & Martin, 2014, Hunstiger et al., 2023). In 13-lined ground squirrels, UCP1 protein and mRNA levels were highest during torpor and IBA compared to active animals in the spring and fall (Ballinger & Andrews, 2018; Hindle & Martin, 2014). In AGS, UCP1 had increased mRNA in non-fasted animals compared to active and torpid animals, but protein levels were increased for non-hibernators compared to torpid animals housed at 5°C (Barger et al., 2006). AGS housed at –10°C had high levels of mRNA in non-fasted non-hibernators and torpid animals, and greater levels of protein expression in hibernators when compared to fasted and non-fasted active animals (Barger et al., 2006). In another study, UCP1 in the BAT of AGS was shown to be highest in animals during late hibernation and directly after terminal arousal in the spring but that protein expression would begin to taper after terminal arousal (Hunstiger et al., 2023). These same animals showed no expression of UCP1 in WAT at any time point during the hibernation season unlike some other hibernating species (Hunstiger et al., 2023). For example, in the Syrian hamster, mRNA for UCP1 increased in WAT pre-hibernation and gradually decreased after prolonged cold exposure (Chayama et al., 2019). In small hibernating mammals, just like their non-hibernating counterparts, UCP1 is recruited during times of increased cold stress and metabolic activity to maintain rheostasis. But is BAT enough to maintain thermoregulation on its own?

4.4 MUSCLE NON-SHIVERING THERMOGENESIS AND THE ROLE OF CALCIUM

4.4.1 Muscle Non-Shivering Thermogenesis

The concept of muscular thermogenesis is almost exclusively related to altered calcium cycling to and from the sarcoendoplasmic reticulum (Pant et al., 2016). The proposed mechanism involves the interruption of SERCA ATP-hydrolyzed symporting of calcium and hydrogen between the cytosol and the SR to reestablish calcium concentration gradients for muscle contraction (Gorski et al., 2017; Pant et al., 2016; Winther et al., 2013). This occurs by the interaction of a small 31AA peptide sarcolipin (SLN) with SERCA, altering the V_{max}, the rate of calcium transport (Gorski et al., 2017). SLN stabilizes the Ca^{2+}-binding sites of SERCA when calcium is bound, so when ATP is hydrolyzed, instead of transporting the calcium across the SR membrane, calcium remains bound and that energy is released as heat (Pant et al., 2016; Winther et al., 2013). In some cases, it has been postulated that phospholamban (PLB) may serve a similar function by altering calcium affinity of SERCA isomers, but other research refutes this postulate, since PLB only interacts with certain SERCA conformations (Gorski et al., 2017; Sahoo et al., 2013; Shaikh et al., 2016). Alternatively, it is suggested that PLB can only serve as a source for thermogenesis when co-expressed with SLN resulting in "super inhibition" by interrupting PLB pentamers, increasing SLN/PLB/ SERCA interactions, and thus increasing uncoupling and energy production (Asahi et al., 2002; Fajardo et al., 2013). In most mammal physiology, expression of SLN is seen in slow-twitch muscle fibers with the SERCA2a isoform, and PLB is expressed in cardiac and fast-twitch muscle fibers with SERCA1a isoform (Fajardo et al., 2013; Gorski et al., 2017). SLN and PLB can interact with both isoforms of SERCA, and thus all muscle fiber types have the opportunity for altered calcium cycling and contributions to thermoregulation of body temperature.

4.4.2 Intracellular Calcium Maintenance During Hibernation

Intracellular calcium concentrations are highly controlled to maintain normal cellular function. Many mechanisms to control Ca^{2+} concentrations are in place, but when these safeguards are not functioning properly, disease and/or cellular death can occur. Hibernators have developed adaptations in these mechanisms to maintain a new definition of "normal function" and have multiple mechanisms to maintain cyclical cytoplasmic Ca^{2+} levels across the hibernation and summer active seasons. These mechanisms utilize the same molecular machinery as non- hibernators but alter which pathways are increased or decreased throughout torpor and active seasons. For example, skeletal muscle of Daurian ground squirrels (*Spermophilus dauricus*) show increasing cytoplasmic calcium levels from early torpor to late torpor time points within a single torpor bout and increase further at arousal (Zhang et al., 2019, Figure 4.3A). However, when reentry into torpor occurs, the cytoplasmic Ca^{2+} concentration is back to early torpor levels (Zhang et al., 2019), suggesting a broader Ca^{2+} rotation needs to be maintained for the torpor-arousal cycling through the hibernation season. To preserve this cycle, intracellular calcium concentration is maintained by four main pathways: store operated calcium entry, ryanodine/SERCA cycling, mitochondrial calcium uptake, and calmodulin free calcium binding.

Ryanodine/SERCA cycling has been of utmost interest recently due to the identification of peptides (sarcolipin and phospholamban) that can directly influence SERCA activity, thus impacting calcium movement in and out of the sarcoplasmic reticulum. When muscle is stimulated for contraction, the L type (voltage gated) calcium channels on the plasma membrane open to allow a small amount of extracellular calcium to come into contact with the ryanodine receptor located in the SR membrane (Felder et al., 2002; Laver 2007a, 2007b). This allows for a calcium-induced calcium release (CICR) from the SR (through the ryanodine receptor), increasing intracellular calcium concentration for the activation of actin myosin cross-bridging (reviewed in Ríos, 2018). A tight regulation of this pathway is particularly important to maintain the ability to shiver, one of the main thermogenesis mechanisms. For relaxation of muscles to occur, the SERCA pump transports the calcium back into the SR for storage (reviewed in Periasamy et al., 2017b). The relaxation steps require ATP (not contraction), and thus it is an energy-intensive process (Periasamy et al., 2017b). Sarcolipin and phospholamban can directly inhibit calcium transport back into the SR, as discussed earlier (Bal et al., 2012). Recent studies in hibernating ground squirrels have shown that SERCA activity levels are maintained relatively high during torpor compared to summer active state to avoid high intracellular calcium levels and prevent muscle injury (Zhang et al., 2019, Figure 4.3B; Guo et al., 2017). Furthermore, SLN levels in hibernating AGS's and 13-lined ground squirrels' skeletal muscle are significantly reduced, allowing for SERCA to function with less inhibition, which could help explain the SERCA activity data seen in Zhang et al. (2019) (Hunstiger et al., 2023; Oliver et al., 2019).

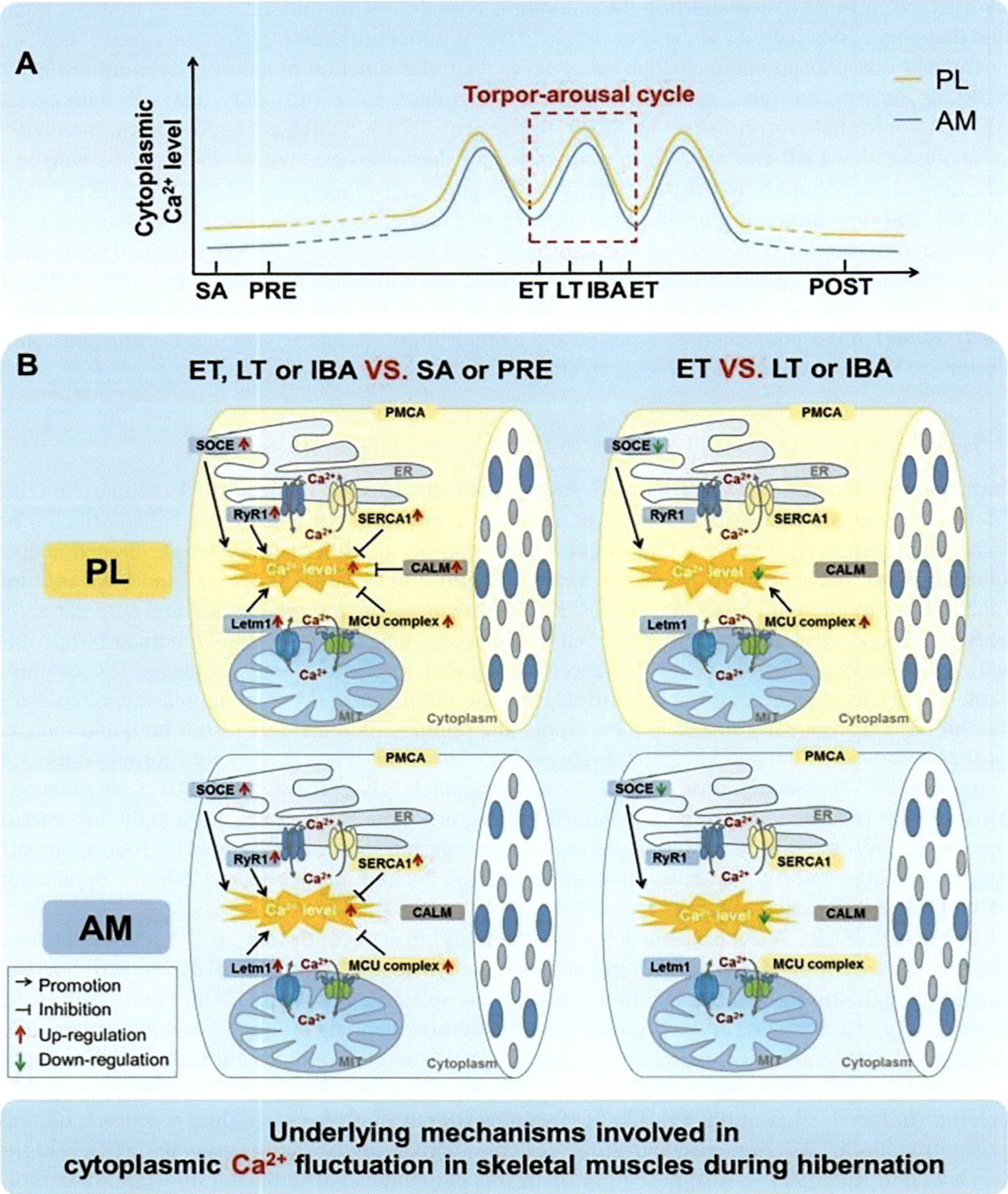

FIGURE 4.3 Graphical summary of Zhang et al.'s (2019) study that focused on the potential roles of major Ca^{2+} channels and proteins in cytoplasmic Ca^{2+} fluctuations in skeletal muscles of Daurian ground squirrels during hibernation. (A) Fluctuation of cytoplasmic Ca^{2+} levels in skeletal muscle fibers during the torpor-arousal cycle. (B) Increased activation probability or protein expression of SOCE, RyR1, and LETM1 may participate in the periodic elevation of cytoplasmic Ca^{2+} during hibernation, whereas the increased expression of SERCA1, MCU, and CALM (only for PL) may be potential mechanisms through which hibernators can attenuate cytoplasmic Ca^{2+} and restore Ca^{2+} homeostasis during hibernation. Compared with that in LT or IBA groups, the decreased activation of SOCE and increased expression of MCU may participate in the partial downregulation of cytoplasmic Ca^{2+}. *AM*, adductor magnus; *ET*, early torpor group; *IBA*, inter-bout arousal group; *LT*, late torpor group; *PL*, plantaris; *POST*, post-hibernation group; *PRE*, pre-hibernation group; SA, summer active group. Reprinted from Zhang et al. (2019) under the Creative Commons Attribution (CC BY) license.

Additionally, dietary lipids have been shown to modulate SERCA activity in cardiac tissue, which has a profound effect on entrance/exit from torpor, suggesting an even larger role of calcium in hibernation mechanisms (Giroud et al., 2013, 2018). Regulation of ryanodine/SERCA calcium cycling can impact the more newly emerging non-shivering thermogenesis mechanism as well as shivering efficiency. For hibernating species, energy conservation is the main requirement, and thus a tight regulation of the calcium cycling pathway could be of significant importance and needs to be studied in more detail.

Intracellular calcium concentration can increase due to ryanodine release from SR stores but also from extracellular influx, through store-operated calcium entry (SOCE). SOCE specific proteins stromal interaction molecule-1 (STIM1) and calcium-release-activated calcium modulator (CRACM1, now known as Orai1) are required for entrance of calcium from extracellular stores (Salido et al., 2009; Várnai et al., 2009; Vig et al., 2006). In hibernators, SOCE activation has been shown in late torpor and during IBAs and is suggested to be one of the major reasons for the increase in cytoplasmic calcium levels, along with ryanodine leakage, seen during these time points (Zhang et al., 2019). Furthermore, the efflux of calcium from mitochondrial stores through the leucine zipper-EF-hand containing transmembrane protein 1 (LETM1) could also account for some of this increase in cytoplasmic calcium levels (Zhang et al., 2019). LETM1 protein levels have been shown to be elevated during hibernation. However, cytoplasmic calcium levels seem to fluctuate with the torpor/IBA cycle, requiring a tight regulation of calcium entry and reuptake (Zhang et al., 2019, Figure 4.3A). For example, the protein levels for the other major mitochondrial calcium transporter, mitochondrial calcium uniporter complex (MCU), are increased during early torpor, which could help alleviate cellular stress caused by higher calcium concentrations (Zhang et al., 2019). Protein expression patterns change from early torpor to late and seem to allow for the torpor/IBA intracellular calcium cycling seen in the study (Zhang et al., 2019, Figure 4.3B). Furthermore, intracellular free calcium binding through calmodulin is elevated during hibernation and could help stabilize calcium concentrations (Zhang et al., 2019). Despite the stresses of chronic cold, low metabolism, and muscle disuse, hibernating animals are able to maintain intracellular calcium levels and may even utilize calcium concentration as a signal for torpor cycling.

4.5 UCP1 AND SLN: BOTH ARE NECESSARY

4.5.1 Non-Hibernators

When comparing the regulation of body temperature observed in mouse models that have altered abundance, expression, or deactivation of UCP1 and SLN, trends start to emerge to help define which thermogenic mechanisms are most needed. Mice with ablated intrascapular BAT defended their body temperature, 35.8 ± 0.4°C, to the same degree as their wild-type counterparts, 36.3 ± 0.2°C, when exposed to chronic cold (Bal et al., 2012). By contrast, $SLN^{-/-}$ mice exposed to 4°C environment experienced drastic decreases in core body temperature from 37°C to 32.2 ± 1.4°C after 4 hours and a continued drop in body temperature to 26.9 ± 1.9°C after 6 hours (Bal et al., 2012, 2017). When mice with both forms of NST disabled, either using UCP1:SLN double knockout mice, or $SLN^{-/-}$ mice with downregulated activity of BAT due to warm exposure, were challenged with acute cold, hypothermia occurred (Bal et al., 2017; Rowland et al., 2015). In these same models, gradual cold conditions resulted in mice being unable to maintain adequate fat stores (Bal et al., 2017; Rowland et al., 2015). In traditional rodent models, SLN uncoupling is essential for maintenance of thermoregulation and body composition.

4.5.2 Hibernators

When compared to novel animal models, there are relatively sparse amounts of data available regarding muscle NST. However, in hibernation models the 13-lined ground squirrel and AGS, SLN,

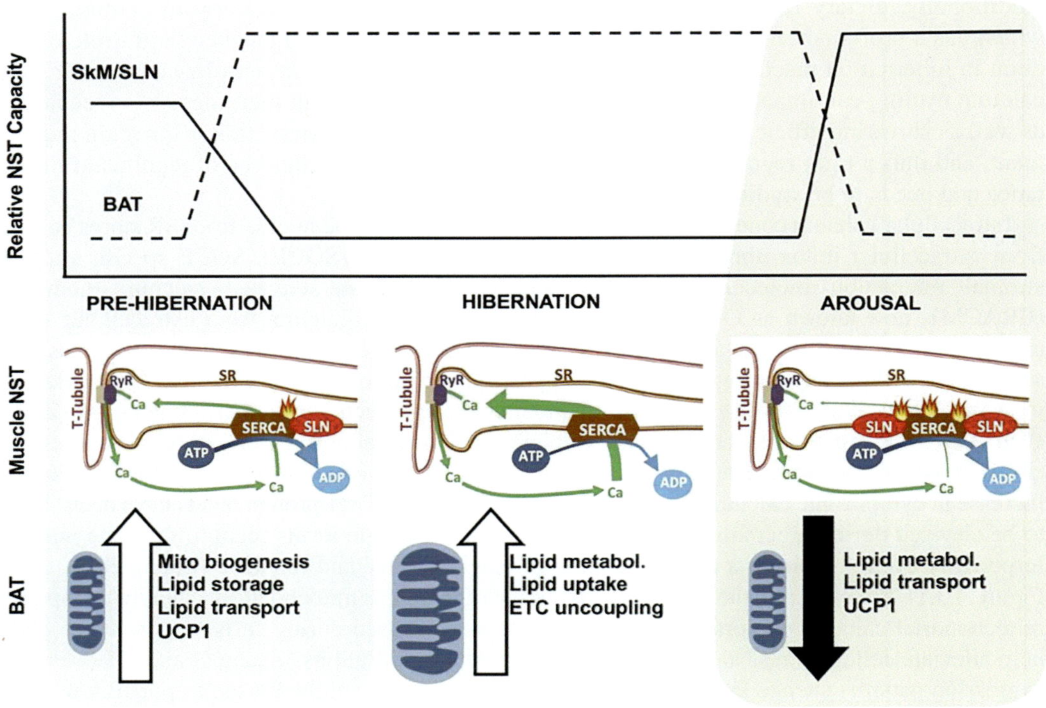

FIGURE 4.4 Conceptual model showing annual profile of non-shivering thermogenic (NST) capacity from skeletal muscle via sarcolipin (SkM/SLN; solid line) and brown adipose tissue (BAT; dashed line). During pre-hibernation (late summer, early fall), BAT undergoes growth, increased lipid content, increased proteins responsible for lipid transport and metabolism, as well as elevated UCP1 expression. These alterations enhance BAT thermogenic capacity and allow for rapid heat production during IBAs. Concurrently, sarcolipin (SLN) mRNA and protein expression are significantly decreased in preparation for hibernation (winter). To reach torpor Tb, SLN protein content must be lowered to decrease skeletal muscle non-shivering thermogenesis and allow for entrance into torpor. SLN expression is increased post arousal (terminal arousal, March–April) while BAT thermogenic mechanisms are decreased. Reprinted from Oliver et al. (2019) with permission from the publisher.

SERCA1a, and 2a proteins have been seen to be expressed and to fluctuate in expression over the duration of the hibernation season (Anderson et al., 2016, Hunstiger et al., 2023; Oliver et al., 2019). In the 13-lined ground squirrel and AGS, mRNA levels of SLN are seen to be present across the hibernation season but downregulated during the spring and summer (Anderson et al., 2016; Fedorov et al., 2014). Additionally, protein levels follow a similar pattern with SLN being decreased during torpor and increased in the spring after terminal arousal (Hunstiger et al., 2023; Oliver et al., 2019). In AGSs experiencing an IBA, SERCA 1a had depressed expression during the hibernation season in diaphragm tissue and SERCA 2a had decreased expression in the diaphragm and *extensor digitorum longus* (Hunstiger et al., 2023). Similarly, in the quadriceps of the 13-lined ground squirrel, both SERCA isoforms appear to have depressed expression during torpor and an increase in expression during arousal (Anderson et al., 2016; Oliver et al., 2019). However, in AGS, UCP1 expression in BAT was at the highest levels during the hibernation season, but muscle NST was suppressed during that time (Hunstiger et al., 2023). After terminal arousal, UCP1 expression in BAT dwindles while muscle NST is enhanced to compensate (Hunstiger et al., 2023) (Figure 4.4). In small rodents, both hibernators and non-hibernators, muscle NST could be a major thermogenic mechanism available for recruitment in the regulation of full-body thermogenesis required for torpor, arousal, and a successful hibernation season.

4.6 HIBERNATORS AS AN IDEAL MODEL FOR THERMOGENESIS STUDIES

In many cases human disease is studied in small rodent mammals, typically rats and mice, as they share 99% of their genomes with humans, are small and easy to house, and have a fast population turnover rate (Vandamme, 2014). Studies can vary from testing therapeutic agents, xenografts, and transgenic models to breeding specific lines or populations to mimic human disease (Rydell-Törmänen & Johnson, 2019; Vandamme, 2014). A drawback of some of these models is that anatomical or physiological differences affect the interpretation of results and the lack of etiology (a physiologically relevant method of disease induction). For instance, in many mice lines, the disease is mimicked by the overexpression or underexpression of certain genes. However, when compared to the human disease of interest, these genes do not reflect an accurate disease progression, symptomology is skewed out of proportion to what is seen in humans, and progression fails to mimic the complexity of the clinical presentation of human disease (Rydell-Törmänen & Johnson, 2019). Some of these drawbacks may be overcome by using novel animal models that naturally present or are adapted to disease-like symptoms without any genetic modifications.

For instance, hibernators from different clades have genomic regions of accelerated evolution that parallel their non-hibernating counterparts (Ferris & Gregg, 2019). These regions disproportionately overlap with human obesity-susceptible genes (Ferris & Gregg, 2019). Hibernation has been recorded in at least seven orders of mammals from monotreme to primates, suggesting that the genes involved are present in both hibernators and non-hibernators, and that the differences could be explained by regulatory changes (Ferris & Gregg, 2019; Table 4.1). It has been suggested that bears tolerate obesity and a sedentary lifestyle during the hibernation season but exit their dens during the spring metabolically healthy and thus has been a useful model to study human obesity (Fröbert et al., 2020). Two smaller species closer in size to mice and rats that mimic obesity models are the 13-lined ground squirrels and the little brown bat, both of which become hyperphagic in the fall, resulting in obesity and insulin resistance, before entering torpor (Ferris & Gregg, 2019; Geiser & Stawski, 2011; Ruf & Geiser, 2015). These postulations would suggest that small hibernators naturally attuned to seasonal oscillations between lean and obese phenotypes, as well as fluctuations in metabolic demands during hibernation, may have genes or subsets of genes that regulate these phenotypes that directly link to the obesity phenotype in humans. There is much we can learn from hibernating species, and new avenues of research are becoming more evident as we better understand the mechanisms underlying this unique phenotype.

REFERENCES

Anderson, K. J., Vermillion, K. L., Jagtap, P., Johnson, J. E., Griffin, T. J., & Andrews, M. T. (2016). Proteogenomic analysis of a hibernating mammal indicates contribution of skeletal muscle physiology to the hibernation phenotype. *Journal of Proteome Research*, *15*(4), 1253–1261. https://doi.org/10.1021/acs.jproteome.5b01138

Asahi, M., Kurzydlowski, K., Tada, M., & MacLennan, D. H. (2002). Sarcolipin inhibits polymerization of phospholamban to induce superinhibition of sarco (endo) plasmic reticulum Ca2+-ATPases (SERCAs). *Journal of Biological Chemistry*, *277*(30), 26725–26728. https://doi.org/10.1074/jbc.C200269200

Bal, N. C., Maurya, S. K., Sopariwala, D. H., Sahoo, S. K., Gupta, S. C., Shaikh, S. A., … & Periasamy, M. (2012). Sarcolipin is a newly identified regulator of muscle-based thermogenesis in mammals. *Nature Medicine*, *18*(10), 1575–1579. https://doi.org/10.1038/nm.2897

Bal, N. C., Singh, S., Reis, F. C. G., Maurya, S. K., Pani, S., Rowland, L. A., & Periasamy, M. (2017). Both brown adipose tissue and skeletal muscle thermogenesis processes are activated during mild to severe cold adaptation in mice. *Journal of Biological Chemistry*, *292*(40), 16616–16625. https://doi.org/10.1074/jbc.M117.790451

Ballinger, M. A., & Andrews, M. T. (2018). Nature's fat-burning machine: Brown adipose tissue in a hibernating mammal. *Journal of Experimental Biology*, *221*(Suppl_1), jeb162586. https://doi.org/10.1242/jeb.162586

Barger, J. L., Barnes, B. M., & Boyer, B. B. (2006). Regulation of UCP1 and UCP3 in Arctic ground squirrels and relation with mitochondrial proton leak. *Journal of Applied Physiology*, *101*(1), 339–347. https://doi.org/10.1152/japplphysiol.01260.2005

Barnes, B. M. (1989). Freeze mvoidance in a Mammal: body temperatures below 0°C in an Arctic hibernator. *Science*, *244*(4912), 1593–1595. https://doi.org/10.1126/science.2740905

Barnes, B. M., & Buck, C. L. (2000). Hibernation in the extreme: burrow and body temperatures, metabolism, and limits to torpor bout length in Arctic ground squirrels. In *Life in the Cold*. Springer Berlin Heidelberg. 65–72. https://doi.org/10.1007/978-3-662-04162-8_7

Barnes, B. M., & Ritter, D. (2019). Patterns of body temperature change in hibernating Arctic ground squirrels. In *Life In The Cold: Ecological, Physiological, And Molecular Mechanisms* (1st ed.). Carey, C., Florant, G.L., Wunder, B.A., & Horwitz, B. (Eds). CRC Press. https://doi.org/10.1201/9780429040931-12

Bell DG, Tikuisis P, & Jacobs I. Relative intensity of muscular contraction during shivering. *Journal of Applied Physiology* (1992). 72(6), 2336–42. https://doi.org/10.1152/jappl.1992.72.6.2336

Bienboire-Frosini, C., Wang, D., Marcet-Rius, M., Villanueva-García, D., Gazzano, A., Domínguez-Oliva, A., & Mota-Rojas, D. (2023). The role of brown adipose tissue and energy metabolism in mammalian thermoregulation during the perinatal period. *Animals*, *13*(13), 2173– 2173. https://doi.org/10.3390/ani13132173

Boyles, J. G., Thompson, A. B., McKechnie, A. E., Malan, E., Humphries, M. M., & Careau, V. (2013). A global heterothermic continuum in mammals. *Global Ecology and Biogeography*, *22*(9), 1029–1039. https://doi.org/10.1111/geb.12077

Buck, C. L., & Barnes, B. M. (1999a). Annual cycle of body composition and hibernation in free-living Arctic ground squirrels. *Journal of Mammalogy*, *80*(2), 430–442. https://doi.org/10.2307/1383291

Buck, C. L., & Barnes, B. M. (1999b). Temperatures of hibernacula and changes in body composition of Arctic ground squirrels over winter. *Journal of Mammalogy*, *80*(4), 1264–1276. https://doi.org/10.2307/1383177

Buck, C. L., & Barnes, B. M. (2000). Effects of ambient temperature on metabolic rate, respiratory quotient, and torpor in an Arctic hibernator. *American Journal of Physiology-Regulatory, Integrative and Comparative Physiology*, *279*(1), R255–R262. https://doi.org/10.1152/ajpregu.2000.279.1.R255

Buck, C. L., Breton, A., Kohl, F., Tøien, Ø., & Barnes, B. (2008). Overwinter body temperature patterns in free-living Arctic ground squirrels Spermophilus parryii. In *Hypometabolism in animals: Torpor, hibernation and cryobiology* Lovegrove B. G., McKechnie A. E., (Eds). University of KwaZulu-Natal, Pietermaritzburg. (pp. 317–326).

Cannon, B., Connolly, E., Obregon, M.J., & Nedergaard, J. (1988). Perinatal Activation of Brown Adipose Tissue. In: Künzel, W., Jensen, A. (Eds) *The Endocrine Control of the Fetus*. Springer, Berlin, Heidelberg. https://doi.org/10.1007/978-3-642-72975-1_27

Cannon B, & Nedergaard J. (2004). Jan, Brown adipose tissue: function and physiological significance. *Physiological Reviews*, 84(1), 277–359. https://doi.org/10.1152/physrev.00015.2003

Cannon, B., & Nedergaard, J. (2012 Feb.). Yes, even human brown fat is on fire! *Journal of Clinical Investigation*, *122*(2), 486–489. https://doi.org/10.1172/JCI60941

Chayama, Y., Ando, L., Sato, Y., Shigenobu, S., Anegawa, D., Fujimoto, T., ... & Yamaguchi, Y. (2019). Molecular basis of white adipose tissue remodeling that precedes and coincides with hibernation in the Syrian hamster, a food-storing hibernator. *Frontiers in Physiology*, *9*, 1973. https://doi.org/10.3389/fphys.2018.01973

Cinti, S. (2006). The role of brown adipose tissue in human obesity. *Nutrition, Metabolism and Cardiovascular Diseases*, *16*(8), 569–574. https://doi.org/10.1016/j.numecd.2006.07.009

Cranford, J. A. (1978). Hibernation in the Western Jumping Mouse (Zapus princeps). *Journal of Mammalogy*, *59*(3), 496–509. https://doi.org/10.2307/1380226

Fajardo, V. A., Bombardier, E., Vigna, C., Devji, T., Bloemberg, D., Gamu, D., ... & Tupling, A. R. (2013). Co-expression of SERCA isoforms, phospholamban and sarcolipin in human skeletal muscle fibers. *PLoS One*, *8*(12), e84304. https://doi.org/10.1371/journal.pone.0084304

Fedorov, V. B., Goropashnaya, A. V., Stewart, N. C., Tøien, Ø., Chang, C., Wang, H., ... & Barnes, B. M. (2014). Comparative functional genomics of adaptation to muscular disuse in hibernating mammals. *Molecular Ecology*, *23*(22), 5524–5537. https://doi.org/10.1111/mec.12963

Felder, E., Protasi, F., Hirsch, R., Franzini-Armstrong, C., & Allen, P. D. (2002). Morphology and molecular composition of sarcoplasmic reticulum surface junctions in the absence of DHPR and RyR in mouse skeletal muscle. *Biophysical Journal*, *82*(6), 3144–3149. https://doi.org/10.1016/S0006-3495(02)75656-7

Ferris, E., & Gregg, C. (2019). Parallel accelerated evolution in distant hibernators reveals candidate cis elements and genetic circuits regulating mammalian obesity. *Cell Reports*, *29*(9), 2608–2620. https://doi.org/10.1016/j.celrep.2019.10.102

Frank, C. L., Karpovich, S., & Barnes, B. M. (2008). Dietary fatty acid composition and the hibernation patterns in free-ranging Arctic ground squirrels. *Physiological and Biochemical Zoology*, *81*(4), 486–495. https://doi.org/10.1086/589107

Fröbert, O., Frøbert, A. M., Kindberg, J., Arnemo, J. M., & Overgaard, M. T. (2020). The brown bear as a translational model for sedentary lifestyle-related diseases. *Journal of Internal Medicine*, *287*(3), 263–270. https://doi.org/10.1111/joim.12983

Geiser, F. (2007). Yearlong hibernation in a marsupial mammal. *Naturwissenschaften*, *94*(11), 941–944. https://doi.org/10.1007/s00114-007-0274-7

Geiser, F., & Stawski, C. (2011). Hibernation and torpor in tropical and subtropical bats in relation to energetics, extinctions, and the evolution of endothermy. *Integrative and Comparative Biology*, *51*(3), 337–348. https://doi.org/10.1093/icb/icr042

Giesbrecht, G. (1994). Shivering thermogenesis during acute hypercapnia. *Canadian Journal of Physiology and Pharmacology*, 72(3) 238–242. https://doi.org/10.1139/y94-037

Giroud, S., Frare, C., Strijkstra, A., Boerema, A., Arnold, W., & Ruf, T. (2013). Membrane phospholipid fatty acid composition regulates cardiac SERCA activity in a hibernator, the Syrian hamster (Mesocricetus auratus). *PLoS ONE*, *8*(5), e63111–e63111. https://doi.org/10.1371/journal.pone.0063111

Giroud, S., Stalder, G., Gerritsmann, H., Kübber-Heiss, A., Kwak, J., Arnold, W., & Ruf, T. (2018). Dietary lipids affect the onset of hibernation in the Garden Dormouse (Eliomys quercinus): Implications for cardiac function. *Frontiers in Physiology*, *9*, 1235. https://doi.org/10.3389/fphys.2018.01235

Gorski, P.A., Ceholski, D.K., Young, H.S. (2017). Structure-Function Relationship of the SERCA Pump and Its Regulation by Phospholamban and Sarcolipin. In: Krebs, J. (Eds) *Membrane Dynamics and Calcium Signaling. Advances in Experimental Medicine and Biology*, vol 981. Springer, Cham. https://doi.org/10.1007/978-3-319-55858-5_5

Grigg, G., Nowack, J., Bicudo, J. E. P. W., Bal, N. C., Woodward, H. N., & Seymour, R. S. (2022). Whole-body endothermy: Ancient, homologous and widespread among the ancestors of mammals, birds and crocodylians. *Biological Reviews*, *97*(2), 766–801. https://doi.org/10.1111/brv.12822

Grigg, G. C., Beard, L. A., & Augee, M. L. (2004). The evolution of endothermy and its diversity in mammals and birds. *Physiological and Biochemical Zoology*, *77*(6), 982–997. https://doi.org/10.1086/425188

Guo, Q., Mi, X., Sun, X., Li, X., Fu, W., Xu, S., Wang, Q., Arfat, Y., Wang, H., Chang, H., & Gao, Y. (2017). Remarkable plasticity of Na+, K+-ATPase, Ca2+-ATPase and SERCA contributes to muscle disuse atrophy resistance in hibernating Daurian ground squirrels. *Scientific Reports*, *7*(1), 10509. https://doi.org/10.1038/s41598-017-10829-6

Haman F. (2006). Shivering in the cold: from mechanisms of fuel selection to survival. *Journal of Applied Physiology, 100*(5), 1702–8. https://doi.org/10.1152/japplphysiol.01088.2005

Harlow, H. J., Lohuis, T., Anderson-Sprecher, R. C., & Beck, T. D. I. (2004). Body surface temperature of hibernating black bears may be related to periodic muscle activity. *Journal of Mammalogy*, *85*(3), 414–419. https://doi.org/10.1644/1545-1542(2004)085<0414:BSTOHB>2.0.CO;2

Heldmaier, G., Ortmann, S., & Elvert, R. (2004). Natural hypometabolism during hibernation and daily torpor in mammals. *Respiratory Physiology & Neurobiology*, *141*(3), 317–329. https://doi.org/10.1016/j.resp.2004.03.014

Herpin, P., Damon, M., & Le Dividich, J. (2002). Development of thermoregulation and neonatal survival in pigs. *Livestock Production Science*, *78*(1), 25–45. https://doi.org/10.1016/S0301-6226(02)00183-5

Hindle, A. G., & Martin, S. L. (2014). Intrinsic circannual regulation of brown adipose tissue form and function in tune with hibernation. *American Journal of Physiology-Endocrinology and Metabolism*, *306*(3), E284–E299. https://doi.org/10.1152/ajpendo.00431.2013

Hohtola, E. (2004). Shivering Thermogenesis in Birds and Mammals. Evolution, Mechanisms, Adaptation, and Application. Twelfth International Hibernation Symposium. Biological Papers of the University of Alaska, number 27. Fairbanks, Alaska, USA: Institute of Arctic Biology, University of Alaska.

Hohtola, E., & González-Alonso, J. (2020). Motor unit function during cold induced thermogenesis in muscle—New perspectives on old concepts. *Acta Physiologica*, *228*(2). https://doi.org/10.1111/apha.13408

Horvath Steven, M., Spurr, G. B., Hutt B. K., & Hamilton, L. H. (1956). Metabolic Cost of Shivering. *Journal of Applied Physiology*, *8* (6), 595–602. https://doi.org/10.1152/jappl.1956.8.6.595

Hunstiger, M., Johannsen, M. M., & Oliver, S. R. (2023). Non-shivering thermogenesis is differentially regulated during the hibernation season in Arctic ground squirrels. *Frontiers in Physiology*, *14*, 1207529. https://doi.org/10.3389/fphys.2023.1207529

Jonasson, K. A., & Willis, C. K. R. (2012). Hibernation energetics of free-ranging little brown bats. *Journal of Experimental Biology*, *215*(12), 2141–2149. https://doi.org/10.1242/jeb.066514

Karpovich, S. A., Tøien, Ø., Buck, C. L., & Barnes, B. M. (2009). Energetics of arousal episodes in hibernating Arctic ground squirrels. *Journal of Comparative Physiology B*, *179*(6), 691–700. https://doi.org/10.1007/s00360-009-0350-8

Kitao, N., & Hashimoto, M. (2012). Increased thermogenic capacity of brown adipose tissue under low temperature and its contribution to arousal from hibernation in Syrian hamsters. *American Journal of Physiology-Regulatory, Integrative and Comparative Physiology*, *302*(1), R118–R125. https://doi.org/10.1152/ajpregu.00053.2011

Kleiber, M. (1932). Body size and metabolism. *Hilgardia*, *6*(11), 315–353. https://doi.org/10.3733/hilg.v06n11p315

Kronfeld-Schor, N., Richardson, C., Silvia, B. A., Kunz, T. H., & Widmaier, E. P. (2000). Dissociation of leptin secretion and adiposity during prehibernatory fattening in little brown bats. *American Journal of Physiology-Regulatory, Integrative and Comparative Physiology*, *279*(4), R1277– R1281. https://doi.org/10.1152/ajpregu.2000.279.4.R1277

Laver, D. R. (2007a). Ca2+ stores regulate Ryanodine receptor Ca2+ release channels via luminal and cytosolic Ca2+ sites. *Clinical and Experimental Pharmacology and Physiology*, *34*(9), 889–896. https://doi.org/10.1111/j.1440-1681.2007.04708.x

Laver, D. R. (2007b). Ca2+ stores regulate ryanodine receptor Ca2+ release channels via luminal and cytosolic Ca2+ sites. *Biophysical Journal*, *92*(10), 3541–3555. https://doi.org/10.1111/j.1440-1681.2007.04708.x

Lee, P., Swarbrick, M. M., & Ho, K. K. Y. (2013). Brown adipose tissue in adult humans: A metabolic renaissance. *Endocrine Reviews*, *34*(3), 413–438. https://doi.org/10.1210/er.2012-1081

Letunic, I., & Bork, P. (2007). Interactive Tree Of Life (iTOL): An online tool for phylogenetic tree display and annotation. *Bioinformatics*, *23*(1), 127–128. https://doi.org/10.1093/bioinformatics/btl529

Lidell, M.E. (2018). Brown adipose tissue in human infants. In: Pfeifer, A., Klingenspor, M., Herzig, S. (Eds) *Brown adipose tissue. Handbook of experimental pharmacology*, vol 251. Springer, Cham. https://doi.org/10.1007/164_2018_118

Lodhi, I. J., & Semenkovich, C. F. (2009). Why We Should Put Clothes on Mice. *Cell Metabolism*, *9*(2), 111–112. https://doi.org/10.1016/j.cmet.2009.01.004

Long, R. A., Martin, T. J., & Barnes, B. M. (n.d.). *Body temperature and activity patterns in free-living Arctic ground squirrels*. https://academic.oup.com/jmammal/article/86/2/314/891514

Mejsnar, J., & Janský, L. (1970). Shivering and nonshivering thermogenesis in the bat (Myotis myotis Borkh.) during arousal from hibernation. *Canadian Journal of Physiology and Pharmacology*, *48*(2), 102–106. https://doi.org/10.1139/y70-017

Montani, J.P., Schutz, Y. & Dulloo, A.G. (2018). The contribution of Swiss scientists to the assessment of energy metabolism. *European Journal of Clinical Nutrition*, *72*, 665–679. https://doi.org/10.1038/s41430-018-0139-5

Morton, M., Maxwell, C., & Wade, C. (1974). Body size, body composition, and behavior of juvenile Belding ground squirrels. *The Great Basin Naturalist*, *34*(2), 121–134.

Muthu, P., Talent, J. M., Gryczynski, I., & Borejdo, J. (2008). Cross-bridge duty cycle in isometric contraction of skeletal myofibrils. *Biochemistry*, *47*(20), 5657–5667. https://doi.org/10.1021/bi7023223

Nedergaard, J., Bengtsson, T., & Cannon, B. (2010). Three years with adult human brown adipose tissue. *Annals of the New York Academy of Sciences*, *1212*(1), E20–E36. https://doi.org/10.1111/j.1749- 6632.2010.05905.x

Nedergaard, J., & Cannon, B. (2018). *Brown adipose tissue as a heat-producing thermoeffector* In: Romanovsky, A. A. (Ed) *Handbook of Clinical Neurology*, vol 156, pp. 137–152 Elsevier. https://doi.org/10.1016/B978-0-444-63912-7.00009-6

Nespolo, R. F., Mejias, C., & Bozinovic, F. (2022). Why bears hibernate? Redefining the scaling energetics of hibernation. *Proceedings of the Royal Society B: Biological Sciences*, *289*(20220456). https://doi.org/10.1098/rspb.2022.0456

Nicholls, D. G. (2021), Mitochondrial proton leaks and uncoupling proteins. *Biochim Biophys Acta Bioenergetics* *1862*(7), 148428. https://doi.org/10.1016/j.bbabio.2021.148428

Obregón, M. J., Jacobsson, A., Kirchgessner, T., Schotz, M. C., Cannon, B., & Nedergaard, J. (1989). Postnatal recruitment of brown adipose tissue is induced by the cold stress experienced by the pups. An analysis of mRNA levels for thermogenin and lipoprotein lipase. *Biochemical Journal*, *259*(2), 341–346. https://doi.org/10.1042/bj2590341

Oliver, S. R., Anderson, K. J., Hunstiger, M. M., & Andrews, M. T. (2019). Turning down the heat: Down-regulation of sarcolipin in a hibernating mammal. *Neuroscience Letters*, *696*, 13–19. https://doi.org/10.1016/j.neulet.2018.11.059

Pant, M., Bal N. C., & Periasamy, M. (2016). Sarcolipin: A Key Thermogenic and Metabolic Regulator in Skeletal Muscle. *Trends in Endocrinology & Metabolism*. *27*(12), 881–892. https://doi.org/10.1016/j.tem.2016.08.006

Periasamy, M., Herrera, J. L., & Reis, F. C. G. (2017a). Skeletal muscle thermogenesis and its role in whole body energy metabolism. *Diabetes & Metabolism Journal*, *41*(5), 327–327. https://doi.org/10.4093/dmj.2017.41.5.327

Periasamy, M., Maurya, S.K., Sahoo, S.K., Singh, S., Reis, F.C.G. & Bal, N.C. (2017b). Role of SERCA Pump in Muscle Thermogenesis and Metabolism. In *Comprehensive Physiology*, R. Terjung (Ed.). https://doi.org/10.1002/cphy.c160030

Porter, C. (2017). Quantification of UCP1 function in human brown adipose tissue. *Adipocyte*, *6*(2), 167– 174. https://doi.org/10.1080/21623945.2017.1319535

Porter, W. P., & Kearney, M. (2009). Size, shape, and the thermal niche of endotherms. *Proceedings of the National Academy of Sciences*, *106*(Supplement_2), 19666–19672. https://doi.org/10.1073/pnas.0907321106

Regan, M. D., Chiang, E., Martin, S. L., Porter, W. P., Assadi-Porter, F. M., & Carey, H. V. (2019). Shifts in metabolic fuel use coincide with maximal rates of ventilation and body surface rewarming in an arousing hibernator. *American Journal of Physiology. Regulatory, Integrative and Comparative Physiology*, *316*(6), R764–R775. https://doi.org/10.1152/ajpregu.00379.2018

Reinhardt, K.D. (2022). Hibernation. In: Vonk, J., Shackelford, T.K. (Eds) *Encyclopedia of Animal Cognition and Behavior*. pp. 3112–3117, Springer, Cham. https://doi.org/10.1007/978-3-319-55065-7_231

Rigano, K. S., Gehring, J. L., Evans Hutzenbiler, B. D., Chen, A. V., Nelson, O. L., Vella, C. A., Robbins, C. T., & Jansen, H. T. (2017). Life in the fat lane: Seasonal regulation of insulin sensitivity, food intake, and adipose biology in brown bears. *Journal of Comparative Physiology B*, *187*(4), 649–676. https://doi.org/10.1007/s00360-016-1050-9

Ríos, E. (2018). Calcium-induced release of calcium in muscle: 50 years of work and the emerging consensus. *Journal of General Physiology*, *150*(4), 521–537. https://doi.org/10.1085/jgp.201711959

Robbins, C. T., Lopez-Alfaro, C., Rode, K. D., Tøien, Ø., & Nelson, O. L. (2012). Hibernation and seasonal fasting in bears: The energetic costs and consequences for polar bears. *Journal of Mammalogy*, *93*(6), 1493–1503. https://doi.org/10.1644/11-MAMM-A-406.1

Rowland, L. A., Bal, N. C., & Periasamy, M. (2015). The role of skeletal-muscle-based thermogenic mechanisms in vertebrate endothermy. *Biological Reviews*, *90*(4), 1279–1297. https://doi.org/10.1111/brv.12157

Ruf, T., & Geiser, F. (2015). Daily torpor and hibernation in birds and mammals. *Biological Reviews*, *90*(3). https://doi.org/10.1111/brv.12137

Ruf, T., Giroud, S., & Geiser, F. (2022). Hypothesis and Theory: A two-process model of torpor-arousal regulation in hibernators. *Frontiers in Physiology*, *13*, 901270. https://doi.org/10.3389/fphys.2022.901270

Rydell-Törmänen, K., & Johnson, J. R. (2019). The applicability of mouse models to the study of human disease. *Mouse cell culture: methods and protocols*, 3–22. Springer Nature. https://doi.org/10.1007/978-1-4939-9086-3_1

Sahoo, S. K., Shaikh, S. A., Sopariwala, D. H., Bal, N. C., & Periasamy, M. (2013). Sarcolipin protein interaction with sarco(endo)plasmic reticulum Ca2+ ATPase (SERCA) is distinct from phospholamban protein, and only sarcolipin can promote uncoupling of the SERCA pump. *Journal of Biological Chemistry*, Mar 8; *288*(10), 6881–9. doi: 10.1074/jbc.M112.436915

Salido, G. M., Sage, S. O., & Rosado, J. A. (2009). Biochemical and functional properties of the store- operated Ca2+ channels. *Cellular Signalling*, *21*(4), 457–461. https://doi.org/10.1016/j.cellsig.2008.11.005

Shaikh, S. A., Sahoo, S. K., & Periasamy, M. (2016). Phospholamban and sarcolipin: Are they functionally redundant or distinct regulators of the Sarco(Endo)Plasmic Reticulum Calcium ATPase? *Journal of Molecular and Cellular Cardiology 91*, 81–91. https://doi.org/10.1016/j.yjmcc.2015.12.030

Sheriff, M. J., Fridinger, R. W., Tøien, Øivind, Barnes, B. M., & Buck, C. L. (2013). Metabolic rate and prehibernation fattening in free-living Arctic ground squirrels. *Physiological and Biochemical Zoology*, *86*(5), 515–527. https://doi.org/10.1086/673092

Staples, J. F. (2016). Metabolic Flexibility: Hibernation, Torpor, and Estivation. In *Comprehensive Physiology*. R. Terjung (Ed.). *6*(2). Wiley. https://doi.org/10.1002/cphy.c140064

Swoap, S. J., Gutilla, M. J., Liles, L. C., Smith, R. O., & Weinshenker, D. (2006). The full expression of fasting-induced torpor requires β3-Adrenergic receptor signaling. *The Journal of Neuroscience*, *26*(1), 241–245. https://doi.org/10.1523/JNEUROSCI.3721-05.2006

Symonds, M. E. (2013). Brown adipose tissue growth and development. *Scientifica*, *2013*, 305763. https://doi.org/10.1155/2013/305763

Symonds M. E., Pope M., & Budge H. (2015). The Ontogeny of Brown Adipose Tissue. *Annual Review of Nutrition 35*, 295–320. doi: 10.1146/annurev-nutr-071813-105330

Tøien, Ø., Blake, J., & Barnes, B. M. (2015). Thermoregulation and energetics in hibernating black bears: Metabolic rate and the mystery of multi-day body temperature cycles. *Journal of Comparative Physiology B*, *185*(4), 447–461. https://doi.org/10.1007/s00360-015-0891-y

Tøien, Ø., Drew, K. L., Chao, M. L., & Rice, M. E. (2001). Ascorbate dynamics and oxygen consumption during arousal from hibernation in Arctic ground squirrels. *American Journal of Physiology- Regulatory, Integrative and Comparative Physiology*, *281*(2). https://doi.org/10.1152/ajpregu.2001.281.2.R572

Trevathan, W. R. (2017). *Human Birth* (W. R. Trevathan, Ed.). Routledge. New York, NY. https://doi.org/10.4324/9780203789599

Vandamme, T. F. (2014). Use of rodents as models of human diseases. *Journal of Pharmacy and Bioallied Sciences*, *6*(1), 2–9. https://doi.org/10.4103/0975-7406.124301

Vargas-Castillo, A., Torres, N., & Tovar, A. R. (2021). Endocrine regulation of brown and beige adipose tissue. In Uloa-Aguirre, A. & Tao, Y.-X. (Eds). *Cellular Endocrinology in Health and Disease*, 247–259. Academic Press. https://doi.org/10.1016/B978-0-12-819801-8.000

Várnai, P., Hunyady, L., & Balla, T. (2009). STIM and Orai: The long-awaited constituents of store- operated calcium entry. *Trends in Pharmacological Sciences*, *30*(3), 118–128. https://doi.org/10.1016/j.tips.2008.11.005

Vig, M., Peinelt, C., Beck, A., Koomoa, D. L., Rabah, D., Koblan-Huberson, M., Kraft, S., Turner, H., Fleig, A., Penner, R., & Kinet, J.-P. (2006). CRACM1 is a plasma membrane protein essential for store-operated Ca2+ Entry. *Science*, *312*(5777), 1220–1223. https://doi.org/10.1126/science.1127883

Wang, L. C. H., & Wolowyk, M. W. (1988). Torpor in mammals and birds. *Canadian Journal of Zoology*, *66*(1). https://doi.org/10.1139/z88-017

Wang, Y., Zhu, T., Ke, S., Fang, N., Irwin, D. M., Lei, M., Zhang, J., Shi, H., Zhang, S., & Wang, Z. (2014). The Great Roundleaf Bat (Hipposideros armiger) as a good model for cold-induced browning of intra-abdominal white adipose tissue. *PLoS ONE*, *9*(11), e112495. https://doi.org/10.1371/journal.pone.0112495

Williams, C. T., Barnes, B. M., & Buck, C. L. (2012). Daily body temperature rhythms persist under the midnight sun but are absent during hibernation in free-living Arctic ground squirrels. *Biology Letters*, *8*(1), 31–34. https://doi.org/10.1098/rsbl.2011.0435

Winther, A. M., Bublitz, M., Karlsen, J. L., Møller, J. V., Hansen, J. B., Nissen, P., Buch-Pedersen, M. J. (2013). The sarcolipin-bound calcium pump stabilizes calcium sites exposed to the cytoplasm. *Nature*, *495*(7440), 265–9. https://doi.org/10.1038/nature11900

Wu, C.-W., Biggar, K. K., & Storey, K. B. (2013). Biochemical adaptations of mammalian hibernation: Exploring squirrels as a perspective model for naturally induced reversible insulin resistance. *Brazilian Journal of Medical and Biological Research*, *46*(1), 1–13. https://doi.org/10.1590/1414-431X20122388

Yan, X., Niu, Q., Gao, X., Shen, S., He, N., Wang, H., Fang, R., Gao, Y., & Chang, H. (2021). Differential protein metabolism and regeneration in gastrocnemius muscles in high-fat diet fed mice and pre-hibernation Daurian ground squirrels: A Comparison between Pathological and Healthy Obesity. *Zoological Studies*, *60*, e6. https://europepmc.org/article/med/34386092

Zhang, J., Li, X., Ismail, F., Xu, S., Wang, Z., Peng, X., Yang, C., Chang, H., Wang, H., & Gao, Y. (2019). Priority strategy of intracellular Ca2+ homeostasis in skeletal muscle fibers during the multiple stresses of hibernation. *Cells*, *9*(1), 42–42. https://doi.org/10.3390/cells9010042

Unit 2

Skeletal Muscle Non-Shivering Thermogenesis

Evolutionary, Mechanistic, and Ecological Aspects

OVERVIEW

Endothermic vertebrates (including some fishes, mammals, and birds) possess internal mechanisms to produce heat that assist maintenance of core body temperature (Tc). Birds and mammals exhibit both shivering thermogenesis (ST) and non-shivering thermogenesis (NST). As discussed in Unit 1, brown adipose tissue (BAT) specific to mammalian clades, including humans, has been demonstrated as a sophisticated tissue of NST. However, all birds and several mammalian clades do not have functional BAT and rely mostly on skeletal muscle for their ST and NST. In practice, however, the visible process of ST ceases after a few hours during a prolonged cold exposure. Whether skeletal muscle has a NST component has been a subject of discussion for quite some time now. Recent data in mammals show that skeletal muscles are the primary source of NST recruited during cold exposure. Another advantage of such recruitment is that skeletal muscle represents ~40–45% of body mass and can be effective in increasing energy expenditure and whole-body metabolism. All contributors to this unit are active researchers in their specific fields on various aspects of skeletal muscle. It is our goal that readers should firmly grasp the content and build a clear sense of how key discoveries were made, what interesting questions remain, and how they may be answered.

Fish were among the earliest vertebrates to evolve, and some fish have a modified skeletal muscle, called a "heater organ," to produce heat, enabling them to achieve regional endothermy. Chapter 5 deals with molecular mechanisms of NST in the heater organ of various fish. Next, the Chapter 6 presents an interesting approach to understanding the whole body, i.e., tachymetabolic endothermy. An extensive survey of the paleontological literature using established proxies is presented in this chapter. Chapter 7 focuses on the understanding of avian mechanisms of muscle NST. Birds are of

DOI: 10.1201/9781003391470-6

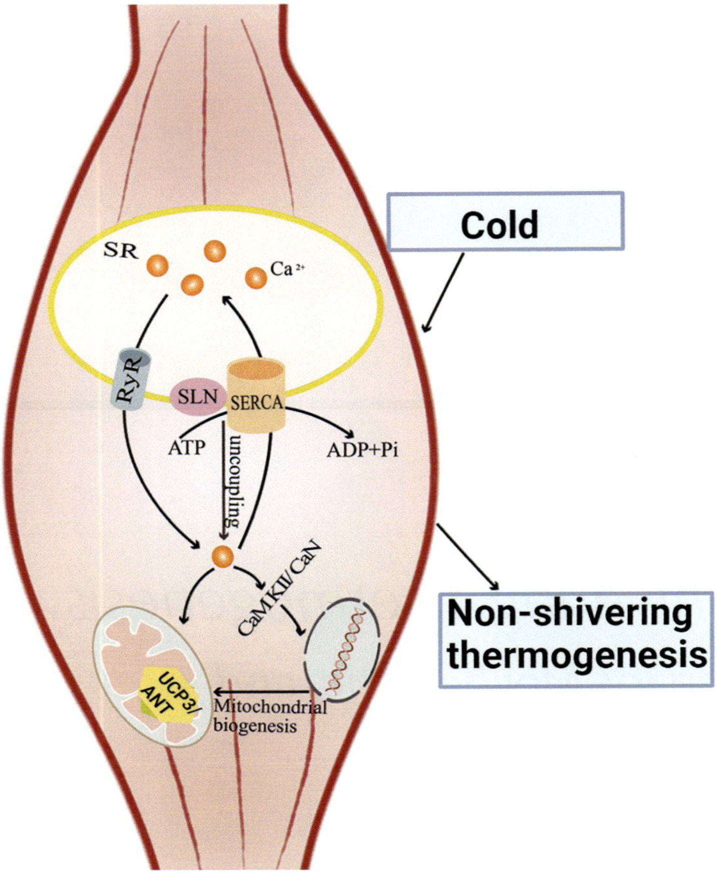

particular interest in studies of NST, as they lack BAT, the major contributor to NST in mammals. Here the authors attempt to provide a better understanding of the mechanistic details of avian muscle NST, especially during cold adaptation, and the process of regulation of the same. Over the past several decades, the major debate has revolved around whether muscle NST can be activated without contraction. Interestingly, research in last few years has shown that in resting phases of the muscle, Ca^{2+} leak through a ryanodine receptor (RyR)s is involved in thermogenesis, setting the activity of the **S**arco **E**ndoplasmic **R**eticulum **C**a^{2+} **AT**Pase (SERCA) and ATP utilization. Chapter 8 aims to show the readers the role of Ca^{2+} leak through RyR1 in resting muscle. Chapter 9 describes in further detail the mechanism of SERCA-based heat production and its recruitment by neural circuits. This chapter presents the allostatic induction of muscle thermogenesis as relevant to energy balance and evidences its relevance in humans. Finally, Chapter 10 describes SERCA and its dynamic regulation by sarcolipin in muscle NST and its interplay with the BAT function.

5 Endothermy in Fish
A Remarkable Story of Convergent Evolutionary Adaptation

Jens P.C. Franck and Sean W.M. Robinson
University of Winnipeg, Winnipeg, Canada

5.1 INTRODUCTION

Less than 0.1% of fish species have the physiological and anatomical capability to generate and conserve metabolic heat (Dickson & Graham, 2004). The challenge of conserving metabolic heat is greater for fish, as the conductive property of water transfers heat from the fish body. The challenge for fish is exacerbated by the direct exposure of the gills to water for oxygen transfer. Most species of endothermic fish are characterized as regional endotherms, such as cranial endothermy (CE), which elevates temperatures that benefit the brain and ocular regions. The cranial endothermy adaptation is considered a minimum requirement for endothermic fish and has evolved independently in several lineages of large pelagic predators such as billfishes (Istiophoridae and Xiphiidae), tunas (Scombridae), lamnid sharks (Lamnidae), opahs (Lampridae), and the common thresher shark (*Alopias vulpinus*) (Alexander, 1998; Bernal & Sepulveda, 2005; Carey & Teal, 1969b; Don Stevens & Fry, 1971; Linthicum & Carey, 1972a; Wegner et al., 2015; Weng & Block, 2004). Increasing and maintaining elevated cranial temperatures benefit visual acuity and central nervous system function when foraging in colder waters below the thermocline, as well as when migrating long distances (Carey et al., 1981; Dickson & Graham, 2004; Friedlander et al., 1976). Swordfish have been documented to experience thermal changes of 19°C within a 2 hour time frame (Carey & Robinson, 1981) and observed diel vertical movements that span 8 hours in sub-thermocline waters (600 m at ~6°C) while maintaining constant cranial temperatures (~28°C) (Carey & Robinson, 1981; Madigan et al., 2021). Other CE species will also make daily dives beneath the thermocline; however, few match the depths and duration reached by the big eye thresher shark (Weng & Block, 2004). In addition to diving below the thermocline, large pelagic predators can perform yearly migrations up to ~8000 km. Much like diving below the thermocline, long-distance migrations result in traversing cool ocean temperatures (~14°C) (Fujioka et al., 2018). Pacific bluefin tuna (*Thunnus orientalis*) perform transcontinental migrations from their spawning grounds in the East Pacific Ocean, specifically the Eastern China Sea and Sea of Japan, to the West Pacific Ocean (Fujioka et al., 2018).

Although endothermic fish share several morphological adaptations that permit them to create and conserve metabolic heat, multiple studies indicate that these are convergent adaptations and are not derived traits. Block and colleagues in an early molecular study mapped the occurrence of endothermy on a phylogenetic tree of the Scombroidei assemblage of fish, which includes the tuna and billfish (Block et al., 1993). This study revealed that within the Scombroidei, endothermy evolved at least three separate times. Cranial endothermy, utilizing the thermogenic heater organ, evolved twice: in the Xiphidae family (swordfish) and in the Istiophoridae family of billfish (marlin). Block and colleagues argue that the multiple occurrences of endothermy in distinct lineages of fishes supports a strong selective pressure favoring niche expansion over increased aerobic capacity. The thermal expansion theory dates back to early studies on Atlantic bluefin tuna by Carey and colleagues (Carey & Lawson, 1973; Carey & Teal, 1969b). The thermal niche expansion theory has been

DOI: 10.1201/9781003391470-7

challenged by others, however, Dickson and Graham argue that endothermy alone is not sufficient to permit continued dives below the thermocline and that the benefit of endothermy is offset by its energetic costs (Dickson & Graham, 2004). In a meta study of 89 published tagging studies of 41 species of fish that included 14 regional endotherms it was found that the endothermic fish do not exploit greater temperature ranges than ectothermic fish (Harding et al., 2021). The same study did find that endothermic fish swim faster than ectotherms, which would potentially benefit these fish for long migratory swims or provide an advantage for prey capture. An *ex vivo* study of muscle preparations from yellowfin tuna (*Thunnus albacares*) and the ectothermic eastern Pacific bonito (*Sarda chiliensis*) showed that the deep red slow-twitch muscle from yellowfin tuna develops 40% more power at 30°C than at 25°C. Bonito slow muscle tissue explants, in comparison, showed impaired performance at higher temperatures (Altringham & Block, 1997). Altringham and Block argue that the higher power output and large amount of slow-twitch muscle mass would permit the yellowfin tuna to maintain higher sustainable speeds for longer durations than the ectothermic bonito. Because endothermic fish are often large pelagic fish, it is inherently difficult to experimentally evaluate competing theories on the physiological advantage conferred by endothermy. For this reason, it is beyond the scope of this review to support or refute the niche expansion theory.

5.2 ANATOMICAL ADAPTATIONS

5.2.1 Heater Organ

The heater organ (HTR) is a highly specialized tissue derived from the extraocular muscle tissue in billfish (Istiophoridae; swordfish, marlin, spearfish, sailfish) and two species of Scombridae, butterfly mackerel, and slender tuna (Dickson & Graham, 2004) (Figure 5.1). While the HTR in all cases originates from the extraocular eye muscle (EM), the specific EM muscle is not uniform across species (Block, 1986; Morrissette et al., 2003). Billfish HTR is derived from the superior rectus extraocular muscle, the butterfly mackerel HTR is derived from the lateral rectus, and slender tuna has the HTR formed from a fusion of all four extraocular muscles (Carey, 1982; Sepulveda et al., 2007). The HTR is differentiated from extraocular muscle by the absence of contractile muscle fibers and the retention of extensive t-tubule and high mitochondria concentration (60% of HTR is comprised of mitochondria) (Morrissette et al., 2003; Tullis et al., 1991).

This cellular composition provides sufficient energy to sustain non-shivering thermogenesis (NST). NST provides organisms with a strategy to increase their basal metabolic rates producing

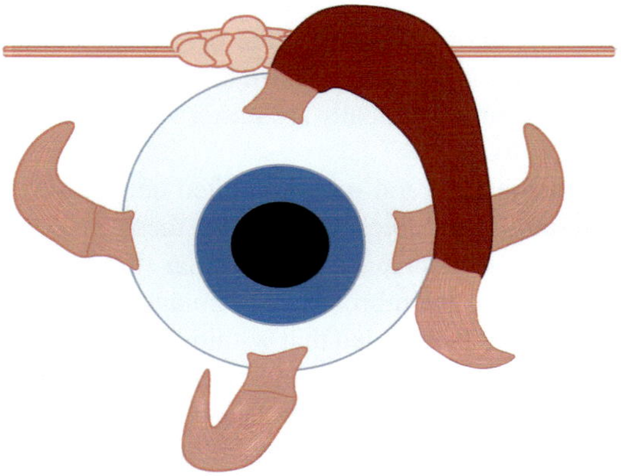

FIGURE 5.1 Heater organ. The billfish heater organ (red) is derived from the superior rectus eye muscle. It is enriched with mitochondria and devoid of contractile proteins.

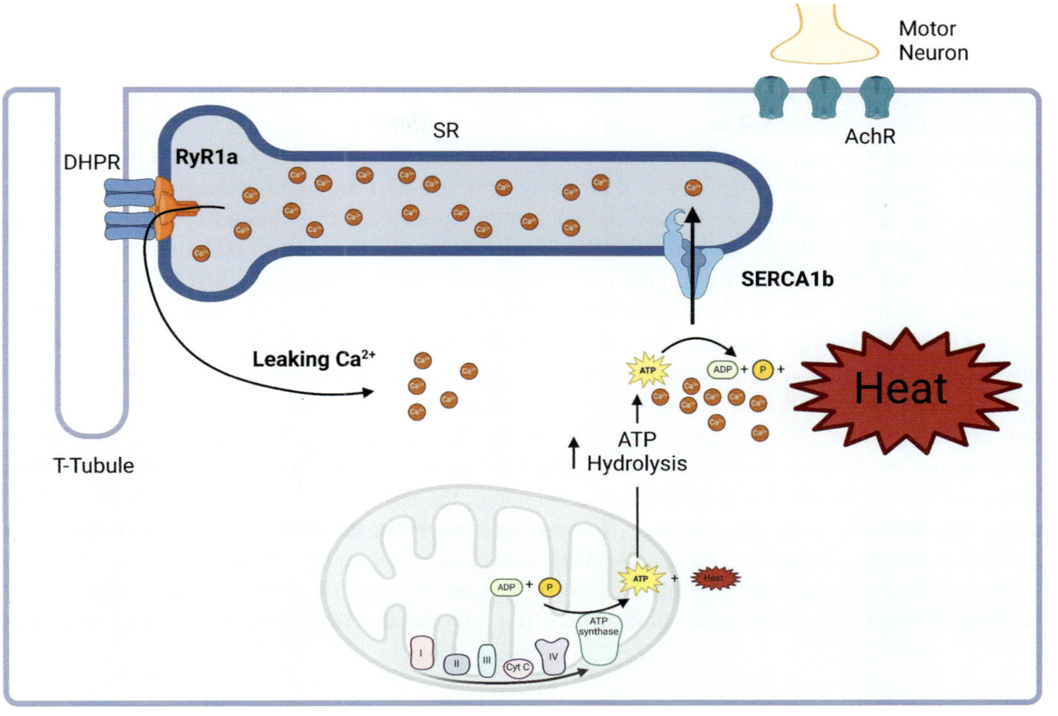

FIGURE 5.2 Excitation-thermogenic coupling model for heater organ tissue. Stimulation of the postsynaptic membrane in heater cells results in DHPR-RyR1a–mediated calcium release. The increased "leakiness" of the RyR1a channel stimulates an increase in calcium transport by SERCA1B resulting in metabolic heat. Adapted from Morrissette et al. (2003).

heat as a by-product. The mechanism of NST within the HTR has been proposed to be due to a "leaky" ryanodine receptor (RyR1a) (Morrissette et al., 2003). RyR1a allows Ca^{2+} to "leak" from the sarcoplasmic reticulum (SR) into the cytosol devoid of central nervous system stimulation. The resulting increased cytosolic Ca^{2+} concentration initiates sarcoplasmic reticulum Ca^{2+} ATPase isoform1b (SERCA1b) to actively transport Ca^{2+} back into the SR, producing heat as a by-product (Figure 5.2). Interestingly, the SERCA1b isoform is distinctly expressed in the HTR and has an enzyme activity like the toadfish swim bladder, a high-frequency contracting muscle known for its rapid calcium-sequestering rate (Morrissette et al., 2003).

In addition, maintaining constant tissue temperatures positively impacts heat production via NST, as calcium pump activity will increase as temperatures increase in the absence of an established Ca^{2+} gradient (da Costa & Landeira-Fernandez, 2009). Taken together, the 'leaky' RyR1a, SERCA1b isoform and abundant mitochondria provide the HTR with the cellular machinery to energetically finance consistent cranial temperatures to combat frigid ocean water. The role for NST in endothermy of fish and evidence for its molecular basis will be discussed in greater detail later in this chapter.

An alternate mode of cranial endothermy in certain fish species, including tunas, lamnid sharks, and manta rays, increases brain and eye temperature by diverting heat generated from skeletal muscle, specifically slow-oxidative red skeletal muscle (RM), used for continuous swimming (Burne, 1923; Carey & Teal, 1969a). To efficiently transport heat, these fish require adaptations to reduce the amount of heat lost from the RM to the cranium. These fish species are continuous swimmers and utilize RAM ventilation, which describes a mode of respiration in which water is continuously passed through the mouth and over the gills (Wegner et al., 2013).

5.2.2 Countercurrent Heat Exchanger (Rete Mirabile)

To prevent heat loss at the site of respiration, a vascular adaptation known as *rete mirabile* ("wonderful nets") is present in many species of endothermic fish. Rete mirabile act as vascular countercurrent heat exchangers and allow for the retention of metabolically generated heat, as warm venous blood will transfer heat to cooler arterial blood before reaching the gills. One of the first studies to characterize the elaborate anatomy of the rete system was by Carey and colleagues (1971). In this study of bigeye tuna (*Thunnus obesus*) the alternating cutaneous veins and arteries are described as well as the presence of arteries and veins that branch off at right angles into the dark red muscle. In the mako shark (*Isurus oxyrinchus*) the red muscle vein, located behind the level of the first dorsal fin, carries warm venous blood to the cranium, derived from the red muscle vein (RM) (Wolf et al., 1988) (Figure 5.3). This vein acts as a drain for the anterior half of the RM. As it makes its way toward the brain and eyes, the red muscle vein contains venules extending laterally that run parallel with arteries, specifically the lateral cutaneous vein and artery, together establishing the first rete-conserving RM heat (Wolf et al., 1988). Close to the last gill slit, the red muscle vein exits the red muscle and enters the neural canal, where it loses the vascular connection with the lateral cutaneous rete (Wolf et al., 1988). Once inside the neural canal, the red muscle vein joins the myelonal vein along the dorsal side of the spinal cord and enters the brain through an enlarged foramen. The myelonal vein then branches into smaller tributaries that cover and warm the brain and eyes (Wolf et al., 1988). To conserve heat in this area, warm venous blood exiting the cranium is drained into a sinus, specifically the suborbital sinus where the suborbital rete is located (Wolf et al., 1988).

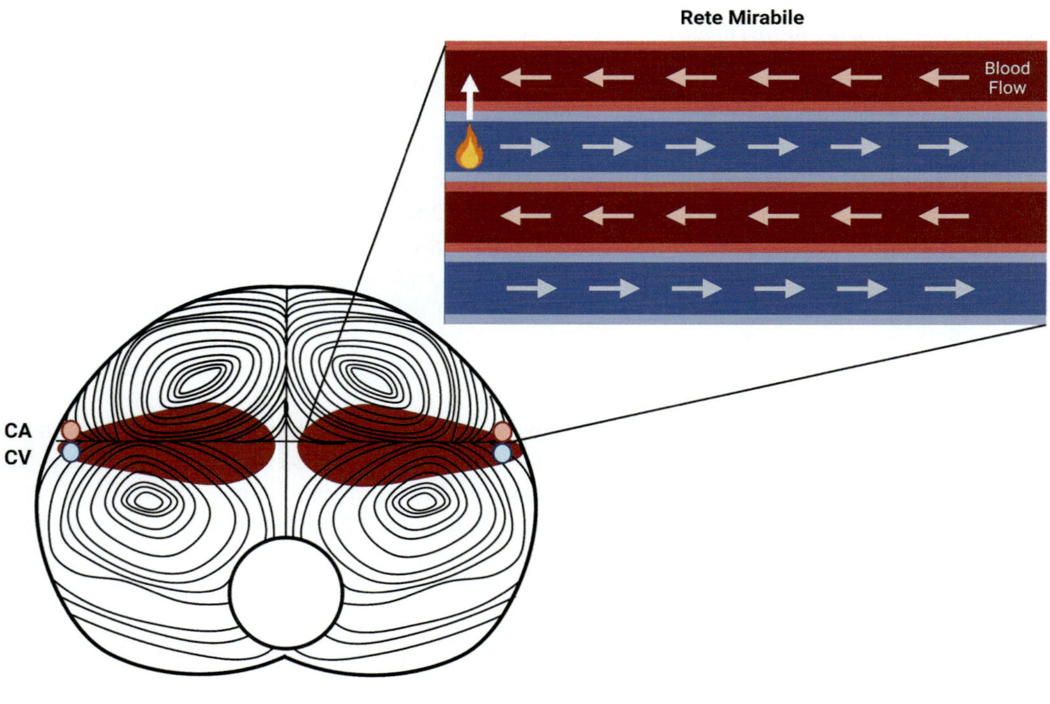

FIGURE 5.3 Mako shark rete mirabile. The cutaneous artery (CA) and vein (CV) run adjacent to each other. Heat transferred from the metabolically active red muscle via the red muscle vein joins the myelonal vein that delivers the warmed blood to a network of veins that surround the brain. In addition, tributaries extend inward and adjacent to one another from the CA and CV along the myotomes, reducing the amount of lateral heat transfer.

Unlike the lateral cutaneous rete, the suborbital rete bathes arteries within the rete before returning to the heart, thus conserving heat within the cranium (Wolf et al., 1988).

It is noteworthy that not all rete are assembled in the same manner. The rete mirabile of opah is an arrangement of alternating afferent (deoxygenated) and efferent (oxygenated) arteries rather than the alternating arteries and veins of the mako shark (Wegner et al., 2015). The opah rete is located in fat-insulated gill arches. The cold oxygenated blood is warmed by the heat of the deoxygenated blood returning to the respiratory exchange surfaces (gills) before being distributed to the fish body.

The presence of retia is not the only anatomical adaptation cranial endotherms have adopted to conserve metabolic heat. Other adaptations include increased fatty adipose tissue surrounding the eyes and brain and medially positioned RM (Linthicum & Carey, 1972b). Increased fatty tissue is used to insulate the eyes and brain by reducing convective heat transfer. Fatty tissue is most abundant in billfishes and Scombrids, with decreasing amounts observed in sharks and manta rays. This is reflective of the cranial temperatures within each species, as an increased number of fatty tissues is indicative of greater cranial temperatures.

5.2.3 Internalized Red Muscle

Medially positioned red muscle tissue, or internalized red muscle, is another anatomical adaptation of endothermic fish (Figure 5.4a). Red muscle in ectothermic fish is typically found in more lateral positions (Figure 5.4b). The red muscle is shifted both internally and more anterior than that of ectothermic fish. Red muscle is required for continuous swimming by fish and is highly aerobic (Carey et al., 1971). For the endothermic shark species the relative anterior position of red muscle varies, with the salmon shark (*Lamna ditropis*) having the most anterior placement (Bernal et al., 2003). Of the three thresher shark species, only the endothermic pelagic thresher (*Alopias vulpinus*) has internalized red muscle similar to other endothermic fish lineages (Bernal & Sepulveda, 2005; Sepulveda et al., 2005). The centralized red muscle location that has been converged on in both lamnid sharks and tunas confers a unique swimming style of stiff-bodied undulatory swimming referred to as *thunniform swimming* (Shadwick & Goldbogen, 2012; Syme & Shadwick, 2011).

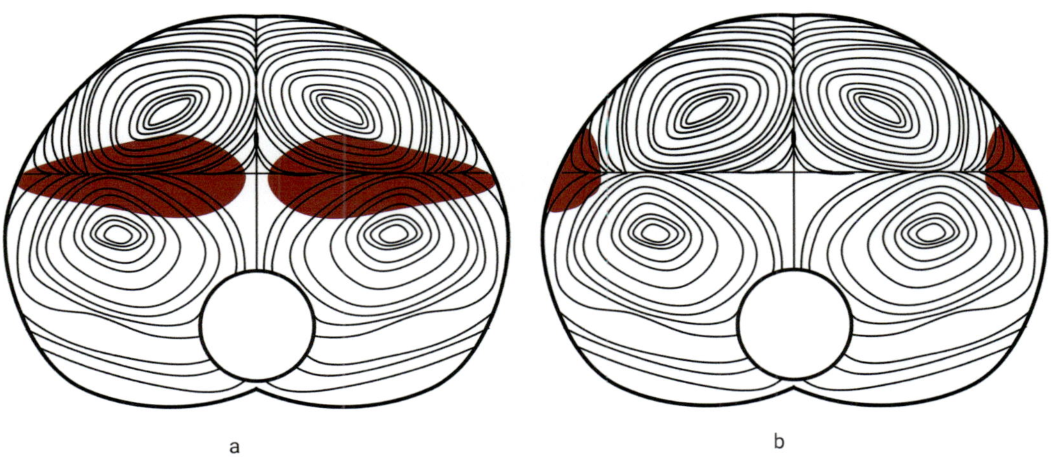

FIGURE 5.4 Red muscle placement. Aerobic red muscle is internalized in endothermic fish (a) compared to the lateral placement typical of ectothermic fish (b).

5.3 MOLECULAR BASIS FOR ENDOTHERMY IN FISH

Research in our laboratory and others identifies slow oxidative red muscle as the primary source of metabolic heat in endothermic fish. In our 2019 study on the endothermic smalleye opah (*Lampris incognitus*) we investigated the potential molecular basis of endothermic heat (Franck et al., 2019). A previous study by Wegner and colleagues showed that smalleye opah produce heat in their deep pectoral muscle, which is used for continuous swimming behavior (Wegner et al., 2015). We focused our study on the potential role of sarcolipin (SLN), a small transmembrane peptide in the sarcoplasmic reticulum which reduces the affinity of SLN for Ca^{2+} resulting in a "slippage" of Ca^{2+} back into the cytosol. The SERCA1 Ca^{2+} ATPase pump increases activity to compensate for the slippage resulting in increased ATP hydrolysis and heat production (Bal et al., 2012; Bal & Periasamy, 2020). Sarcolipin transcript levels in the dark red pectoral muscle of opah were found to be threefold higher than transcripts in light red muscle and white muscle supporting the role of futile calcium cycling as the source of metabolic heat in opah. Wu and colleagues conducted a genomic analysis of genes associated with the evolutionary adaptation of endothermy in selected species including mackerel tuna (*Euthynnus affinis*), Pacific bluefin tuna (*Thunnus orientalis*), sailfish (*Istiphorus platypterus*), and opah (*Lampris incognitus*) (Wu et al., 2023). Based on a comparative RNA-seq analysis, genes associated with NST, such as *sln*, were found to be upregulated in the heater tissue of sailfish but lower expressed in tuna red muscle. Contrarily, genes associated with shivering thermogenesis (ST) are upregulated in the red muscle of tuna, arguing that ST has a greater role in thermogenesis in tuna red muscle and futile calcium cycling a greater role in the heater tissues of billfish. In another comparative study Wang and

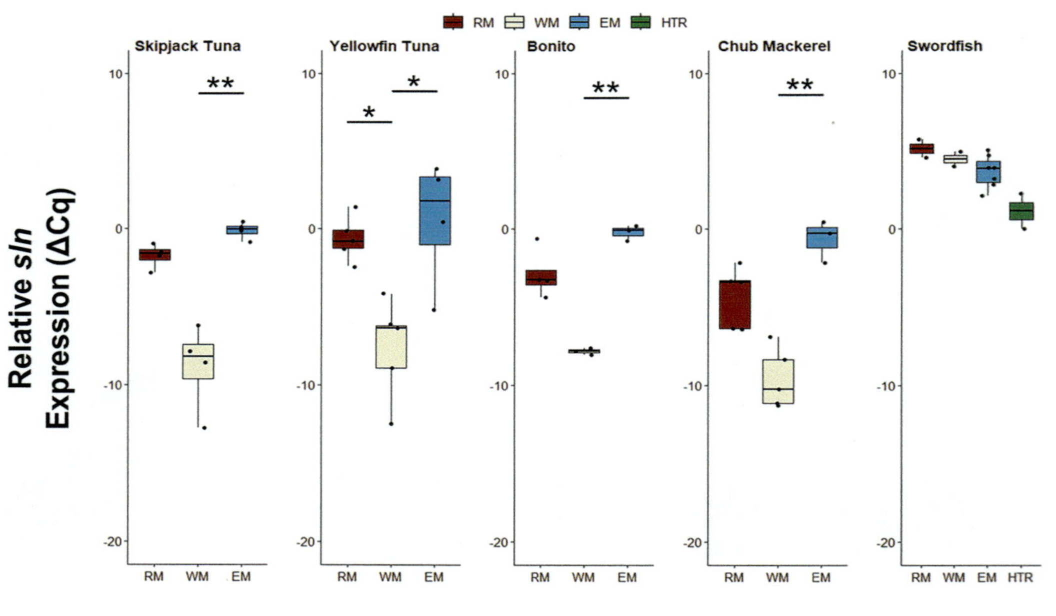

(a)

FIGURE 5.5 (a) *Sln* transcript levels measured relative to housekeeping gene *ef1α* (ΔCq: housekeeping – target gene) for slow-oxidative red skeletal muscle (RM), fast-glycolytic white skeletal muscle (WM), extraocular eye muscle (EM), and heater organ (HTR; exclusive to swordfish) in skipjack tuna (*Katsuwonus pelamis*: 1A), yellowfin tuna (*Thunnus albacares*: 1B), Eastern Pacific bonito (*Sarda chiliensis*: 1C), Pacific chub mackerel (*Scomber japonicus*: 1D), and swordfish (*Xiphias gladius*: 1E). $*P < 0.05$, $**P < 0.01$ (Robinson et al., 2024b).

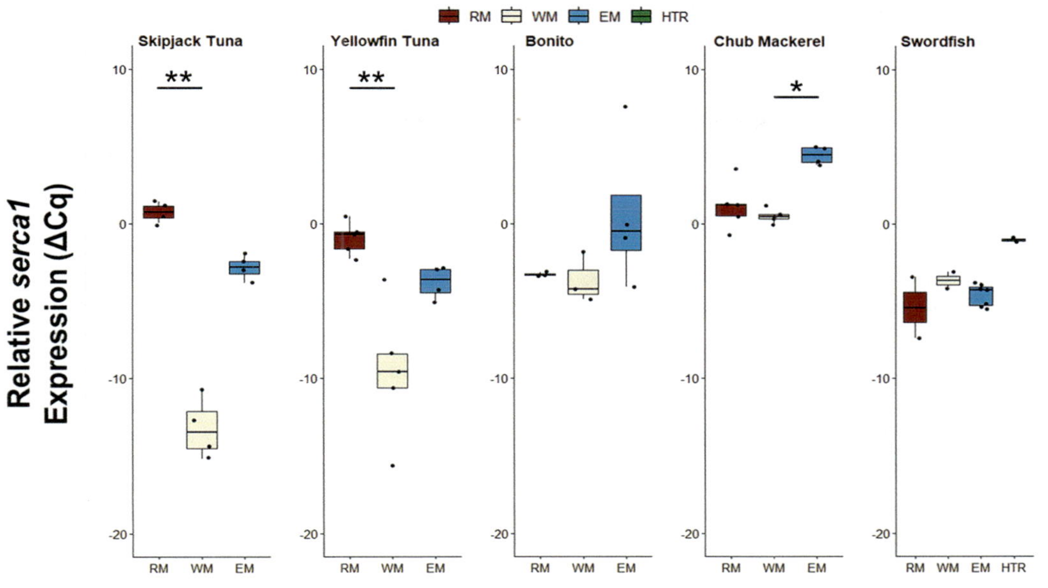

(b)

FIGURE 5.5 (b) *Serca1* transcript levels measured relative to housekeeping gene *ef1α* (ΔCq: housekeeping – target gene) for slow-oxidative red skeletal muscle (RM), fast-glycolytic white skeletal muscle (WM), extraocular eye muscle (EM), and heater organ (HTR; exclusive to swordfish) in skipjack tuna (*Katsuwonus pelamis*), yellowfin tuna (*Thunnus albacares*), Eastern Pacific bonito (*Sarda chiliensis*), Pacific chub mackerel (*Scomber japonicus*), and swordfish (*Xiphias gladius*). *$P < 0.05$, **$P < 0.01$ (Robinson et al., 2024b).

colleagues selected five endothermic species (opah, turkey, chicken, mouse, and human) to identify genes that show convergent evolution for the endothermy phenotype (Wang et al., 2022). Genes related to primary metabolic processes were enriched in the analysis, and further analysis of genes with convergent amino acid substitution identified a mitochondrial Na^+/Ca^{2+} exchanger that was restricted to whole-body endothermic mammals, birds, and opah and did not appear in the regional endothermic fish. Thirty-six genes were identified with adaptive mutations in the endothermic fish analyzed with many of the genes related to mitochondrial function.

We extended our studies to investigate *sln* and *serca1* transcript levels in muscle tissues derived from two endothermic scombrid species, skipjack tuna (*Katsuwonus pelamis*) and yellowfin tuna (*Thunnus albacares*), as well as two closely related ectothermic scombrids, eastern pacific bonito (*Sarda chiliensis*) and Pacific chub mackerel (*Scomber japonicus*) (Robinson et al., 2024b). We additionally investigated *sln* and *serca1* transcript levels in the muscle tissues of swordfish (*Xiphias gladius*), a cranial endotherm. We measured transcript levels in slow oxidative red muscle (RM), glycolytic white muscle (WM), and eye muscle (EM). We additionally measured transcript levels in the specialized heater tissue (HTR) of swordfish. Interestingly, the *sln* transcript levels in all four scombrid species examined trended or was significantly higher in the RM and EM tissues compared to WM (Figure 5.5). *Sln* transcript levels for RM, WM, EM, and HTR were not significantly different.

Transcript levels for the SERCA1 pump were significantly higher in RM compared to WM in both endothermic species and trended higher in the EM tissues. In contrast, *serca1* levels in the ectothermic scombrid species did not significantly differ between tissues except for EM *serca1* transcript expression in Pacific chub mackerel, which was significantly higher when compared to WM tissues (Figure 5.5b).

5.3.1 Key Regulatory Site Amino Acid Substitution in Sarcolipin

Sarcolipin amino acid sequences were aligned from representatives of Lampridae (opah), Xiphiidae (swordfish), and Scombridae (Figure 5.6). The fifth amino acid, threonine (T), is substituted by a nonpolar alanine (A) in the tunas and with a valine (V) residue in opah. Threonine 5 (T5) is a key regulatory site in sarcolipin. Phosphorylation of sarcolipin T5 relieves the inhibitory (uncoupling) effects of SLN on the SERCA pump (Bhupathy et al., 2009; Toyoshima et al., 2013). Rearrangement of the N-terminus upon phosphorylation of the T5 residue reduces the energy barrier for passage of ions (Cao et al., 2016). Deletion of the N terminal residues of SLN also results in loss of the uncoupling function (Sahoo et al., 2013). The loss of the T5 phosphorylation regulatory site for red muscle endotherms, tunas and opah, would presumably result in SLN having a larger uncoupling effect on the SERCA pump. Conversely, the presence of the T5 regulatory site in swordfish would reduce the uncoupling effect of SLN on the pump upon phosphorylation, further supporting the role of a leaky RyR1a calcium channel in the futile cycling mechanism as originally proposed by Morrissette and colleagues (Morrissette et al., 2003). The leaky RyR1a channel is also the likely primary mechanism for heat production in the swordfish cranial heater organ.

5.3.2 Transcriptional Machinery Required for Endothermy Conserved in Ectothermic Fishes

Recent meta studies on the evolution of endothermy highlight the fact that the SERCA-based NST in fish may be the ancestral mechanism of thermogenesis that predates UCP1-based uncoupling of the mitochondria. In this review we have emphasized the importance of anatomical adaptations, such as the countercurrent heat exchanger and internalized red muscle, of endothermic fish. Along

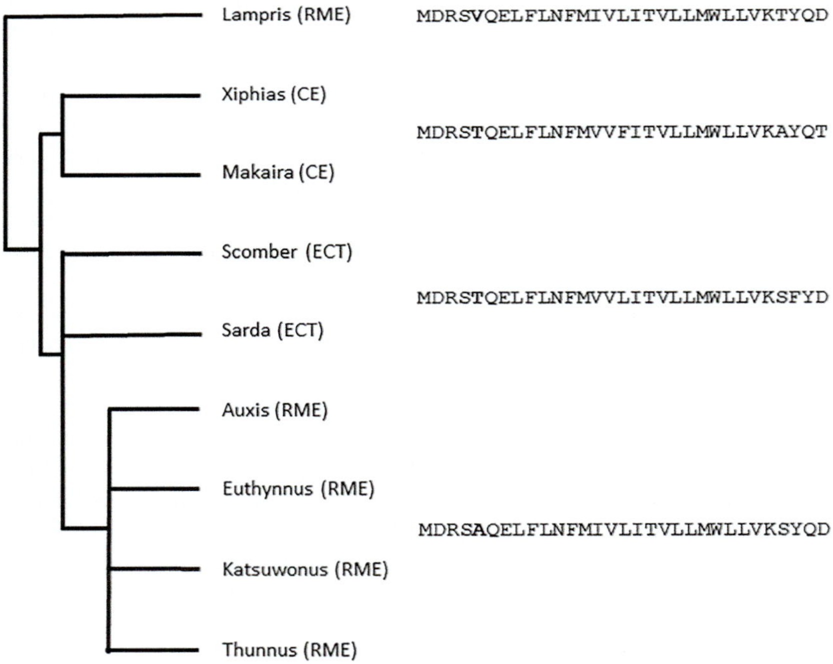

FIGURE 5.6. Sarcolipin amino acid substitution across diverse assemblage of endothermic and ectothermic species. The fifth amino acid residue, threonine (T), is substituted with valine (V) in opah (Lampris) and alanine (A) in the skipjack (Katsuwonus), which are both red muscle–based endotherms (RME). The swordfish (Xiphias), a cranial endotherm (CE), retains the threonine residue, as do the ectothermic Scomber and Sarda species (ECT). Phylogeny topology adapted from Davesne et al. (2018).

with others, we have shown that ectothermic fish that lack these adaptations still possess the transcriptional machinery that is upregulated upon cold challenge. Specifically, we showed that cold-adapted medaka (*Oryzias latipes*) increase the expression of both *sln* and *serca1* transcripts, both critical components of the futile calcium cycling mechanism (Robinson et al., 2024a). Medaka, cold-adapted to 8°C, increased *serca1* transcript expression approximately fourfold (Figure 5.7a) and *sln* transcript expression approximately sevenfold compared to fish held at 24°C (Figure 5.7b). In another ectothermic fish model species, zebrafish, Little and Seebacher (2013), studying the role

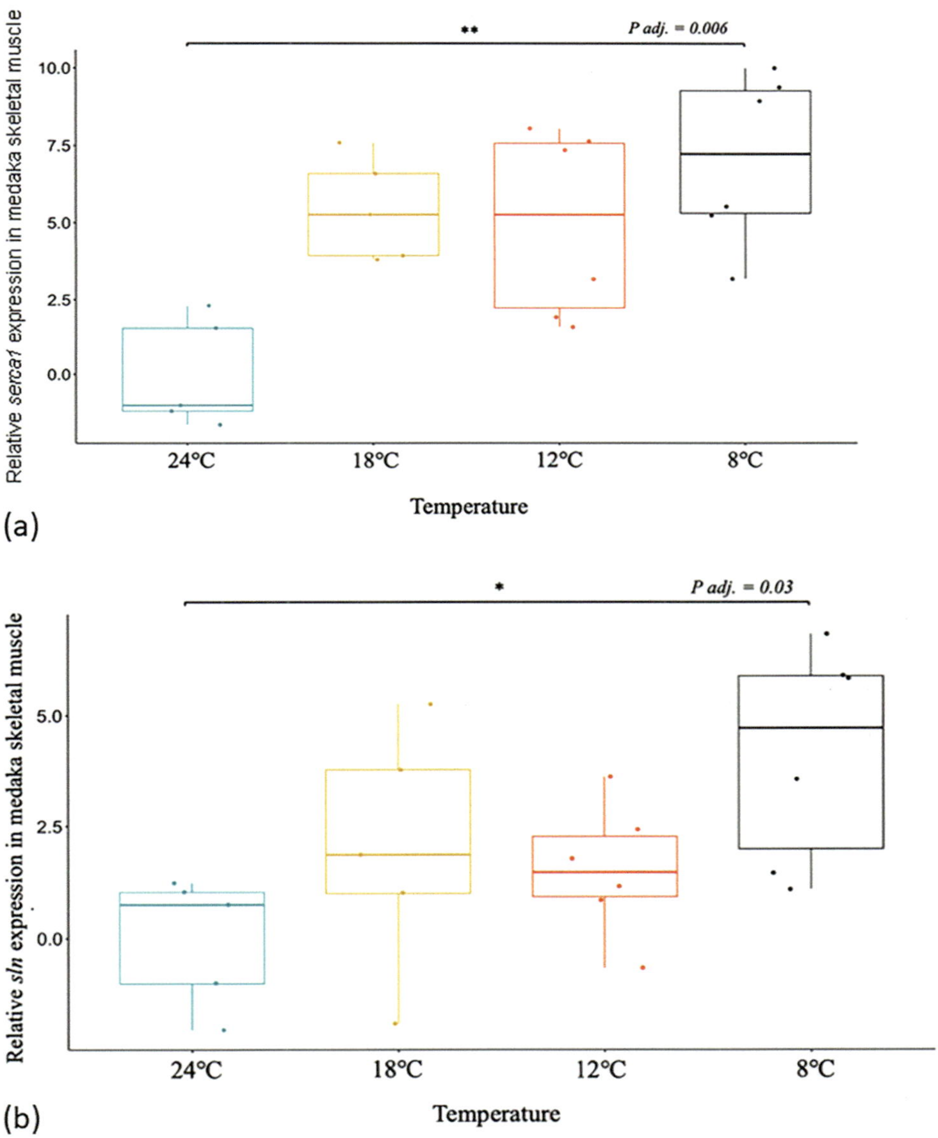

FIGURE 5.7 (a) *Serca1* transcript expression relative to housekeeping gene *ef1α* in Japanese medaka (*Oryzias latipes*) skeletal muscle acclimated to 24°C, 18°C, 12°C, and 8°C. Kruskal-Wallis was used, and results are expressed as boxplots and compared to 24°C. n = 5–6 biological replicates per group; $^{**}P < 0.01$. (b) Sln transcript expression relative to housekeeping gene *ef1α* in Japanese medaka (*Oryzias latipes*) skeletal muscle acclimated to 24°C, 18°C, 12°C and 8°C. Kruskal-Wallis was used and results are expressed as boxplots and compared to 24°C; n = 5-6 biological replicates per group $^{*}P < 0.05$. (Robinson et al., 2024b).

of the thyroid hormone, showed that *serca1* transcript expression increased significantly in cold-acclimated fish. Little and Seebacher suggest that the observed increase in SERCA1 would improve locomotor performance at low temperatures but also acknowledge the potential role for thermogenesis.

5.4 CONCLUDING REMARKS

The adaptation of endothermy in fish is a remarkable example of evolutionary convergence. Anatomical features such as internalized slow twitch red muscle are common in endothermic fish but are not monophyletic traits; rather, they have been converged on in multiple fish lineages. Another heat-conserving adaptation, the countercurrent vascular heat exchangers known as rete mirabile, have evolved independently. Likewise, specialized heat-generating tissues such as the heater organ have been adapted from different extraocular eye muscles. A common through line for endothermic fish is the fact that almost all are large pelagic fish that often dive to great depths and swim long distances. This fact has led to the proposal that selective pressures favoring endothermy are related to niche expansion. While the evolution of endothermy is an example of convergent evolution, the underlying genetic framework is conserved with a wider assemblage of endotherms including mammals. For example, the futile NST calcium cycle proposed for the billfish heater organ (Morrissette et al., 2003) has recently been shown to be a heat-producing mechanism employed by mammals by increasing their basal ATP turnover (Meizoso-Huesca et al., 2022). Additionally, the molecular evidence that sarcolipin and the sarcoplasmic reticulum ATPase pump are upregulated in cold-adapted ectothermic fish highlights the fact that the predisposition for the evolution of endothermy in fish is widely conserved (Little & Seebacher, 2013; Robinson et al., 2024a). The loss of the Threonine 5 (T5) regulatory site in the SLN sequence for opahs and tunas supports a greater role of SLN in red muscle–based NST, as the amino acid substitution would result in the inability to inhibit (regulate) the uncoupling mechanism of SLN on the SERCA1 pump.

REFERENCES

Alexander, R. L. (1998). Blood supply to the eyes and brain of lamniform sharks (Lamniformes). *Journal of Zoology*, *245*(3), 363–369. https://doi.org/10.1111/j.1469-7998.1998.tb00111.x

Altringham, J. D., & Block, B. A. (1997). Why do tuna maintain elevated slow muscle temperatures? Power output of muscle isolated from endothermic and ectothermic fish. *Journal of Experimental Biology*, *200*(20), 2617–2627. https://doi.org/10.1242/jeb.200.20.2617

Bal, N. C., Maurya, S. K., Sopariwala, D. H., Sahoo, S. K., Gupta, S. C., Shaikh, S. A., Pant, M., Rowland, L. A., Bombardier, E., Goonasekera, S. A., Tupling, A. R., Molkentin, J. D., & Periasamy, M. (2012). Sarcolipin is a newly identified regulator of muscle-based thermogenesis in mammals. *Nature Medicine*, *18*(10), 1575–1579. https://doi.org/10.1038/nm.2897

Bal, N. C., & Periasamy, M. (2020). Uncoupling of sarcoendoplasmic reticulum calcium ATPase pump activity by sarcolipin as the basis for muscle non-shivering thermogenesis. *Philosophical Transactions of the Royal Society B: Biological Sciences*, *375*(1793), 20190135. https://doi.org/10.1098/rstb.2019.0135

Bernal, D., Sepulveda, C., Mathieu-Costello, O., & Graham, J. B. (2003). Comparative studies of high performance swimming in sharks I. Red muscle morphometrics, vascularization and ultrastructure. *Journal of Experimental Biology*, *206*(16), 2831–2843. https://doi.org/10.1242/jeb.00481

Bernal, D., & Sepulveda, C. A. (2005). Evidence for temperature elevation in the aerobic swimming musculature of the common thresher shark, alopias vulpinus. *Copeia*, *1*, 146–151.

Bhupathy, P., Babu, G.J., Ito, M., Periasamy, M., 2009. Threonine-5 at the N-terminus can modulate sarcolipin function in cardiac myocytes. *Journal of Molecular and Cellular Cardiology 47*, 723–729. https://doi.org/10.1016/j.yjmcc.2009.07.014

Block, B. A. (1986). Structure of the brain and eye heater tissue in marlins, sailfish, and spearfishes. *Journal of Morphology*, *190*(2), 169–189. https://doi.org/10.1002/jmor.1051900203

Block, Barbara A., Finnerty, John R., Stewart, Alexandre F.R., & Kidd, Jessica. (1993). Evolution of Endothermy in Fish: Mapping Physiological Traits on a Molecular Phylogeny. *Science*, *260*(9), 210–214.

Burne, R. H. (1923). *VII.-Some peculiarities of the blood-vascular system of the porbeagle shark (lamna cornubica).* https://doi.org/10.1098/rstb.1924.0007

Cao, Y., Wu, X., Wang, X., Sun, H., Lee, I., 2016. Transmembrane dynamics of the Thr-5 phosphorylated sarcolipin pentameric channel. *Archives of Biochemistry and Biophysics 604*, 143–151. https://doi.org/10.1016/j.abb.2016.06.016

Carey, F. G. (1982). A brain heater in the swordfish. *Science*, *216*(4552), 1327–1329. https://doi.org/10.1126/science.7079766

Carey, F. G., & Lawson, K. D. (1973). Temperature regulation in free-swimming bluefin tuna. *Comparative Biochemistry and Physiology Part A: Physiology*, *44*(2), 375–392. https://doi.org/10.1016/0300-9629(73)90490-8

Carey, F. G., & Robinson, B. H. (1981). Daily patterns in the activities of swordfish, Xiphias gladius, observed by acoustic telemetry. *Fishery Bulletin*. https://www.semanticscholar.org/paper/Daily-patterns-in-the-activities-of-swordfish%2C-by-Carey-Robinson/f5a34c322d5f9483d48b25bf88d324c3f4e9fe44

Carey, F. G., & Teal, J. M. (1969a). Mako and porbeagle: Warm-bodied sharks. *Comparative Biochemistry and Physiology*, *28*(1), 199–204. https://doi.org/10.1016/0010-406X(69)91335-8

Carey, F. G., & Teal, J. M. (1969b). Regulation of body temperature by the bluefin tuna. *Comparative Biochemistry and Physiology*, *28*(1), 205–213. https://doi.org/10.1016/0010-406X(69)91336-X

Carey, F. G., Teal, J. M., & Kanwisher, J. W. (1981). The visceral temperatures of mackerel sharks (Lamnidae). *Physiological Zoology*, *54*(3), 334–344. https://doi.org/10.1086/physzool.54.3.30159948

Carey, F. G., Teal, J. M., Kanwisher, J. W., Lawson, K. D., & Beckett, J. S. (1971). Warm-bodied fish. *American Zoologist*, *11*(1), 137–143. https://doi.org/10.1093/icb11.1.137

da Costa, D. C. F., & Landeira-Fernandez, A. M. (2009). Thermogenic activity of the Ca2+-ATPase from blue marlin heater organ: Regulation by KCl and temperature. *American Journal of Physiology-Regulatory, Integrative and Comparative Physiology*, *297*(5), R1460–R1468. https://doi.org/10.1152/ajpregu.90993.2008

Davesne, D., Meunier, F.J., Friedman, M., Benson, R.B.J., Otero, O., 2018. Histology of the endothermic opah (Lampris sp.) suggests a new structure-function relationship in teleost fish bone. *Biology Letters 14*, 20180270. https://doi.org/10.1098/rsbl.2018.0270

Dickson, K. A., & Graham, J. B. (2004). Evolution and Consequences of Endothermy in Fishes. *Physiological & Biochemical Zoology*, *77*(6), 998–1018. https://doi.org/10.1086/423743

Don Stevens, E., & Fry, F. E. J. (1971). Brain and muscle temperatures in ocean caught and captive skipjack tuna. *Comparative Biochemistry and Physiology Part A: Physiology*, *38*(1), 203–211. https://doi.org/10.1016/0300-9629(71)90109-5

Franck, J. P. C., Slight-Simcoe, E., & Wegner, N. C. (2019). Endothermy in the smalleye opah (Lampris incognitus): A potential role for the uncoupling protein sarcolipin. *Comparative Biochemistry and Physiology Part A: Molecular & Integrative Physiology*, *233*, 48–52. https://doi.org/10.1016/j.cbpa.2019.03.024

Friedlander, M. J., Kotchabhakdi, N., & Prosser, C. L. (1976). Effects of cold and heat on behavior and cerebellar function in goldfish. *Journal of Comparative Physiology*, *112*(1), 19–45. https://doi.org/10.1007/BF00612674

Fujioka, K., Fukuda, H., Tei, Y., Okamoto, S., Kiyofuji, H., Furukawa, S., Takagi, J., Estess, E., Farwell, C. J., Fuller, D. W., Suzuki, N., Ohshimo, S., & Kitagawa, T. (2018). Spatial and temporal variability in the trans-Pacific migration of Pacific bluefin tuna (Thunnus orientalis) revealed by archival tags. *Progress in Oceanography*, *162*, 52–65. https://doi.org/10.1016/j.pocean.2018.02.010

Harding, L., Jackson, A., Barnett, A., Donohue, I., Halsey, L., Huveneers, C., Meyer, C., Papastamatiou, Y., Semmens, J. M., Spencer, E., Watanabe, Y., & Payne, N. (2021). Endothermy makes fishes faster but does not expand their thermal niche. *Functional Ecology*, *35*(9), 1951–1959. https://doi.org/10.1111/1365-2435.13869

Linthicum, D. S., & Carey, F. G. (1972a). Regulation of brain and eye temperatures by the bluefin tuna. *Comparative Biochemistry and Physiology Part A: Physiology*, *43*(2), 425–433. https://doi.org/10.1016/0300-9629(72)90201-0

Linthicum, D. S., & Carey, F. G. (1972b). Regulation of brain and eye temperatures by the bluefin tuna. *Comparative Biochemistry and Physiology Part A: Physiology*, *43*(2), 425–433. https://doi.org/10.1016/0300-9629(72)90201-0

Little, A. G., & Seebacher, F. (2013). Thyroid hormone regulates muscle function during cold acclimation in zebrafish (Danio rerio). *The Journal of Experimental Biology*, *216*(Pt 18), 3514–3521. https://doi.org/10.1242/jeb.089136

Madigan, D. J., Richardson, A. J., Carlisle, A. B., Weber, S. B., Brown, J., & Hussey, N. E. (2021). Water column structure defines vertical habitat of twelve pelagic predators in the South Atlantic. *ICES Journal of Marine Science*, *78*(3), 867–883. https://doi.org/10.1093/icesjms/fsaa222

Meizoso-Huesca, A., Pearce, L., Barclay, C. J., & Launikonis, B. S. (2022). Ca2+ leak through ryanodine receptor 1 regulates thermogenesis in resting skeletal muscle. *Proceedings of the National Academy of Sciences*, *119*(4), e2119203119. https://doi.org/10.1073/pnas.2119203119

Morrissette, J. M., Franck, J. P. G., & Block, B. A. (2003). Characterization of ryanodine receptor and Ca2+-ATPase isoforms in the thermogenic heater organ of blue marlin (*Makaira nigricans*). *Journal of Experimental Biology*, *206*(5), 805–812. https://doi.org/10.1242/jeb.00158

Robinson, S., Hechter, D., Almoumen, F., & Franck, J. P. C. (2024a). Sarcolipin (sln) and Sarcoplasmic Reticulum calcium ATPase pump (serca1) expression increase in Japanese medaka (Oryzias latipes) skeletal muscle tissue following cold challenge. *Comparative Biochemistry and Physiology Part A: Molecular & Integrative Physiology*, *287*, 111534. https://doi.org/10.1016/j.cbpa.2023.111534

Robinson, S., Wegner, N.C., Sepulveda, C.A., & Franck, J.P.C. (2024b). Relative Sarcolipin (SLN) and Sarcoplasmic Reticulum Ca2+ ATPase (SERCA1) Transcripts Levels in Closely Related Endothermic and Ectothermic Scombrid Fishes: Implications for Molecular Basis of Futile Calcium Cycle Non-Shivering Thermogenesis (NST). Comparative Biochemistry and Physiology Part A, in Press. https://doi.org/10.1016/j.cbpa.2024.111667

Sahoo, S. K., Shaikh, S. A., Sopariwala, D. H., Bal, N. C., Periasamy, M. (2013). Sarcolipin Protein Interaction with Sarco(endo)plasmic Reticulum Ca2+ATPase (SERCA) Is Distinct from Phospholamban Protein, and Only Sarcolipin Can Promote Uncoupling of the SERCA Pump. *The Journal of Biological Chemistry*, *288*(10), 6881–6889. https://doi.org/10.1074/jbc.M112.436915

Sepulveda, C. A., Dickson, K. A., Frank, L. R., & Graham, J. B. (2007). Cranial endothermy and a putative brain heater in the most basal tuna species, Allothunnus fallai. *Journal of Fish Biology*, *70*(6), 1720–1733. https://doi.org/10.1111/j.1095-8649.2007.01446.x

Sepulveda, C. A., Wegner, N. C., Bernal, D., & Graham, J. B. (2005). The red muscle morphology of the thresher sharks (family Alopiidae). *Journal of Experimental Biology*, *208*(22), 4255–4261. https://doi.org/10.1242/jeb.01898

Shadwick, R. E., & Goldbogen, J. A. (2012). Muscle function and swimming in sharks. *Journal of Fish Biology*, *80*(5), 1904–1939. https://doi.org/10.1111/j.1095-8649.2012.03266.x

Syme, D. A., & Shadwick, R. E. (2011). Red muscle function in stiff-bodied swimmers: There and almost back again. *Philosophical Transactions of the Royal Society B: Biological Sciences*, *366*(1570), 1507–1515. https://doi.org/10.1098/rstb.2010.0322

Toyoshima, C., Iwasawa, S., Ogawa, H., Hirata, A., Tsueda, J., Inesi, G., 2013. Crystal structures of the calcium pump and sarcolipin in the Mg2+-bound E1 state. *Nature 495*, 260–264. https://doi.org/10.1038/nature11899

Tullis, A., Block, B. A., & Sidell, B. D. (1991). Activities of Key Metabolic Enzymes in the Heater Organs of Scombroid Fishes. *Journal of Experimental Biology*, *161*(1), 383–403. https://doi.org/10.1242/jeb.161.1.383

Wang, X., Qu, M., Liu, Y., Schneider, R. F., Song, Y., Chen, Z., Zhang, H., Zhang, Y., Yu, H., Zhang, S., Li, D., Qin, G., Ma, S., Zhong, J., Yin, J., Liu, S., Fan, G., Meyer, A., Wang, D., & Lin, Q. (2022). Genomic basis of evolutionary adaptation in a warm-blooded fish. *The Innovation*, *3*(1), 100185. https://doi.org/10.1016/j.xinn.2021.100185

Wegner, N. C., Sepulveda, C. A., Aalbers, S. A., & Graham, J. B. (2013). Structural adaptations for ram ventilation: Gill fusions in scombrids and billfishes. *Journal of Morphology*, *274*(1), 108–120. https://doi.org/10.1002/jmor.20082

Wegner, N. C., Snodgrass, O. E., Dewar, H., & Hyde, J. R. (2015). Whole-body endothermy in a mesopelagic fish, the opah, Lampris guttatus. *Science*, *348*(6236), 786–789. https://doi.org/10.1126/science.aaa8902

Weng, K. C., & Block, B. A. (2004). Diel vertical migration of the bigeye thresher shark (Alopias superciliosus), a species possessing orbital retia mirabilia. *Fishery Bulletin*, *102*(1), 221–229.

Wolf, N. G., Swift, P. R., & Carey, F. G. (1988). Swimming muscle helps warm the brain of lamnid sharks. *Journal of Comparative Physiology B*, *157*(6), 709–715. https://doi.org/10.1007/BF00691001

Wu, B., Gao, X., Hu, M., Hu, J., Lan, T., Xue, T., Xu, W., Zhu, C., Yuan, Y., Zheng, J., Qin, T., Xin, P., Li, Y., Gong, L., Feng, C., He, S., Liu, H., Li, H., Wang, Q., Wang, K. (2023). Distinct and shared endothermic strategies in the heat producing tissues of tuna and other teleosts. *Science China Life Sciences*, *66*(11), 2629–2645. https://doi.org/10.1007/s11427-022-2312-1

6 Whole-Body Endothermy and Non-Shivering Thermogenesis (NST) in Today's Birds and Mammals and in Their Past

Gordon Grigg
University of Queensland, Brisbane, Australia

6.1 INTRODUCTION

Shivering thermogenesis (ST) is usually regarded as the primary response of whole-body endotherms to cold exposure, but much of the energy delivered thereby for heat production goes to futile work. Non-Shivering thermogenesis (NST) is heat production without muscle contraction. It is considered more efficient because the contraction-relaxation cycle is turned off and more of the energy can be released as heat. Strictly speaking, the background metabolic processes, those going on within an endotherm's thermoneutral zone, also produce heat without shivering, but NST usually refers only to the facultative "available on demand" processes. Hence, the term is often qualified using "adaptive" or "regulatory" to remove any ambiguity. Both of those qualifying terms, however, are less than ideal. The former has evolutionary implications, and the latter describes only part of its function. A much more suitable word would be "facultative" as used by de Meis et al. (2005): "facultative" NST, in contrast to the obligate NST associated with BMR. Facultative NST is presumed to be activated to supplement/replace shivering in response to exposure to cold as well as to drive arousals to normothermia after bouts of short- or long-term torpor (hibernation), either alone or as a supplement to a source of passive warming. Whether it may be used to elevate metabolic rate in other circumstances is unknown. In this chapter the main focus will be on NST generated within skeletal muscle. I will follow previous practice and refer to it as "muscle NST" until a better term (MNST? SkMNST?) is found.

This is a particularly interesting time to be reviewing facultative NST as an important component of whole-body endothermy because two long-standing assumptions are under critical reconsideration. The first is the long-held assumption that BAT is the only source of NST in mammals (Cannon et al. 2000; Cannon and Nedergaard 2004). This is now challenged. Since the early 2000s there has been a slowly growing recognition, although still controversial, that placental mammals have an additional source of NST in skeletal muscle. This has been reviewed, and its significance given more flesh, in two milestone publications, Rowland et al. (2015a) and Nowack et al. (2017). What's more, a skeletal muscle NST source has been proposed as being more ancient than BAT (Rowland et al. 2015a, 2015b). This is consistent with monotremes and marsupials being very competent whole-body endotherms despite lacking BAT and, perhaps, having muscle NST to rely on (more on this later in the chapter). The second assumption is that whole-body (tachymetabolic) endothermy evolved quite independently in mammals and birds. This is commonly asserted in the first few lines of any paper dealing with the evolution of their endothermy. However, a recent study observed that the similarity of the mechanism driving muscle NST in birds and (probably) all three extant clades of mammals suggests their common ancestry as long ago as the Late Carboniferous when the

DOI: 10.1201/9781003391470-8

Sauropsida ("reptile-bird" lineage) and Synapsida ("mammal" lineage) are thought to have separated (see Section 6.2.1; Grigg et al. 2022). This prompted a survey of paleontological literature for evidence of whole-body endothermy in the earliest sauropsids and synapsids. Could whole-body endothermy really be so ancient? It seems so. We found abundant evidence of whole-body endothermy in both these lineages as far back as the Late Carboniferous (synapsids) and Middle Permian (sauropsids); Section 6.2.2 provides a brief review of that study. Such great age is consistent with endothermy in birds and mammals having common ancestry, and it provides indirect support. Taking both parts of our study together, there is a strong case for whole-body endothermy in birds and placental mammals being not only ancient but also homologous (Grigg et al. 2022).

Hence, it is timely to be discussing facultative NST and the evolutionary history of whole-body endothermy because it can be considered within the context of these new(ish) understandings and interpretations. Briefly, what emerges from the combination of biochemical, comparative physiological, and paleontological information summarized in the following sections, from my perspective as a comparative physiologist, is that muscle NST is likely to have been a component of whole-body endothermy since its earliest expressions in both sauropsids and synapsids, with BAT evolving more recently as an additional source of NST in eutherian mammals.

Having said that, there have been opinions expressing doubt that skeletal muscle is yet proven to be an important source of NST (Hohtola (2004); Campbell and Dicke (2018); Lømo et al. (2020); Hohtola and González-Alonso (2020); Jastroch et al. (2021) and the significance of muscle NST remains controversial in the three groups of mammals and, to a lesser extent, in birds. Part of the difficulty may be that, unlike in BAT, the muscle NST source is diffuse, not readily identifiable as a discrete "organ," and we have not been accustomed to thinking of the response to cold exposure being so "whole of body." However, a comparison of the thermoregulatory capacities of all four groups (Section 6.3) will provide strong evidence that all of them must have a good capacity for NST, if not from BAT or a BAT analogue and not from skeletal muscle, then from some as yet unknown source. Because NST from skeletal muscle is the best-supported contender so far among the non-BAT–based suggestions, I will continue to refer to that process, or processes, as muscle NST. Much of what I have to say will be equally applicable if additional source(s) of NST turn out to be elsewhere or if multiple sources are identified within skeletal muscle, which seems likely. Also, keeping an open mind, the possibility that a discrete functional BAT analogue is found in birds and non-placental mammals cannot yet be ruled out.

6.1.1 Chapter Outline

Some definitions are necessary, and some background will be helpful as well (Section 6.1). I will then summarize the case made by Grigg et al. (2022) for whole-body endothermy being ancient and homologous in monotremes, marsupials, placental mammals, and birds (Section 6.2). That will be followed by a brief review of the thermoregulatory biology seen in each of the four extant groups of vertebrate endotherms, including a summary of what is known about facultative NST in each group (Section 6.3). Because most readers are likely to be more familiar with eutherian (placental) than other mammals, I have included more detail about monotremes and marsupials. Following that is a discussion summarizing the debate that led to the conclusion that whole-body endothermy in birds and mammals is plesiomorphic and the likely role of NST in its evolution (Section 6.4). A short section suggesting future directions concludes the discussion (Section 6.5).

6.1.2 Definitions

Most invertebrates, fish, amphibians, and extant reptiles are *ectotherms* ("outside heat") with a low metabolic rate (*bradymetabolism*), too low to influence body temperature (Tb) and with little thermal insulation. Many regulate Tb behaviorally, some very effectively. Some ectotherms use interesting strategies to maintain warmth in parts of their body. Familiar examples are tuna relying on countercurrent perfusion of their lateral musculature to retain by-product warmth from swimming, and leatherback turtles retaining their by-product heat by having good insulation and a globose body shape. These are examples of ectotherms showing *regional endothermy*, and their metabolism is bradymetabolic. In contrast, extant birds and mammals are *whole-body (tachymetabolic) endotherms* ("high metabolic intensity, inside heat") and, except when in torpor or hibernation, their basal metabolic rate is 5–10 times higher than in ectotherms, producing enough "waste" heat from the aerobic metabolism of their internal organs to maintain Tb within their *thermoneutral zone* (TNZ). This is a range of ambient temperature, Ta, over which Tb can be regulated by controlling the rate of heat loss or gain with minimal energy cost, making adjustments to integumentary insulation, surface blood flow, and evaporation. At low ambient temperatures (Ta) below their TNZ, "normal" body temperature (Tb) can be maintained by exercise, by *shivering thermogenesis* (ST), and/or by *facultative non-shivering thermogenesis* (NST). This is a "regulatory" or "adaptive" heat source, able to be upregulated quickly in response to cold or for arousal from torpor or hibernation. It is often called *adaptive NST* (a term that has the drawback of its evolutionary usage) and sometimes *regulatory NST* (better but potentially constraining), but, as discussed above, the word used by De Meis et al. (2005), *facultative NST*, would be more appropriate. The term is generic; it can refer to either BAT-generated or skeletal-muscle–based NST (***muscle NST*** in this chapter) and, potentially, to other facultatively implemented heat sources fitting this description, some undoubtedly still remaining to be recognized. Sometimes, mostly in medical literature, the by-product heat from processes contributing to resting metabolic rate is also included in the (also technically correct) term NST, so the qualifiers such as "regulatory," "adaptive," or (better) "facultative" are useful. Also, the term *endothermy* used by itself can be ambiguous; without any qualifier it can be applied to regional endothermy, so it is good practice to use a qualifier at least in the early stages of an article. It is quite common for probably bradymetabolic vertebrates showing regional endothermy to be reported as being "endothermic," possibly leaving uncertainty about the author's meaning. The evolution of whole-body endothermy should not be seen as a strategy to gain and maintain a high and stable Tb. Rather, its tachymetabolism permits a high-energy lifestyle with enough energy to provide also for the higher temperature required for that; the two factors are interdependent. Many tachymetabolic endotherms remain *homeothermic* indefinitely, but many show periods of *heterothermy* as well, going into *daily torpor* or multi-day torpor (hibernation). These two terms will be further defined in the sections that follow.

6.1.3 Some Background

My interest in facultative NST grew out of a field study of short-beaked echidnas in the Australian Alps beginning in 1987. Short-beaked and long-beaked echidnas and platypus are the only surviving Monotremata, the oldest of the three extant mammalian groups, and they are inevitably of evolutionary interest. Our study was prompted by reports from cross-country skiers of echidnas wandering around in the snow. Echidnas were already known to be endothermic, but also to have very labile Tb, and some authors considered their thermoregulatory capacities dubious. Hence, we wondered if they were lost, just "visiting," or actually resident above the snowline (then thought very unlikely). As will be discussed in more detail later in the chapter, we soon found that they were indeed resident above the winter snowline and—very exciting at that time—we found they spent much of the winter in hibernation (Grigg et al. 1989). This was the first unambiguous report of hibernation in a non-placental mammal. Significantly, in hibernation they undertake the same pattern of periodic arousals to normothermia that characterizes hibernation in eutherian (placental) mammals. At that time there was no reason to doubt that the arousals would be driven by UCP1 and brown adipose tissue (BAT). However, reports in the early to mid-1990s that monotremes and marsupials lacked both of those accepted requirements for NST raised significant questions. Did shivering alone account for a hibernating echidna's periodic arousals or did they have some unknown and different source of facultative NST? During the early to mid-1990s, skeletal muscle was identified as a possible facultative heat source in marsupials (more about this later in the chapter), and although I was unaware initially of similar hints emerging from placentals, my then colleagues and I became convinced that the two older mammalian groups must have a different source of NST, probably in muscle. We noted in a speculative paper about the evolution of endothermy, "It seems likely that a widespread, ancient, and controllable mechanism of regulatory NST will be found in the skeletal muscle of mammals" (Grigg et al. 2004). That would explain the rapid arousals we had seen in hibernating echidnas, but we had all three mammalian groups in mind. Knowing that the source of avian NST is thought to be in skeletal muscle, we also noted that "If it turns out to be the same mechanism in both birds and mammals, that could imply that endothermy in these two groups stems from a common origin." As will be discussed in the following sections, current understanding suggests both of those predictions seem likely to be correct.

This was pretty much the way things were when I retired in 2007; the "echidna arousal" puzzle remained, and I got involved with other things (Grigg and Kirshner 2015). But in 2017, serendipity delivered a paper to my desk for review (to become Nowack et al. 2017), and I was delighted to discover that there was by then a growing recognition that skeletal muscle is a major source of facultative NST in placental mammals, herein referred to as muscle NST (Rowland et al. 2015a, 2015b; Nowack et al. 2017), in addition to BAT. Stemming from the two predictions made in 2004, two questions suggested themselves straightaway. First, how similar to the muscle NST in birds is the now recognized facultative muscle NST in placental mammals? Strong similarity would be consistent with their endothermies having an ancient common origin, as we conjectured in 2004. Common ancestry would imply endothermy being as old as the differentiation of the Sauropsida (reptile, bird line) and Synapsida (mammalian line) within the Amniota in the Late Carboniferous. So, the second question was: Could a search of the paleontological literature using proxies for whole-body endothermy find evidence of it in fossils dating back in time, even to the Late Carboniferous? Both these questions will be discussed in the next section.

6.2 THE CASE FOR WHOLE-BODY ENDOTHERMY BEING ANCIENT AND HOMOLOGOUS IN MAMMALS AND BIRDS

Given the predictions about the possibility of common ancestry made by Grigg et al. (2004) and following on from the review and interpretations of muscle NST in placentals by Nowack et al. (2017), it was logical to assemble a small team of collaborators with expertise in biochemistry, comparative physiology, and paleontology. The aims were to compare the then-known biochemical mechanisms driving muscle NST in birds with those driving muscle NST in placental mammals and likely, on the basis of their comparable endothermy, to be operational in monotremes and marsupials as well. Similarity would imply homology of the mechanisms driving NST in birds and mammals, and an ancient origin at least as long ago as when sauropsids and synapsids differentiated. The second aim was to search the paleontological literature for evidence of whole-body endothermy in the fossil record.

The open-access paper reporting the results in full, and our interpretation of them, is easily accessible (Grigg et al. 2022), so only an abbreviated summary is presented here.

6.2.1 How Similar Are the Molecular Mechanisms Driving Muscle NST in Mammals and Birds?

The mechanism of the most commonly accepted form of facultative NST within skeletal muscle is generally regarded as resulting from modification of the activity of the sarcoplasmic reticulum Ca^{2+}-ATPase pump (SERCA), which normally pumps Ca^{2+} back into the sarcoplasmic reticulum following contraction, allowing the muscle to relax. The modification comprises the uncoupling of the pump, now thought to be by the protein sarcolipin (SLN), which allows the hydrolysis of ATP, releasing heat (Bal et al. 2012, 2016). However, it cannot be assumed that the molecular detail of this process and its regulation is the same in all three mammal groups, as well as in birds. The details are best known in placental mammals (Eutheria), much less known in birds, and less still—so far—in the non-eutherian mammals. In our comparison of the individual molecular processes then known to contribute an increase in skeletal muscle NST in response to cold across these four groups, similarities rather than differences emerged (Grigg et al. 2022). From 13 known muscle-NST–related responses by eutherians to cold exposure, all but one were identified in birds too and six in non-eutherian mammals (information lacking for the other seven). After millions of years of separation, some difference in the muscle NST biochemistry between birds and mammals would not be surprising. Importantly, the single response not identified as increasing in birds after cold exposure was SLN. In birds the SLN is slightly different structurally, and its role in NST is uncertain (Bal et al. 2018). Furthermore, the relevance of SLN as the key trigger to initiate the release of heat in mammals has recently been questioned, with the focus shifting instead to regulation of Ca^{2+} leak through the ryanodine receptors (RyR1) to generate heat at the SERCA Ca^{2+} pump in resting muscle (Meizoso-Huesca et al. 2022). Clearly the fine detail of how all this works remains to be resolved. In a follow-up paper by the same research group, evidence was presented for the extent of the Ca^{2+} leak through RyR1 being regulated via the sympathetic nervous system (Singh et al. 2023). That may address what has been an uncertainty; clearly the effectiveness of a diffusely distributed agent, such as SERCA-based NST located in skeletal muscle, depends on having a very effective control mechanism.

In comparing what was then known about muscle NST in mammals and birds and considering the huge time gulf since the separation of synapsids and sauropsids, we judged the similarity sufficient to be consistent with the hypothesis that their whole-body endothermies have ancient common ancestry, though not, of course, providing solid proof. This conclusion should be revisited when there is more information about the fine details of the mechanics and control of muscle NST in birds and placental mammals and, presumably, in monotremes and marsupials too, in which muscle NST is posited but not yet confirmed (see discussion later in the chapter). However, the similarity was certainly enough to justify a search in the fossil record for evidence of whole-body endothermy; as summarized in the next section, we found plenty of it.

6.2.2 Is Whole-Body Endothermy Ancient? Indications of it Are Found Quite Early in the Amniote Fossil Record

Because the sauropsids and synapsids are thought to have separated in the Late Carboniferous, our muscle NST-based hypothesis that endothermy in birds and mammals has a common ancestry implies its origin at least as old as that. If our hypothesis is correct, we could hope to find evidence of endothermy that far back. Accordingly, we conducted a survey of paleontological literature in search of information about extinct amniotes in which attributes indicative of endothermy had been reported.

6.2.2.1 Proxies for Whole-Body Endothermy

Although physiological and biochemical evidence of whole-body endothermy cannot be obtained directly from fossilized skeletal material, such material often provides strong indirect evidence that can be used as a proxy for it. In comparison to ectothermy, whole-body endothermy is characterized by higher and sustained growth rate, higher levels of activity, higher arterial blood pressure, higher rates of blood flow, and a capacity for maintaining a higher body temperature, and all these attributes can leave evidence in their skeleton directly or via competent reconstructions of body size and posture. We considered six proxies, and full explanations can be found in the original paper (Grigg et al. 2022), but I will summarize them here. We were most comfortable about judgments when there were at least two proxies in support for any taxon under consideration. Among nearly 50 taxa in which we found relevant information we relied on a single proxy in only six, all using bone histology. This proxy is the most commonly available, from numerous studies of growth rate made by examining thin sections of bone under polarized light. Relying on that alone is not ideal (Schucht et al. 2021), but the presence of fibrolamellar bone and primary and secondary osteons over a size range implying rapid growth throughout most of ontogeny can be indicative (Farlow et al. 1995; Padian and de Ricqlès 2020). Large body size and erect stature can be taken as a good indicator of endothermy because ectothermic metabolic intensities could not provide the high systemic arterial blood pressures and four-chambered hearts necessary to adequately perfuse the top of the body. This criterion is satisfied if heart-to-top-of-body distance (commonly the head) exceeds about 50 cm (Seymour 1976, 2016, elaborated further by Grigg et al. 2022). Likewise, large nutrient foramina in long bones indicate capacity for the high rate of bone perfusion necessary to support rapid somatic growth and for repair of microfractures caused by high levels of activity as seen in extant whole-body endotherms (Seymour et al. 2012, 2019). Obligatory bipedality was common in many extinct sauropsids. To assign a bipedal taxon as endothermic we accepted a study showing that any bipedal sauropsid or synapsid >10 kg body mass would be unable to move swiftly unless it was tachymetabolic, and any biped >200 kg would need to be tachymetabolic even to maintain a walking pace (Pontzer et al. 2009). We treated with caution any Tb measurements from very large aquatic sauropsids derived by paleothermometry, recognizing that achieving it by gigantothermy was a possibility. However, after a substantial study modeling large mosasaurs as a representative example of such beasts, and with reference to field data on the achievement of gigantothermy by crocodiles over a large size range (Grigg et al. 1998; Seebacher et al. 1999), we were able to show that endotherm-like Tbs measured by paleothermometry in the low to mid-30s°C in very large plesiosaurs, ichthyosaurs, and mosasaurs (Bernard et al. 2010; Harrell et al. 2016) could not have been the result of gigantothermy (Brice and Grigg, 2023). Accordingly, Tb measured by paleothermometry in such cases became our fifth proxy, validated now across the size range. The sixth proxy, presence of respiratory turbinal bones in the nasal passages, is a well-established proxy for whole-body endothermy (Hillenius, 1992, 1994; Ruben, 1995). We accepted its presence as being diagnostic but considered that absence does not necessarily imply ectothermy.

6.2.2.2 Results from the Fossil Record

More detailed results are easily available in the original paper (Grigg et al. 2022). Briefly, a survey of paleontological literature representing a wide range of about 50 mostly higher-level taxa of sauropsids and synapsids using these proxies provided abundant evidence of occurrences of whole-body endothermy. Its taxonomic diversity was sufficiently representative to suggest that whole-body endothermy was likely the "default situation" throughout both lineages, extending back in time to the Middle Permian and Late Carboniferous, well before the Permian-Triassic mass extinction event (PTME). A nearly contemporary paper by Benton (2020) on the evolution of endothermy mentioned whole-body endothermy prior to the PTME as if that was of little importance and featured the Mesozoic as the time when endothermy emerged with real significance. Because our focus was on the oldest expressions of whole-body endothermy in each clade, I will not discuss our results for the Mesozoic taxa beyond emphasizing that endothermy was clearly widespread in both groups then and thereafter (Benton 2020; Grigg et al. 2022). Instead, I will summarize findings only from amniotes prior to the PTME. Given the great geological upheavals associated with that event, it is not surprising that there are fewer remains of the biota from that time. However, fossilized pareiasaurs (Mid-Late Permian Sauropsida) and "pelycosaurs" (Late Carboniferous Synapsida) were diverse and fairly well represented, and we found solid evidence of whole-body endothermy in both groups. Each of these two groups will now be summarized briefly.

6.2.2.2.1 Pre-Triassic Sauropsids

Among pre-PTME (pre-Triassic) sauropsids the earliest we found with whole-body endothermy were the Pareiasauria. These were large stocky herbivores (500–1,000kg) with semi-erect to upright limb posture. They appeared in the Mid-Permian, diversified and became distributed globally, but did not survive the PTME. Bone structure was interpreted in several studies covering seven genera to indicate fast growth rates and elevated metabolism. This is congruent with their semi-erect to upright posture (sprawling gait is often taken as an indication of ectothermy). These indications of endothermy are confirmed by larger species having heart-to-head/top-of-body vertical distances larger than could be serviced by ectothermic metabolic intensity, an important proxy established by Seymour (1976, 2016). Additionally, stable oxygen isotope values from phosphates in teeth and bone from *Bradysaurus* and some unidentified pareiasaurs clustered with those from coeval tachymetabolic endothermic synapsids, indirectly suggesting endothermy. The pareiasaurs were the oldest group for which we could find relevant studies for consideration. They satisfied at least two of the proxies for whole-body endothermy. (Full analysis in Grigg et al. 2022, Supplementary Material.)

6.2.2.2.2 Pre-Triassic Synapsids

We reviewed the incongruently grouped "pelycosaurs" (Caseasauria, Ophiacodontidae, Edaphosauridae, and Sphenacodontia) and several pre-Triassic Therapsida (Dinocephalia, Gorgonopsia, Dicynodontia, Therocephalia, and Cynodontia). All four groups of "pelycosaurs" arose in the Late Carboniferous and none survived past the mid-Permian, unless the Therapsida (among which are our own ancestors) descended from the Sphenacodontia as has been suggested Benson (2012). Of the four, strong evidence for whole-body endothermy emerged in the latter two, edaphosaurids and sphenacodontids, both of which have a tall "sail" along the back, like that in the well-known *Dimetrodon* (a sphenacodontid). In both groups the combination of long-bone nutrient foramina and the large distance between the heart and the top of their sails (1.5 m in *Dimetrodon*) speak strongly for them having been whole-body endotherms. Among therapsids the Dinocephalia, Gorgonopsia, Dicynodontia, Therocephalia, and Cynodontia all arose in the mid-Permian but only the last three survived the PTME. Whole-body endothermy was indicated by data on osteology (in four of the five groups), large size of long bone foramina (in three groups), presence of respiratory turbinals (in two groups), large heart-to-head/top-of-body (in two groups), and in one group there was a confirmation by paleothermometry. In short, there can be little doubt that Therapsida were and are characteristically tachymetabolic endotherms and were characterized by it before the end of the Palaeozoic. (Full analysis in Grigg et al. 2022, Supplementary Material.)

6.2.2.2.3 Summary

Concerning the paleontological data (Section 6.2.2), the earliest groups for which we found sufficient relevant information were the pareiasaurs (Mid-Late Permian Sauropsida) and edaphosaurs and sphenacodontids (Late Carboniferous Synapsida). We found convincing evidence of endothermy in these three groups prior to the PTME, and in many groups subsequently not reviewed herein but detailed in the original paper. Our proxies provided unequivocal support for whole-body endothermy being widely expressed in both lineages, so, assuming the proxies are valid, there is little room for doubt that whole-body endothermy is very, very old, with expressions of it nearly as far back as the time when the sauropsids and synapsids separated in the Late Carboniferous. These findings are consistent with our hypothesis about a common ancestry for endothermy in mammals and birds and therefore provide indirect support for it.

6.2.3 A More Recent Study Supporting the Common Ancestry of Birds and Mammals

Interestingly, a study by Cherlin (2023), starting from a very different perspective, came to conclusions very similar to those of Grigg et al. (2022). Published in Russian and with an English summary, its title is "Preadaptivity of non-contractive thermogenesis in the evolution of warm-bloodedness in vertebrates." It seems his starting point was essentially intuitive. Cherlin noted that skeletal muscle is present in all vertebrates and posited that its primary functional characteristics for contraction and relaxation also preadapted it as a heating organ and that its being repurposed for "non-contractive thermogenesis" (or NST) was essentially inevitable. He discussed and considered that possibility comprehensively, citing an enormous number of research publications including our own, and, as reported in the English summary, made a case that

> the presence of skeletal muscles in all vertebrates and the common biochemical foundations of the contraction–relaxation cycle represent a single preadaptive property of [are pre-adapted for] the manifestation of non-contractive thermogenesis in all vertebrates, starting with fish, which is the basis for the evolution of warm-bloodedness. Therefore, it is understandable and unsurprising [that] modern data [show] that the first terrestrial vertebrates were most likely animals with high levels of both metabolism and body temperature.

By implication, he also concluded that whole-body endothermy was plesiomorphic for sauropsids and synapsids, writing that "tachymetabolic endothermia in mammals (monadromous, marsupials, and placentals) and birds [is] homologous."

6.2.4 Why Are All the Extant Non-avian Sauropsids Ectothermic? (It May Have Led Early Researchers Astray)

Readers may have noticed that the case made above for whole-body endothermy being widely distributed and characteristic of extinct sauropsids is at odds with the ectothermic physiology of the living sauropsids (except birds). If the above conclusions are essentially correct, how could that be? The explanation for crocodylians is straightforward: they are certainly sauropsids, and there is very good evidence for them having endothermic ancestry (Seymour et al. 2004). This is commonly referred to and seems to have become well accepted. Concerning other extant reptiles, the Lepidosauria (Rhynchocephalia [tuatara] and Squamata [snakes and lizards]), one possibility is that they too have reverted to ectothermy from endothermic ancestry. There are, after all, many benefits of ectothermy (Pough 1980), and, as well as in crocodylians, likely reversions from endothermy to ectothermy have been recognized in phytosaurs (de Ricqlés et al., 2003; Legendre et al. 2016) and notosuchians (Cubo et al. 2020; Cubo et al. 2023; de Araújo Sena et al., 2023), and there may be many more. A case could possibly be made for varanid lizards having an endothermic ancestry. Their femoral nutrient foramina are in the size range of mammals (Seymour et al. 2012, Knaus et al. 2021), they have multicameral lungs (Lambertz et al. 2015), and winter arousals of *Varanus rosenbergi* have been reported under conditions unfavorable for passive rewarming (Rismiller and Mckelvey, 2000) (Section 6.4.1), all of which could be interpreted as evidence for endogenous heat production. In this context it is interesting to note that the enormous aquatic and air-breathing mosasaurs, certainly whole-body endotherms (Brice and Grigg 2023), are also classified in Squamata. If not reversion, another possible explanation could be that grouping non-archosauromorph "reptiles" in Sauropsida may not even be appropriate. Very early amniote fossils are few and far between, typically poorly preserved, and their phylogenetic context is often uncertain. It is not certain that only two long-lived lineages emerged from Amniota, and the ancestors of extant non-archosaurian reptiles may not necessarily have had a capacity for endothermy. Padian and De Ricqlès (2020) lamented that, "Palaeobiologists are historically bound to a dichotomous set of terms developed long ago to describe the relatively depauperate living vertebrate fauna," and saw that as a constraint for researchers speculating about the physiology of extinct vertebrates. They were not thinking of Sauropsida and Synapsida in particular, but there does seem to be an assumption that apparent descendants of "stem amniotes" should fit neatly into one or the other. Finding a place for turtles in the family tree has been challenging (Lee 1997, 2013; Lyson and Bever 2020); could they comprise a third lineage? One thing is certain: if all extant Lepidosauria are placed correctly as members of Sauropsida, they have certainly been very poor representatives of their ancestry! The ectothermy of extant "reptiles" probably misled the early palaeontologists to interpret dinosaurs and all the other big extinct "reptiles" as cold-blooded—and who could blame them for that?

6.3 THERMAL RELATIONS OF THE EXTANT MONOTREMES, MARSUPIALS, EUTHERIANS, AND BIRDS, WITH A FOCUS ON TORPOR, HIBERNATION, AND NST

As part of a chapter speculating about the likely contribution of NST (including BAT) to the evolution of whole-body endothermy, it is logical to review the thermoregulatory biology characteristic of the four extant groups, with a particular focus on heterothermies such as torpor and hibernation. These indicate capability for recovery to normothermia, likely to involve a form of facultative NST. The extent of their similarity is relevant to whether mammalian and avian endothermies have a common ancestry. Only a brief summary is necessary because excellent comprehensive recent reviews of occurrence of torpor and hibernation are available (Geiser 2021; Geiser and Cooper 2023; Withers et al. 2016).

6.3.1 Monotremes

I will focus particularly on these, the only surviving egg-laying mammals, providing a lot more detail about monotremes than the other two mammalian groups and the birds because monotremes are often overlooked. They are sometimes dismissed as primitive second-rate mammals and less relevant, and their thermoregulatory biology is often misunderstood and sometimes even misrepresented (e.g., Grady et al. 2014; Martinez et al. 2023). They are not the easiest to work on, often hard to get permits to do research on, and far distant from where most researchers live. They are easily forgotten. However, as the oldest group among the surviving Mammalia, there is good reason to suggest that monotremes have something to contribute to discussions about the evolution of endothermy—and they do. I have a particular interest because, after their apparent lack of BAT was recognized (Hayward and Lisson 1992; Rose et al. 1999; Kabat et al. 2003a, 2003b; Polymeropoulos et al. 2012; Jastroch et al. 2021), I puzzled for years about what was driving the spectacular periodic arousals we had seen in hibernating echidnas in 1987. Nicol (2017) has provided a very comprehensive review of many aspects of the body temperature, metabolism, and the thermoregulatory physiology of monotremes.

It is desirable to do physiological research in the field whenever possible, and perhaps that is particularly true for studies of thermal biology. For monotremes there are field data from long-term studies of platypus (*Ornithorhynchus anatinus*) and the short-beaked echidna (*Tachyglossus aculeatus*). I am unaware of field data on Tb of Papua-New Guinea's long-beaked echidnas, *Zaglossus bartoni*, but there are long-term data from two held at Taronga Zoo in Sydney.

6.3.1.1 Platypus

Field data showed that platypus are very good whole-body endotherms. After finding hibernation in short-beaked echidnas in the Australian Alps during the winter of 1987 (see Section 6.3.1.2), we were almost immediately able to look at platypus because there was a good population conveniently close to our echidna study area. Water temperatures in the snowmelt-fed Thredbo River range from about 3°C in winter—and sometimes much lower—to 20°C in summer. In a previous study in the same river, Grant (1983) had monitored Tb in a single male with an attached radio transmitter and a thermistor probe inserted 8 cm into the posterior alimentary tract. He reported that it maintained its Tb at about 32°C while swimming in water of 0°C in the partly frozen river for at least five hours. Platypus lack a continuous insulating fat layer over the whole body and must rely on fur and the air trapped within it for insulation. The bill, webbed feet, and the area under the tail are naked, comprising about 30% of the body surface, and heat loss via that route in cold water may be minimized by specialized cardiovascular arrangements (Grant and Dawson 1978a). Would platypus show torpor or enter hibernation in winter? We monitored Tb data from five adult platypus (725–1,500g) with implanted temperature-sensitive radio transmitters as they ranged freely, foraging daily in the river, between April and October in 1988. Of these, one male and one female were tracked continuously for 160 and 180 days, respectively, all through winter. All five returned a typical pattern of homeothermic endothermy, averaging about 32°C, cooler than most placentals, and ranging from 29°C to 34°C in their normal diurnal activity cycle. There are anecdotal reports of captive platypus entering torpor for several days, but I can find no published data. It would not be greatly surprising, but none of the individuals in our study showed any sign of torpor and they did not hibernate (Grigg et al. 1992b).

Because monotremes often seem to be regarded as rather second-rate mammals, operating at lower Tb than we do (and in echidnas with much more variability), I thought a comparison of platypus performance with that of sea otters (*Enhydra lutris*) would be worth including here. Both sea otters and platypus tolerate very cold water, and an adult platypus weighs about the same as a neonate otter (approx. 1.5 kg). Sea otters in a study by Wright et al. (2021) ranged in body size from neonate to adult, 1.4–44.5 kg. Their pups are born at any time of year in Southern California but mostly in Spring in the Gulf of Alaska. Like platypus, sea otters lack blubber and rely on air trapped within thick fur, but platypus have more naked skin exposed. At the northern end of their range otters encounter winter sea surface temperatures of about 2°C, close to the 3°C winter water temperatures encountered by platypus in the Thredbo River. Admittedly, modal platypus Tb is 4–5°C lower than sea otter's but maintaining Tb in the low to mid-30s°C in 3°C water implies a well-developed capacity for tachymetabolic endothermy, especially with about 30% of the body surface naked.

6.3.1.1.1 NST in Platypus?

Neither UCP1 nor BAT has been found in monotremes (see earlier discussion), and apart from their demonstrated capacity for homeothermic endothermy in very cold water, the best evidence for any form of facultative thermogenesis in platypus are data that show an increase in resting metabolic rate at lower temperature. It has been measured as nearly doubling after exposure to air at 5°C and tripling in water at 5°C (Grant and Dawson 1978b). In the respirometer vessel the animals either rested or swam gently. Whether shivering occurred was not reported, and it would be interesting to know if NST was contributing. As is typical of whole-body endotherms, Tb rises quite noticeably during high activity, and platypus swim very actively as they forage, so the by-product heat from that will contribute to countering losses to cold water. Based on their demonstrated competence as whole-body endotherms, the presence of effective NST would seem to be a certainty, and, in the absence of BAT, muscle NST seems to be the most likely candidate.

6.3.1.2 Short-Beaked Echidna

The most studied echidna of the four extant species is the short-beaked echidna, *Tachyglossus aculeatus*. It occurs in suitable habitat throughout mainland Australia and in Tasmania and in parts of New Guinea. Since the late 1980s, substantial field studies of its thermal relations and other aspects have been undertaken in southern Tasmania, in the Australian Alps in New South Wales, in milder high country in SE Queensland, and in hot, arid western Queensland. There is now a substantial literature, far too much to do it justice here. Echidnas lack sweat glands, have subcutaneous insulation, and the hair between their spines is much thicker in the cool Tasmanian climate than in warmer climes. It is almost completely lacking in the hottest parts of their range (Brice et al. 2002a, 2002b).

6.3.1.2.1 Body Temperature

The modal Tb of short-beaked echidnas is close to 32°C, as in the other monotremes (Grigg et al. 1989, 1992b, 1992a; Nicol et al. 2004), but they are relaxed about maintaining a stable Tb, with typical daily cycles of 2–5°C and often more (Figure 6.1), correlating with their daily activity. In the hot, dry climate of central Queensland Tb of 35–36°C is common (Brice et al. 2002a, 2002b). However, unlike platypus, and unlike committed homeothermic endotherms such as humans, echidnas often show a much more substantial daily cycle, with Tb lower during the (usually overnight) resting period. Interestingly, the daily cycle in Tb can differ between seasons and between habitats with different climatic characteristics. For example, in the cool climate at Lovely Banks in Tasmania, Tb shows daily cycles of about 3°C (Nicol et al. 2004); in the warmer climate west of Stanthorpe in SE Queensland, cycles of 3–4°C were recorded in nonhibernating animals in winter (Figure 6.1a) but 3–7°C in the same animal in summer (Figure 6.1b) (Grigg et al. 2004).

The substantial daily cycles are worth considering further. Echidnas have a well-developed capacity for maintaining activity over a wide range of Tb and usually have an apparent lack of interest in maintaining a "normal" Tb. Unfortunately, this was misinterpreted in early experiments as incapacity to maintain a high Tb, thus questioning their thermoregulatory capabilities (Martin 1903). Supposedly more familiar with "normal" mammals, Martin didn't realize that an echidna placed in a respirometer is likely to either settle down and cool down or struggle vigorously to escape. It is actually easy to understand their daily heterothermy, which is not necessarily daily torpor, in strategic terms. Its amplitude increases in hotter and drier environments, and it also varies seasonally (Figure 6.1) Added to their typical mammalian hibernation in cool climates (see discussion later in the chapter), this daily

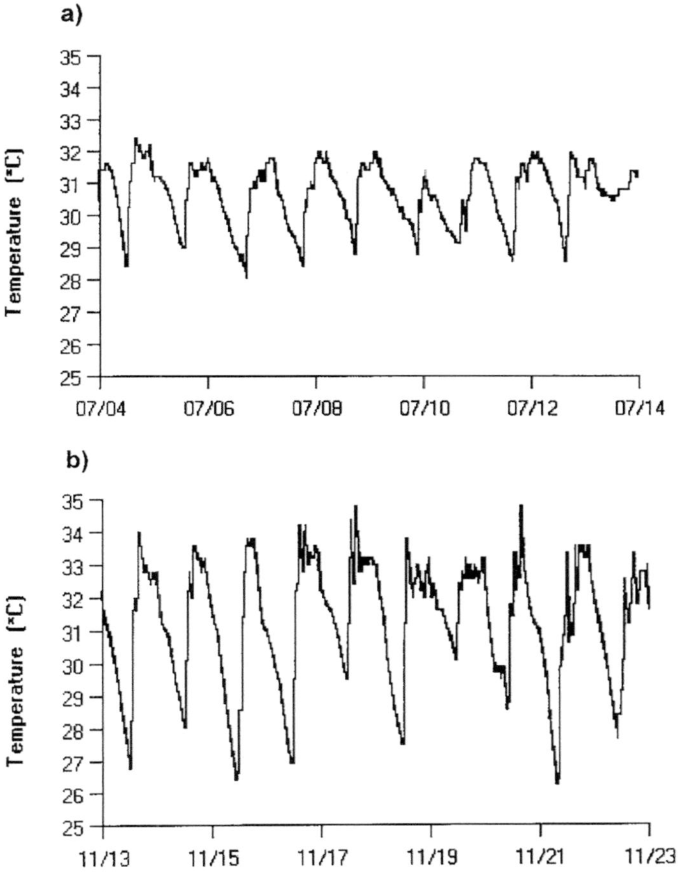

FIGURE 6.1 Daily cycles in body temperature in a 4-kg male echidna in our southeast Queensland study site, which has a warm temperate climate. (a) July (southern Winter). (b) November (late southern Spring). Reprinted from Grigg et al. (2004) with permission from The University of Chicago Press.

heterothermy must provide substantial energy economy, crucial in unproductive habitats, and probably explains their Australia-wide distribution (Brice 2009). Such heterothermy is not unusual. Many endothermic birds and other mammals employ daily heterothermy strategically to reduce energy requirement or to save water loss (Geiser 2021), and it may be more common in nutrient-poor and/or arid environments (Gyhrs et al., 2022; Ostrowski and Williams 2006). There is no doubt that short-beaked echidnas can maintain homeothermy when required. The most striking example must be females maintaining a Tb of 32°C with commendable precision while incubating an egg (Grigg et al. 2004; Nicol and Andersen 2006), even in a shallow burrow above the snowline (Beard et al. 1992). It is thought the ancestors of echidnas were aquatic and platypus-like (Phillips et al. 2009) and quite likely to have been more homeothermically inclined. Their relaxed attitude to maintaining homeothermy may have evolved as a strategy for coping with such a wide range of habitats in their (now) terrestrial habitat. The flexibility echidnas show in their thermoregulatory behavior prompted Grigg et al. (2004) to suggest that short-beaked echidnas may provide a useful living model of what an advanced "proto-endotherm" may have been like (see Section 6.4.1).

6.3.1.2.2 Torpor

Torpor is seldom easy to define and is even more difficult in the heterothermic short-beaked echidna. The following description (modified slightly from the original) may be useful in a general sense:

> [M]any endotherms undergo adaptive hypothermia or torpor from which they can arouse to normothermia, independent of external heat sources, using endogenous metabolic heat production. Torpor is not an inability to thermoregulate; rather it is an adaptive response that reduces energy expenditure(Withers et al. 2016).

That does not, however, help much with echidnas. "Daily heterothermy" seems to fit better for an animal that can be active routinely over a wide range of Tb.

6.3.1.2.3 Hibernation

Short-beaked echidnas are among the largest of all known hibernators, 2–7 kg compared with marmots, 2–8 kg across several species. In contrast to daily torpor, in all three mammalian groups and the one known avian hibernator, deep multiday torpor (hibernation) usually lasts for many days or months, with Tb falling close to ambient levels. In all four groups it is punctuated by periodic but so far unexplained brief arousals to normothermia (Grigg et al. 1989, 1992a). Behavior interpreted as hibernation by echidnas was reported very early in captive echidnas (Martin 1903), but those claims were reviewed and subsequently dismissed (Augee 1978). There is, however, no longer any ambiguity about whether short-beaked echidnas hibernate. As explained in Section 6.1.3, their hibernation was discovered in a study prompted by reports of skiers seeing echidnas in the snow as they rode chairlifts in the Australian Alps (Grigg et al. 1989). We observed winter hibernation initially in five echidnas, three above and two below the snowline. All entered hibernation early in the winter, dropping Tb close to substrate values, and then showing a series of periodic and rapid arousals of at least 20°C to reach their normal Tb in a few hours, exactly matching the hibernation patterns known already in so many placental hibernators (Grigg et al. 1989, 1992a). Subsequent continuous recordings from two males below the snowline convey a good rendition of Tb before, during, and after what we came to realize were typical hibernation events (Figures 6.2 and 6.3).

Additionally, debunking any claims that they are unable to maintain a stable Tb (see discussion earlier in the chapter), females incubating an egg maintain a stable Tb of 32°C for 2–3 weeks (Figure 6.3), even in a shallow burrow above the snowline (Beard et al. 1992; Nicol and Andersen 2006). In Tasmania, echidnas hibernate every year (Nicol and Andersen 2006), but they commonly enter hibernation in mild climates too, perhaps because once they have stored sufficient energy they can take advantage of the looming cooling and "put themselves on ice" until the spring breeding season

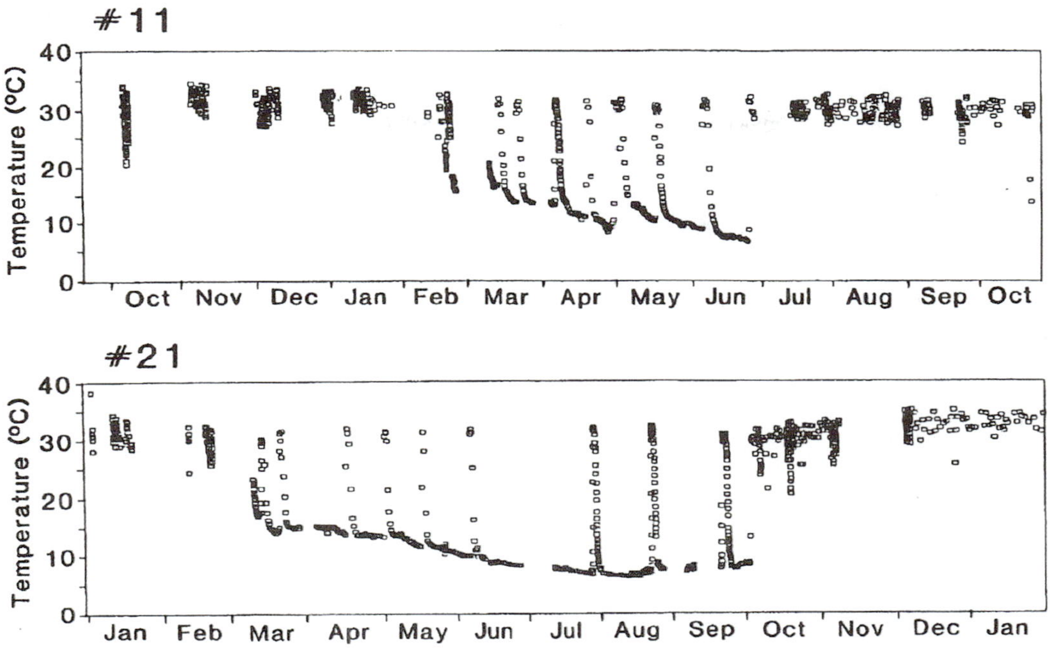

FIGURE 6.2 Body temperature of two 4–5-kg echidnas monitored for a year on the lower slopes of the Australian Alps, near Jindabyne, NSW. Cool temperate/alpine climate. Upper panel, September 1987–October 1988; lower panel January 1988–January 1989. As an historical note, these data predate the availability of implantable data loggers and were gained from calibrated temperature-sensitive radio transmitters implanted in the peritoneum and monitored from a distance at predetermined time intervals with a VHF receiver and cassette tape recorder. Pulse intervals indicating Tb were read later from the tape recorder. It was laborious but effective. Reprinted from Grigg et al. (1992a) with permission from the Royal Zoological Society of New South Wales.

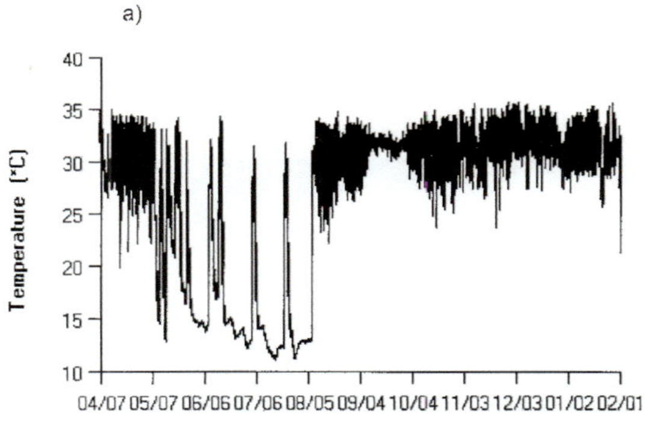

FIGURE 6.3 A record of Tb in a female echidna, April 2000–February 2001, in the warm temperature climate of our southeast Queensland study site. Note the conspicuous period of homeothermy in late September during the incubation of her egg, showing the capacity for tight regulation of Tb.

(Grigg and Beard 2000; Grigg et al. 2004). In milder climates they are facultative hibernators; some will hibernate and others will not (Kuchel 2003). Mating occurs in late winter, even above the snowline, maximizing the single young's chance of growing sufficiently before the following winter (Nicol and Andersen, 2002). Males typically emerge early and disturb females, which may or may not reenter hibernation after mating. The full story of the behavior and timing of these events and the sexual mores of echidnas is well worth reading (Nicol et al. 2019).

6.3.1.2.4 NST in Short-Beaked Echidnas?

When hibernation by short-beaked echidnas was confirmed in 1987, it would have been logical to assume they relied on UCP1 and BAT for their periodic arousals. So, when reports emerged in the mid-1990s suggesting that BAT is lacking in platypus, echidnas, and all studied marsupials (see earlier discussion), this created a puzzle. Their arousals from hibernation looked just as spectacular as those for which a marmot or a ground squirrel would be relying on BAT. Given the prodigious capacity of echidnas to rewarm rapidly during their repetitive arousals during hibernation (Figures 6.2 and 6.3), it seemed likely that they must have some other source of NST, unless it was all attributable to shivering. Also during the 1990s, hints were emerging that there is a second important source of NST in skeletal muscle. I picked up on only some of these, but enough to be convinced in the early 2000s that the skeletal muscle of marsupials and monotremes must be a source for NST (Grigg et al. 2004). A chance observation in September 1988 (Grigg et al. 1992a), whose significance was overlooked initially, does seem to provide good evidence of a source of NST in short-beaked echidnas, presumably unrelated to BAT.

Briefly, on a cold, bleak spring day, we dug a 2.2-kg male echidna (Figure 6.4) out of hibernation to remove its temperature-sensitive transmitter and then continued to log its Tb and behavior for the rest of the day. It warmed from 12°C to 21°C without any movement "except occasional slow movements of the limbs and body" and, above 18°C, "occasional brief twitches" until shivering commenced at about 22°C, after which the warming was more rapid (Figure 6.5). The ambient temperature was 14–15°C and remained so for the rest of the day, too low to account for either stage of the observed rewarming, which plateaued at about 35°C. Logically, the warming we observed prior to the commencement of shivering provides good evidence for them having NST but no information about its source.

SERCA-based NST from skeletal muscle has to be regarded as a possibility, albeit with lower heating capacity in comparison to BAT-based NST. Nicol et al. (2009) compared rates of rewarming during arousals from hibernation in echidnas and marmots (*Marmota marmota*). It was a field study conducted over a single season in each case, with data from 10 individuals of each species, size matched (approximately 2.5–5.0 kg in each group). It was an ambitious project and must have been a fascinating one. Arousals occurred approximately every 13 days in both species, and marmots warmed at about twice the rate of echidnas. The difference between the two would have been expected, but the authors were cautious about making any general statement about the relative capabilities of mammals with and without BAT. They referred to various factors that may have led to

FIGURE 6.4 A 2.2-kg male echidna just after being dug from its hibernaculum in a rotting log. Photo Gordon Grigg.

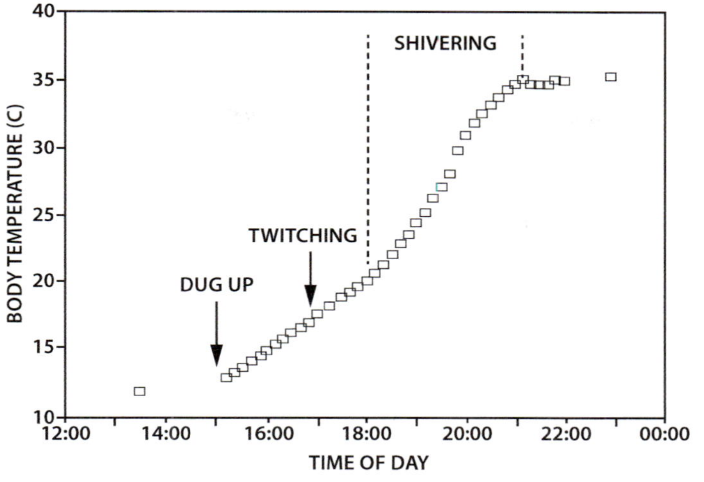

FIGURE 6.5 Increase in Tb of the echidna shown in Figure 6.4 following removal from its hibernaculum. It warmed from 12°C to 21°C without any movement "except occasional slow movements of the limbs and body" and, above 18°C, "occasional brief twitches" until shivering commenced at about 22°C, after which the warming was more rapid. The ambient temperature was 14–15°C and remained so for the rest of the day, too low to account for either stage of the observed rewarming, which plateaued at about 35°C. We interpreted the early phase of the warmup, at least, as possibly being the result of some form of NST. Reprinted from Grigg et al. (1992a) with permission from the Royal Zoological Society of New South Wales.

over- or underestimation of warming rates. Nevertheless, I took two things from it. One is that their conclusions support what I think many of us have considered likely for quite a long time: that BAT-driven NST evolved in placental mammals as a result of selection pressures favoring a capacity for even higher metabolic intensity, and it is tempting to suggest this may have evolved in response to coping in colder climates. Second, I was fascinated to see that they found the highest rates of rewarming in echidnas in the latter half of the arousal phase, matching the shivering phase of the arousal in the echidna that we observed by chance (Figure 6.5).

6.3.1.3 Long-Beaked Echidnas from Papua New Guinea

Field data on Tb seem to be lacking so far, but there has been one long-term study of two captive *Zaglossus bartoni* long held at Taronga Zoo in Sydney. Their Tb was monitored for 3.5 years using implanted data loggers (Grigg et al. 2003). Modal Tb was 31°C with an activity-related daily cycle of 2–4°C. This is like the pattern seen in nonhibernating short-beaked echidnas in mild climates and in platypus, suggesting that this Tb is characteristic of the extant monotremes. They showed no sign of torpor or hibernation, but occasionally there was a day on which no activity-generated Tb increase occurred, presumably indicating a "day off" for some reason. It is worth noting that although captive short-beaks do not readily undertake hibernation, the two in the same pen as the *Zaglossus* showed short periods of torpor, and in winter, both went into hibernation.

6.3.2 Marsupials

Conveniently, there have been two recent and very extensive reviews of both torpor and hibernation in mammals and birds during the last few years, the later one focusing particularly on marsupials (Geiser 2021; Geiser and Cooper 2023). There has also been an excellent review of the ecology and environmental physiology of mammals, with a strong focus on marsupial thermoregulation (Withers et al. 2016). Much of what follows is drawn from these excellent resources, and to meet the purpose of this chapter I need to make only a few points.

Marsupials are the second oldest group of mammals, and unlike monotremes, their status as whole-body endotherms has not been in question. They comprise a very diverse group, with about 225 species in Australia and Papua New Guinea, about 120 in Central and South America, and one, the American Opossum, in the USA and Canada. They are the "pouched mammals," giving birth to hairless young that are subsequently nourished in the mother's pouch. They range in adult size from approximately 5 g to 90 kg, and an average Tb of 35°C in animals at rest in thermoneutral environments is usually accepted as representative, a couple of degrees lower than placentals. However, between different groups of marsupials the range of Tb is much larger, from about 31°C in marsupial moles to about 37°C in kangaroos. There is quite an overlap with Eutherian (placental) mammals at the top end, which is often overlooked. Their metabolic rate is lower than is typical of placental mammals by about 30%. They are typical mammalian whole-body endotherms in the sense of modifying the time of day they are most active and their heat exchange with the environment by seeking sun or sheltering from it as appropriate, adjusting blood flow to their periphery, having hair, losing excess heat by sweating, panting, and, in the case of kangaroos and other Macropodidae, moistening their forearms with saliva to facilitate evaporative cooling. Many of them show striking heterothermy and, like monotremes, they lack UCP1 and BAT, both features of current interest. My aim here is to provide a brief overview of occurrences of torpor and hibernation and discuss the likelihood of marsupials having a capacity for NST.

6.3.2.1 Heterothermy, Torpor, and Hibernation in Marsupials

In addition to their capacity for homeothermic endothermy, more than 40% of marsupials use daily torpor or multiday torpor and in some that extends into hibernation. The benefits from these departures from "normal" Tb differ but usually relate to saving energy and water. Torpor is no longer recognized as an adaptation to living in cold climates, and "torpidating" can occur in whole-body endotherms across a wide diversity of habitats, mainly related to energy or water management. Torpor is almost always characteristic of the smaller animals, 5 g to about 1 kg, although quolls, sometimes called tiger cats, get much heavier and often show daily torpor. Because of the way different expressions of heterothermy grade into each other, definitions are difficult, but for marsupials it was found useful to define daily torpors as lasting for less than a day and in which Tb remains above 10°C (Geiser 2021). That is a useful distinction because Tb continues to be regulated actively during torpor but not so much in hibernation. Multiple days in torpor at a lower Tb can be regarded as hibernation events (Geiser and Cooper 2023). In hibernation, Tb normally falls to substrate temperature and is usually not actively regulated. Practicing torpor or hibernation may enable an animal to survive adverse harsh environmental conditions or shortages of food or water, and the capacity for it may considerably extend the geographic range of a species. Larger marsupials such as kangaroos can also tolerate several degrees of fluctuation in Tb while remaining active, particularly during hot conditions when water is not readily available.

Hibernation or multiday torpor at low Tb is known in several marsupials: two opossums in South America; *Lestodelphys halli*, the Patagonian opossum (order Didelphimorphia) and *Dromiciops gliroides*, the Monito del Monte (order Microbiotheria); and in Australia five pygmy possums (Burramyidae) and the feathertail glider (Acrobatidae) in the Order Diprotodontia) (Geiser 2021). Many are too small for implanting temperature-sensitive transmitters that will transmit for many months, so most is known about their thermal relations from laboratory studies in simulated environmental temperatures. Among the five Burramyids the mountain pygmy possum (Diprotodontia) hibernates in the subnivean space during winter in the Australian Alps, and in captivity Tb fell to 2–3°C in the simulated winter, and adults hibernated successfully for up to seven months, arousing periodically between torpor bouts of up to 20 days, shorter early and late in the season (Geiser and Broome 1991). Körtner and Geiser (1998) undertook a successful field study, endorsing and expanding the results in captivity. Interestingly, a captive study of *Lestodelphys halli* in autumn and early winter with food available showed occasional short torpor, but exposure to cold with food available for most of the time, induced repetitive longer torpors of 2–4 days with several periodic arousals, like a short hibernation pattern expressed over just a few days. The authors put this forward as a possible ancestral stage, intermediate between daily torpor and hibernation (Geiser and Martin 2013), but it could also be an expression of flexibility. *Dromiciops gliroides* was the subject of an interesting study that demonstrated a more direct method of measuring energy saved by hibernation, using quantitative magnetic resonance to determine body composition (fat and lean meat content) before and after hibernating in naturalistic surroundings. Its multiple arousals during a 31-day hibernation period equated to a saving in energy of 88%, comparing favorably with a less direct and more time-consuming process (Mejías et al., 2022). It is worth noting the basic similarity between these two orders of South American marsupials and the Diprotodontia in Australia, despite separation when Gondwana broke up about 140 million years ago.

6.3.2.2 NST in Marsupials?

A comprehensive review left this question open (Jastroch et al. 2021), and I could not find more recent studies focused on it. Many of the early studies quite reasonably sought evidence for UCP1 and BAT, looking for increased metabolic rate and/or Tb rise after an injection of noradrenaline, particularly in cold-adapted subjects, those being the usual assays for BAT-generated NST in eutherian mammals. However, during the 1990s it became clear, as had been suggested by Augee (1978), that marsupials lack UCP1 and BAT (Hayward and Lisson, 1992; Rose et al., 1999; Kabat et al., 2003a, 2003b; Polymeropoulos et al. 2012; Jastroch et al. 2021), and this seems now to have become accepted. There had been, however, a lot of effort put into looking for NST while it was assumed that if marsupials had facultative NST it would be from UCP1-BAT, and much of that research was reviewed recently (Jastroch et al. 2021). The results of those early studies were mixed, which, in retrospect, is not surprising. The results from macropods' responses to noradrenaline were encouraging. A thermogenic response to NA was found in *Potorous tridactylus, Bettongia gaimardi, Perameles gunnii*, and *Macropus rufogriseus* (Ikonomopoulou and Rose 2006; Loudon et al. 1985; Nicol 1978; Rose et al. 1999), but responses to NA between marsupial groups differed. No response was measured in *Monodelphis domestica, Antechinus stuartii, Macropus rufogriseus*, and *Trichosurus vulpecula* (Nicol et al. 1997; Reynolds and Hulbert 1982). Other studies sought a morphological equivalent of BAT in interscapular tissue, but also without convincing success. As improving methodologies were developed, a substantial search for UCP1 in interscapular adipose tissue confirmed a UCP1 analogue in adipose tissue of the South American *Monodelphis domestica* and the Australian dunnart, *Sminthopsis crassicaudata* (Jastroch et al. 2008), but whether it could function to uncouple mitochondrial respiration was uncertain. That finding prompted a substantial further study (Polymeropoulos et al. 2012), which, among other results, found similar adrenergically induced thermogenic responses in both cold (12°C) and warm (24°C) adapted dunnarts and no enhanced response in the 12°C animals. That is, the UCP orthologue did not stimulate the interscapular tissue to an increased thermogenic response, which would have been expected if dunnarts had a classical, eutherian-like NST. Polymeropoulos et al. (2012) concluded that adaptive non-shivering thermogenesis does not occur in this marsupial, despite the cold recruitment of oxidative capacity and UCP1 in the interscapular fat deposit. They observed that the presence of NST in marsupials was controversial and suggested that without NST the arousals from torpor in marsupials might instead be effected by shivering, activity, and basking.

In the apparent absence of BAT and possibly because marsupials do seem to be getting on very well as whole-body endotherms without it, Jastroch et al. (2021) turned their attention to other possible facultative heat sources. They reviewed a few possibilities, including various factors known to augment metabolic rate in response to cold exposure, and speculated about an enhanced capacity for shivering. They also noted that the basal proton leak of isolated skeletal muscle mitochondria in the marsupial *Antechinus flavipes* did not change in cold acclimation (Jastroch et al. 2009). It was suggested that NST might be more difficult to identify in marsupials because they live in warmer climates, but many marsupials live routinely in cold habitats, many show daily torpor, and some hibernate, showing the same overall Tb pattern as do placentals (Geiser 2021;

Geiser and Cooper 2023). The possibility that marsupials could have an independent source of facultative NST in skeletal muscle was not embraced, citing only a paper negative to this view (Campbell and Dicke 2018). However, Jastroch et al. (2021, p. 1092) usefully drew attention to the technical difficulties of "untangling the contribution of muscle shivering vs. non-shivering when both processes are located in the same cell" and constructively acknowledged the possibility that some primary source of NST could be present in marsupials "given their ability to maintain high body temperatures in the cold." Looking ahead, the possibility that the putative source would be similar to that in Eutheria was kept alive by suggesting further work looking for thermogenic adipocytes in marsupials, and of course that possibility remains. Additionally, the review advised that, "the contribution to adaptive NST of other metabolically active organs, such as skeletal muscle, requires better mechanistic experimentation" (p. 1093).

From my perspective, as the periodic and rapid arousals associated with hibernation in placental hibernators involve BAT-based NST (Cannon et al., 2000; Cannon and Nedergaard, 2004), the just as impressive though less rapid arousals during hibernation in marsupials (Nicol et al. 2009) and the singleton hibernating monotreme and bird hibernators will almost certainly be found in due course to be just as reliant on a source of facultative NST, probably in skeletal muscle, although the possibility of a BAT analogue or some other source cannot yet be ruled out.

6.3.3 Eutherian (Placental) Mammals

Body temperatures are typically 36–37°C, but different taxonomic groups range from about 33°C in golden moles and pangolins to 39°C in hares and rabbits (Withers et al. 2016). There is considerable overlap with marsupials, more than seems to be generally recognized. Indeed, one of the conclusions from an analysis of "resting" Tb data collected from nearly 600 species of mammal (and nearly 500 species of bird) was that the "traditional ranking" of Tb increasing from monotremes to marsupials to placentals is somewhat misleading (Clarke and Rothery 2008). Interestingly, after allowing for phylogenetic effects, that same study found no striking relationships between body mass and Tb.

6.3.3.1 Torpor and Hibernation

These terms were defined previously, and patterns of both show a greater diversity of expression than in non-placental mammals, perhaps a consequence of having a greater diversity of taxa. In this section, too, the recent review by Geiser (2021) is very useful. As in marsupials, there are multiple occurrences of placental mammals undertaking departures from the maintenance of a high and stable Tb with many entering shallow daily torpor or long-term (multiday) deep torpor, i.e., hibernation. Estimates of the proportion of placentals expressing one of these heterothermies vary, but 30% is probably conservative. Much more research has been undertaken on placentals than either marsupial or monotreme heterotherms, so these events are much better understood in placentals. Nevertheless, the physiology of daily torpor and of hibernation between all three mammalian groups, particularly marsupials and placentals, appears to be very similar (Geiser 2021, Withers et al. 2016). Althoughonce thought to be adaptations to cope with cold, both torpor and hibernation can be expressed facultatively in mild conditions and most occurrences can be recognized primarily as providing opportunities for saving energy and/or water.

6.3.3.2 NST in Eutheria (Placental Mammals)

Until comparatively recently only one source of facultative NST was recognized: brown adipose tissue (BAT), activated by the uncoupling protein UCP1 (Cannon et al. 2000; Cannon and Nedergaard 2004). As mentioned earlier, there is growing acceptance that there is at least one additional independent source of facultative NST in placental mammals, probably within skeletal muscle, likely to be SERCA based (Bal et al. 2012; Bal et al. 2016; Nowack et al. 2017; Nowack et al. 2019), and included herein as muscle NST. Based on their likely distributions through time, the current hypothesis is that muscle NST in mammals is ancient, having supported synapsid whole-body endothermy throughout its existence, at least broadly similar to the muscle NST in extant birds supporting sauropsid endothermy, and that BAT-based NST evolved as an additional source in the Eutheria (Grigg et al. 2022; see Section 6.3.3). This interpretation is consistent with a genomic study finding that BAT-based NST evolved after the divergence between eutherians and marsupials (Saito et al. 2008).

The topic of facultative NST in eutherians is being addressed elsewhere in this book, so I will mention only a few points here that are relevant to the main topic: muscle NST and the evolution of whole-body endothermy. The SERCA-based source of heat was first described in rabbits, and the suggestion was made soon afterward that it could provide a mechanism for "facultative thermogenesis" (De Meis, 2001; De Meis et al. 2005). Subsequently it was shown in mice that the mechanism is upregulated in response to cold challenge and triggered by the protein sarcolipin (SLN) (Bal et al. 2012, 2016). That came at a time when it was known that marsupials and monotremes lack BAT (Hayward and Lisson 1992; Rose et al. 1999; Kabat et al. 2003a; Kabat, et al. 2003b; Jastroch et al. 2008) but are effective whole-body endotherms nevertheless (Sections 6.3.1, 6.3.2). Many eutherian mammals also lack BAT (Gaudry and Campbell, 2017), particularly larger mammals, or have it only in small amounts and seemingly manage without being compromised. In this, muscle NST is likely involved, but its extent may be challenging to quantify. By 2012, as mentioned earlier, it was also well known that a similar eye muscle–based heat source had been described in some fishes (Section 6.4.1.3). Several subsequent papers were pivotal in bringing muscle NST into a broader biological context, extolling the likelihood of its great age, likely occurrence in all mammalian groups, and suggesting an evolutionary significance (Nowack et al. 2017; Rowland et al. 2015a, 2015b).

The presence of BAT in many eutherian mammals makes it challenging to assess any contribution from muscle NST. Hence, the lack of BAT in the early development of wild-type boar piglets provided a significant opportunity for a particularly compelling study (Nowack et al. 2019). Compared to controls, these authors demonstrated that, as development of the piglets proceeded, heat production increased in response to cold exposure and shivering decreased. Further, this was accompanied by increased expression of SERCA1a and SLN as well as SERCA activity. The authors concluded that, "muscle NST may be the primary mechanism of heat production during cold stress in large mammals lacking BAT, strengthening the hypothesis that muscle NST has likely played an important role in the evolution of endothermy" (p. 1).

Further work since then seems to have further endorsed skeletal muscle as a source for NST, almost all of it on eutherians, which have BAT as well. In one such study, after the surgical removal of as much as possible of the interscapular BAT (about half the total BAT) from Brandt's vole (*Lasiopodomys brandtii*), followed by cold exposure at 4°C for 28 days, resected voles maintained the same Tb and metabolic response to cold as the sham-operated ones. They also had a higher expression of UCP1 in white adipose tissue and of SLN and SERCA in skeletal muscle (Liu et al. 2023). The authors interpreted the results as showing recruitment of skeletal muscle–based thermogenesis in compensation for the reduction in BAT, suggesting a functional interaction between the two thermogenic processes.

In addition to fundamental research, muscle NST has not been overlooked in the medical area. There is quite a lot of interest in treating obesity in humans by exposure to cold, that is, boosting metabolic rate by stimulating NST, obviating the need for losing weight via exercising. Adult humans have little or no BAT and, according to a recent assessment, there is so little of it that it can make only a small contribution to the overall energy expenditure generated by cold exposure. The suggestion is that the increased metabolic rate is from skeletal muscle, not just from shivering but from SERCA-based NST and other muscle-based processes (Townsend et al. 2023). However, in response, the size of the BAT deposit can increase (van der Lans et al. 2013). Attempts are afoot to find pharmacological agents that will stimulate muscle NST directly, thus avoiding discomfort to a patient having to spend long periods in the cold.

Finally, to a couple of papers reporting experimental work at the level of a single muscle fiber. In a study by Meizoso-Huesca et al. (2022), the release of heat from a single myofilament preparation was measured using a thermoluminescent dye under a range of experimental circumstances. This allowed detection of temperature changes as either ryanodine receptor 1 (RyR1) leak or SR Ca^{2+} pump activity was altered experimentally. Mouse (endothermic) and cane toad (ectothermic) filaments were compared. Results showed in both preparations that Ca^{2+} leak through ryanodine receptor 1 regulates thermogenesis in resting skeletal muscle. The leak was much greater in the mouse, and the researchers suspected that there was something about the properties of RyR1 and the SR Ca^{2+} pump in the mammalian skeletal muscle that is critical for it to become a heat generator. Although SLN has long been regarded as the agent uncoupling the sarcoplasmic reticulum Ca^{2+}-ATPase pump, it was not necessary in the muscle type used (*tibialis anterioris*), although it may be in others. Intrigued, I suspect, by the apparent similarity between the microanatomy of the myofibril and its RyR1 receptors and SERCA pumps in mouse and toad, the next study addressed a comparison between them and added human myofibrils as well (Singh et al. 2023). The conclusion was that an evolutionary loss from mammalian muscle of a Ca^{2+}-sensitive RyR isoform allowed the amplification of a futile cycle of Ca^{2+} through the sarcoplasmic reticulum, providing heat for mammals without a need for shivering. Again, the occurrence of thermogenesis without SLN involvement was noted, so that conundrum remains to be resolved. Another outcome from the study is evidence suggesting that the regulation of thermogenesis via muscle NST is likely to be under the control of the sympathetic nervous system. With muscle NST widely dispersed, adequate regulation is obviously crucial. To me these single myofibril experiments seem to provide a significant opportunity for making further comparison between muscle NST in mammals and birds (see Section 6.5).

6.3.4 Birds

The status of birds as whole-body endotherms is never questioned. Their normal body temperatures at rest are higher than in placental mammals, most about 42°C, except the ratite birds (ostrich, emu, cassowary, rhea), whose Tb averages about 39°C (Clarke and Rothery 2008). Their higher Tb is maintained by a metabolic rate higher than that of placental mammals and it is typically regulated quite tightly. They are typically homeothermic whole-body endotherms except that many use daily torpor and there is one known hibernator.

6.3.4.1 Torpor and Hibernation

Nearly half of the 30 or so taxonomic orders of birds have species that abandon their tight regulation at a set point and show daily torpor, defined in birds as a drop in Tb of more than 5°C (Geiser 2021). Most torpor is nocturnal, and warming occurs before the next activity period, effected by shivering and/or (assumed) muscle NST and commonly assisted passively by ambient conditions.

Concerning hibernation, only one species, the western North American common poorwill (*Phalaenoptilus nuttallii*), is known to extend its period of torpor long enough to qualify as hibernation. Woods and Brigham (2004) reported 13 occasions of birds remaining inactive for 10 or more days, the longest being 25 days. In a later study, Woods et al. (2019) sought to identify whether warming was passive or active, and, because poorwills hibernate in exposed sites, they shaded seven birds from the sun to eliminate or minimize passive rewarming. All shaded birds rewarmed nevertheless, one of them strikingly remaining in hibernation for 45 days, during which it showed eight arousal events (Figure 6.6). The authors noted the similarity of this pattern to that seen in hibernating mammals (!). They concluded that an active process was involved but could not resolve whether this was by shivering, NST, or a combination. With more than 10,000 bird species, there would be little surprise if more examples of hibernation were discovered. The occurrence of arousals in this sauropsid as well as in the synapsids provides strong support for the homology of hibernation and whole-body endothermy.

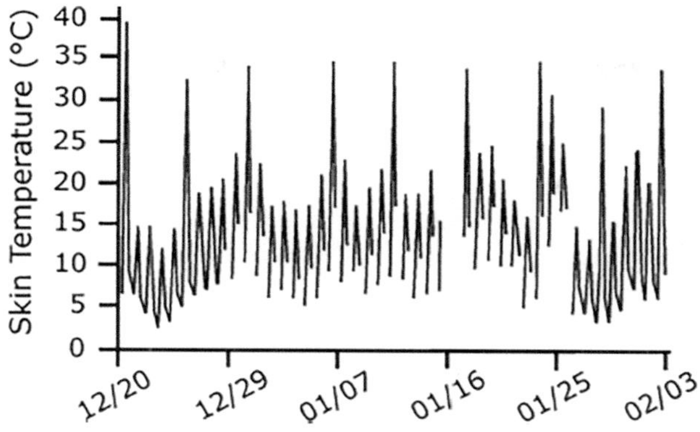

FIGURE 6.6 Body temperature of a common poorwill (*Phalaenoptilus nuttallii*) in southern Arizona, shaded to prevent passive solar warming, showing a 45-day period in hibernation during which it remained inactive but undertook eight brief arousals to normothermia. Because poorwills hibernate in more exposed sites than most hibernators, the daily cycles in Tb are quite large. Reprinted from Woods et al. (2019) with permission from Taylor and Francis.

NST in birds. This is the subject of another chapter, so there is only a brief entry here. By analogy with other whole-body endotherms and the many torpidators (Geiser 2021) and the periodic arousals to normothermia undertaken by the only known avian hibernator (Woods et al. 2019), birds must have a capacity for effective NST from at least one source. As they lack both UCP1 and BAT and have large amounts of skeletal muscle, it is no surprise that muscle has been looked to as the source. Foundational research into a potential mechanism was published around the turn of the century, focused on skeletal muscle (Bicudo et al. 2001, Bicudo et al. 2002). As discussed in Section 6.2.1, the biochemists on our team (Grigg et al. 2022) undertook a comparison between the putative biochemical characteristics of SERCA-based muscle NST in birds and mammals, as then understood, and found considerable similarity between them. The conclusion was that the extent of similarity was consistent with our hypothesis that their whole-body endothermy has a common ancestry but by itself is not proof, of course (Section 6.2). However, although skeletal muscle is the most likely source, there seems not to have been a huge amount of effort into detailing the actual mechanism since those papers in 2001 and 2002, until recently, and much uncertainty still surrounds it. Some of the reason for comparatively little attention may stem from recognition of the difficulty, mentioned by Jastroch et al. (2021), of unraveling the three processes coexisting within the muscle: locomotion, shivering, and, presumably, muscle NST. However, a timely review of the physiology of avian thermoregulation was published recently (Pani and Bal, 2022), discussing the nature and occurrence of shivering, including microshivering, and a review of muscle NST, the latter based on experience gained through researching muscle NST in placental mammals. It reported that the most promising of several potential molecular mechanisms proposed as potential sources of NST from avian skeletal muscle is SERCA based, with another potential muscle mechanism being adenine nucleotide translocase (ANT), a possible mechanism drawn attention to also by Walter and Seebacher (2009).

More recently, and departing from normal approaches using captive subjects, Pani et al. (2023) undertook an analysis of structural and biochemical changes in breast muscle between cool and warm seasons in field-captured pigeons (*Columba livia*). They found structural differences between seasons, perhaps in response to increased shivering in winter, and gene expression analysis showed winter upregulation of SERCA, SLN, and RyR, interpreted as evidence of additional recruitment of muscle NST. These findings are consistent with the current understanding that birds, as well as the three extant mammalian groups, respond to cold with a combination of shivering thermogenesis (ST) and NST, assumed to be from muscle NST in this case. Additionally, they found different results in different muscles, suggesting the possibility of a "division of labor" between muscles and, thereby, the capacity for independent control of ST and muscle NST, functioning independently or together.

6.4 NST AND THE EVOLUTION OF WHOLE-BODY ENDOTHERMY

To assess where NST "fits" in the evolution of whole-body endothermy requires knowledge about whether endothermy's multiple occurrences within amniote lineages arose by convergent evolution or from sharing a common ancestry dating from before the synapsids and sauropsids separated, probably in the Late Carboniferous. As stated in hundreds or even thousands of articles, scientific and those intended for public consumption, the former is accepted as correct: endothermy evolved independently in mammal and birds, and that has become the prevailing opinion. It is now a tenet. Further, the tenet apparently extends to include even endothermy's occurrences in their extinct ancestors in the synapsid and sauropsid lineages (e.g., Faure-Brac et al. 2024). However, much evidence accumulated in the last 30 years or so challenges the tenet.

This section reviews the recent growth of understanding that seems to be leading inevitably to a rejection of that tenet and an acceptance that whole-body endothermy's occurrences in amniotes have arisen from a common ancestry. In considering the possibility of two or more organisms having a common ancestry (i.e., being plesiomorphic), biologists usually have only extant individuals available on which to make such a call. Unusually, in this case we have some knowledge also of the history of whole-body endothermy in both synapsids and sauropsids. We also know of the occurrences of some of its components even earlier in the family tree. Accordingly, the section also reviews various features of extant fish and reptiles that probably carried over to the earliest amniotes, and whose recognition may make a common ancestry for endothermy easier to understand and accept. These features likely contributed to the evolution of whole-body endothermy when, on land and free from the thermal and respiratory constraints of living and breeding in water, selection pressures could have favored the evolution of what we now identify as whole-body endothermy.

Against that background the section will discuss how and where NST "fits" in the evolution of whole-body endothermy.

6.4.1 Independent Origins of Whole-Body Endothermies or a Common Ancestry?

As mentioned earlier, the prevailing tenet about the evolution of whole-body endothermy in mammals and birds is that it evolved independently in each—that is, by convergent evolution. However, both of the two pieces of evidence sometimes mentioned in support of its convergent evolution are under serious question. As discussed in Section 6.2, it can no longer be taken as read that NST in mammals relies only on BAT, whereas BAT is lacking in birds. And it is no longer considered that endothermy in each group arose quite late in the amniote family tree. The whole story has become much more complicated. Evidence for whole-body endothermy in both sauropsids and synapsids has been pushed further and further back in time, and this was well underway, as reviewed by Benton (2020) and others, before Grigg et al. (2022) focused more its pre-Mesozoic occurrences. Also, muscle-based sources of NST have been found in placental mammals as well as in birds, and muscle NST is coming to be recognized as the ancient source of NST in amniote whole-body endothermy, with evidence mounting that BAT in placentals (but not monotremes and marsupials) is an evolutionary "late-comer." This section builds on material presented earlier, presents additional material, and strengthens the more parsimonious case for whole-body endothermy in mammals and birds and their extinct sauropsids and synapsid ancestors having common ancestry.

6.4.1.1 Common Ancestry and Evidence for It from Extant Mammals and Birds

The thermoregulatory biology of each of the four clades of extant whole-body endotherms was summarized in Section 6.3. As that showed, the similarity across all four (monotremes, marsupials, eutherians, and birds) is quite striking. All have a capacity for maintaining homeothermy, and in all groups some species do so all the time. All groups have one or more species that abandon homeothermy periodically and strategically for energy and/or water management by undertaking daily torpor, and all have at least one species undertaking periods of hibernation, all of which include making periodic arousals. Responding to suggestions that torpor and hibernation are essentially the same process but at opposite ends of a continuum, Ruf and Geiser (2015) undertook a comprehensive physiological analysis and concluded that they are functionally quite distinct.

As we shall see, the departures from homeothermy during torpor or hibernation can be quite informative about whole-body endothermy's ancestry, perhaps in particular the occurrence of periodic arousals during hibernation and the similarity of this pattern across the three groups of mammals and the single bird hibernator (Section 6.3 and later in the chapter). The biological significance of the arousals is uncertain, but they are very intriguing, and there is no leading hypothesis emerging from a flotilla of suggestions. Interestingly, similar winter arousals are known among extant reptiles too (Section 6.4.1.3). I would wager that if any convincing explanation for them emerges in one group of hibernators, it will fit them all, including any reptiles so inclined.

From my perspective, the growth in understanding that has led to recognizing the strong likelihood that whole-body endothermy in amniotes has common ancestry goes back to a short paper published in 1992 by Michael Augee and Brett Gooden. Much of it can be traced to the emergence of the capacity to undertake studies of animals ranging freely in their natural habitat with body temperatures being monitored for long periods by radiotelemetry or, later, by data loggers. Prior to about 1990 it was widely held that occurrences of both torpor and hibernation were apomorphic, each having arisen independently to cope with cold and/or food shortage or whatever. Most of the earlier debate was focused on occurrences in the different groups of placental mammals. However, shortly after hibernation was described in a monotreme and a marsupial living in the Australian Alps (Grigg et al. 1989; Geiser and Broome 1991; Sections 6.1.3, 6.3.1, 6.3.2), Augee and Gooden (1992) noted its similarity in all three groups and pondered the multiple occurrences of such similar but very complex physiological and behavioral events. They concluded that the chances were very small that such similarity between the complex processes involved could have evolved independently in three distinct groups of mammals separated by millions of years. Referring to the occurrences of hibernation in placentals, monotremes and marsupials, they observed, "For these to be separate derivations of a complex phenomenon such as hibernation pushes the concept of convergence beyond its usefulness" (p. 175). The parsimonious explanation must be, they asserted, that hibernation in mammals is plesiomorphic. Subsequently, independently but for similar reasons, Malan (1996) noted similarities in the Tb patterns expressed by hibernators to those seen seasonally in extant reptiles and came to the same conclusion. He offered that identical mutation of all the genes underlying hibernation was so unlikely in multiple taxonomically distinct groups that the suite of relevant genes must have been ancestral. Referring to research on the European green lizard (*Lacerta viridis*), he interpreted its seasonal changes in thermoregulatory biology as a possible model for ancestral mechanisms underlying hibernation. Additionally, Lovegrove et al. (1999) interpreted the occurrence of torpor in elephant shrews (Macroscelidae) as support for a plesiomorphic origin of heterothermy. Pursuing echidna work further, Grigg (2004) and Grigg et al. (2004) developed a stepwise model for the evolution of whole-body endothermy within the context of it being plesiomorphic and we saw it as evolving via heterothermy. With short-beaked echidnas in mind (see Section 6.3.1), we proposed in 2004 that the earliest protoendotherms would have been facultatively endothermic, retaining their ectothermic capacity for "constitutional eurythermy" (*sensu* Eisentraut 1960), noting the significance of torpidating or hibernating mammals having tolerance for low Tb. Lovegrove (2012) embraced and extended that hypothesis and, much later, it was pleasing to see that in his extraordinary 2019 book *Fires of Life*, Barry Lovegrove, by then the doyen of endothermy evolution, referred to that concept as "compelling" (Lovegrove 2019). Adding to his earlier contribution, André Malan found further support for the plesiomorphy of mammalian hibernation in a paper dealing with acid-base physiology (Malan 2014).

In short, since the Augee and Gooden paper was published in 1992, there has been considerable support for the homology of mammalian occurrences of torpor and hibernation in mammals, sharing common ancestry. It is a small step to recognize that homology between the complex functions supporting hibernation in a monotreme, several marsupials, and many placentals is simultaneously proof of their endothermies also being homologous.

So what about birds? As related in Section 6.2, the recognition of skeletal muscle–based NST in placental mammals as well as birds (Rowland et al. 2015, Nowack et al. 2017) ignited speculation about the possibility that their endothermies too could have common ancestry with the three mammalian groups. To summarize briefly, following up that possibility, Grigg et al. (2022) reported sufficient biochemical similarity between NST in birds and what was then known about the muscle NST in each of the three mammalian groups to be consistent with them having a common ancestry. However, and very regrettably, at the time of writing what would have been the relevant part of our 2022 paper, I was unaware of the work on an avian hibernator, the common poorwill. This was discussed in Section 6.3.4, so I will be brief here. The only known avian hibernator is the common poorwill (*Phalaenoptilus nuttallii*), a type of nightjar (Family Caprimulgidae) living in western North America (Canada and USA). For many years it has been known to undergo periods of torpor, and Woods and Brigham (2004) recorded torpors of varying lengths up to 25 days, enough to be called hibernation, but lacking data on Tb. Subsequently, Woods et al. (2019) used temperature-sensitive radiotransmitters to monitor roost and skin temperature and found the Tb pattern during bouts of hibernation were very similar to those seen in hibernating mammals (Figure 6.6). Excitingly, for me anyway, the birds rewarmed periodically and briefly to normothermia just as occurs in mammalian hibernators. Whether this was achieved by shivering or NST was not able to be determined. One of their study birds was in hibernation for 45 days, during which it had eight brief, spontaneous arousals. But the similarity of the hibernation processes between this avian hibernator and mammalian hibernators and the combination of this and the similarities in muscle NST provide very strong evidence for whole-body endothermy in sauropsids and synapsids sharing common ancestry, probably in the Late Carboniferous. The findings of multiple occurrences of whole-body endothermy in both Sauropsida (to at least the Middle Permian with no earlier relevant literature available) and Synapsida (to the Late Carboniferous) are consistent with and provide indirect support for its implied ancient origin (Grigg et al. 2022).

Many features of the common ancestry of sauropsids and synapsids obviously date to the stem amniotes and earlier, so their capacity for evolving whole-body endothermy did not spring from nowhere. As we will now review, and in the next subsection, extant fishes and reptiles provide strong evidence that some of its endothermy-relevant features were very likely already in place within the vertebrate lineage much earlier than that.

6.4.1.2 Endothermy-Relevant Features in Extant Fishes

The early evolution of whole-body endothermy in Amniota (Grigg et al. 2022) is much easier to understand when we realize that one of its components, thought to be it SERCA-based NST, is known also in modern fishes. Certainly Cherlin (2023) came to that opinion (Section 6.2.3). The high thermal conductivity of water and the close association between blood and the water at the gills make achieving whole-body endothermy in water almost impossible. However, several groups of extant fishes show regional endothermy, indicating that responses to selection pressures favoring a warmer body occurred in vertebrates long before the amniotes. For example, tuna and mako sharks have a countercurrent arrangement of the blood vascular system within their lateral musculature that effects retention of the by-product heat from swimming (Carey and Teal 1966, 1969). Whether that is augmented by any SERCA-based or other thermogenic source is unknown so far. Interestingly, though, the metabolic rate of Pacific Bluefin tuna (*Thunnus orientalis*) is known to increase in cooler water (Blank et al. 2007), hinting at possible thermogenesis in response to cold (Ciezarek et al. 2019). SERCA-based thermogenesis has become well known in swordfish, marlin, and butterfly mackerel, which have modified muscle tissue adjacent to their eyes (Block and Finnerty 1994), which may improve vision (Block, 1994; Fritsches et al. 2005). It could be argued that this barely counts because the benefit assists food capture, not a regional increase in Tb that can benefit physical performance. However, opahs (*Lampris* sp.), a quite large actinopterygian fish, maintain much of their body a few degrees warmer than seawater by a variety of tricks (Wegner et al. 2015), and hints have emerged that SERCA-based thermogenesis may be operational in deep muscle tissue (Franck et al. 2019). The benefit to opah is thought to be in having an enhanced capacity to forage in deeper, colder water. Perhaps surprisingly, elements of SERCA-based thermogenesis have been found in much smaller fishes too, such as Japanese medaka (*Oryzias latipes*), in which the expression of SLN and SERCA1 transcripts increases in response to cold challenge (Robinson et al. 2024). The authors concluded that,

> the ability of medaka to respond to a cold challenge with an increase in key calcium cycling components, specifically the calcium pump and sarcolipin suggest that teleost fish share a conserved transcriptional program in response to cold stimuli with fish species that possess the requisite anatomical adaptations to conserve metabolic heat. (p. 111534).

I found that a thought-provoking result. Nevertheless, regional endothermy as mentioned in these examples does not confer whole-body endothermy status, and these fish remain bradymetabolic ectotherms. However, the several occurrences do show that, despite the constraints imposed by gill breathing, selection pressures for warmth and higher energy capacity do, and apparently did, exist even in the aquatic environment, and that SERCA-based thermogenesis has been recruited. It seems likely that many more examples will become recognized, particularly if an easy assay for its operation becomes available. Apart from that, it is interesting that selection pressures for warmth and its energetic benefits were operational in water despite the constraints. Bearing in mind the opinion that the elements characterizing skeletal muscle for contraction and relaxation also pre-adapted it for heat production (Cherlin 2023), and combined with its operation in several taxonomically distinct fish taxa, it is a fairly safe bet that SERCA-based thermogenesis was in the genomic toolkit available within stem amniotes.

6.4.1.3 Endothermy-Relevant Features in Extant Reptiles

Malan (1996) saw the mechanisms underlying torpor/hibernation in both mammals and birds as stretching back to their common ancestry with extant reptiles, and many features contributing to whole-body endothermy are already present and functional in living ectothermic reptiles. The full scope of these cannot be covered here, but, for example, as in birds and mammals, the modulation of peripheral blood flow allows control over rates of heat gain and loss in ectothermic reptiles, and there are multiple studies showing very sophisticated behavioral and physiological contributions to thermoregulation (Grigg and Seebacher 1999; Seebacher and Franklin 2005). Skeletal muscle, behavioral thermoregulation, and control over blood flow and thus heat flow through the periphery are characteristic of all terrestrial vertebrates, including frogs and salamanders, as well as whole-body endotherms, and it seems safe to assume that these attributes would have characterized the stem amniotes too. As discussed in Section 6.2.4, the ectothermy of today's crocodylians is accepted as a consequence of reversion from endothermic ancestry (Seymour et al. 2004), and they provide a good example of sophisticated behavioral and physiological thermoregulation (Grigg et al. 1998; Seebacher et al. 1999). Grigg et al. (2022) posited that the ectothermy of modern Squamata may be explained the same way or, alternately, that grouping non-archosauromorph "reptiles" in Sauropsida may not even be appropriate anyway; very early amniote fossils are few and far between, typically poorly preserved, and their phylogenetic context is often uncertain. Either way, their thermoregulatory biology is characteristic of typical vertebrate ectotherms. Interestingly, one squamate is known to show facultative endothermy, and it is hypothesized for at least two more. Some female pythons show facultative endothermy, coiling around their eggs and warming them by shivering (Hutchison et al. 1966; Harlow and Grigg, 1984). This comes close to qualifying as "facultative whole-body endothermy" driven by shivering. Whether there is a component of facultative NST as well is unknown. Also, a South American lizard, the tegu (*Salvator merianae*), shows transient evidence of endothermic capability during its reproductive season, most of which it spends in its burrow. Body temperatures up to 10°C above burrow temperature have been observed (Tattersall et al. 2016). This has been provocative, but a possible explanation has emerged recently. Hervas et al. (2024) found in both males and females that, during Spring, skeletal muscle fibers "increased oxidative phosphorylation" (p. e14162) and also "enhanced uncoupled respiration and heat production via adenine nucleotide translocase (ANT), but not via uncoupling proteins (UCP)" (p. e14162). They concluded that their study "indicates that the unique feature of transitioning to endothermy through non-shivering thermogenesis during the reproductive phase may be facilitated by higher mitochondrial density, function, and uncoupling within the skeletal muscle" (p. e14162). Furthermore, and perhaps the most interesting of all, is *Varanus rosenbergi*, which in Tasmania becomes dormant in winter but in late winter and early spring undertakes periodic, often daily, very brief warm-ups approaching or reaching its normal operating Tb before re-equilibrating to burrow temperature (Rismiller and Mckelvey, 2000). Some of the warm-ups could not be explained by passive warming, provoking tantalizing questions about an endothermic heat source, and the superficial similarity of these arousals to those seen during mammalian hibernation is impossible to ignore (Section 6.3). The authors observed that, "*V. rosenbergi* use mechanisms independent of but in addition to ambient temperature to regulate Tb" (p. 63). Periodic emergences from winter dormancy also occurred in captive South American tegus (*Salvator merianae*) (Sanders et al. 2015), and how commonly this may occur in other reptiles during brumation in the wild remains to be learned. Other reptiles too have been reported to undertake short arousals from dormancy, probably mostly passively, but it is a thin database; how many herpetologists monitor reptiles in the field throughout winter? However, the passive warm-ups in *V. rosenbergi* too are interesting. Overall, the lizard's annual Tb cycle in winter resembles that of a hibernating mammal and, like mammals, it does not use the brief period at normothermia for foraging. Hibernation biologists have speculated for years about the significance of the arousals in mammals, but so far there is no satisfactory explanation. Is it possible that the practice of a making brief returns to normothermia seen in all mammalian hibernators is expressing some sort of ancient requirement?

6.4.1.4 Common Ancestry and a Glimpse Back in Time

Sections 6.4.1.2 and 6.4.1.3 reviewed the endothermy-relevant attributes of extant fishes and reptiles as a basis for deducing what attributes the stem amniotes (ectotherms) may have had. Not unexpectedly, and if extant fishes and reptiles provide any guide, many of the physiological and behavioral components of thermoregulatory biology in endotherms would probably have been components in the ectothermic stem amniotes too, including a capacity for SERCA-based thermogenesis, whether expressed or not. Walter and Seebacher (2009) noted that metabolic processes are highly conserved among vertebrates and offered that, "the transition from ectothermy to endothermy would likely involve quantitative changes in (existing metabolic) pathways rather than de novo structures or processes" (p. 2328). Considering the points made in the current section and given the extent of homologies throughout the vertebrates (think of Shubin, 2008), it is not hard to see that, as ectotherms (Grigg et al. 2022, Faure-Brac et al. 2024), the stem amniotes would have had on board a platform of capabilities onto which whole-body endothermy could have been built comparatively easily. Cherlin (2023) observed that it was "understandable and unsurprising" that "the first terrestrial vertebrates (amniotes) were most likely animals with high levels of both metabolism and body temperature."(p. 392). I would not have gone that far—I have always thought of the very early amniotes as ectothermic—but I do agree that it is indeed understandable and unsurprising that the first terrestrial amniotes very likely had much of the genomic and other machinery giving them that capacity before the Late Carboniferous separation (Reisz 1972) of the two major lineages, Sauropsida and Synapsida, and that it could be expressed in response to appropriate selection pressures.

Concerning the first expressions of whole-body endothermy, we judged that the earliest amniotes were "probably ectothermic" and cited two papers in support (Grigg et al. 2022). We also noted, "We cannot tell whether the first expressions of endothermy pre-dated or followed the separation of sauropsids and synapsids" (p. 789). My inclination is toward "pre-dated," but both possibilities are consistent with the common ancestry hypothesis.

Unfortunately, Faure-Brac et al. (2024) mistakenly assumed we were asserting that even the stem amniotes were endotherms. Acting on that false premise, they conducted osteological analyses on about a dozen each of pre-amniotes and stem or early amniotes and found they were characterized by ectothermy. Although this confirmed our expressed opinion about the earliest amniotes, they used their findings to reject our common ancestry hypothesis, claiming their results supported the tenet of independent origins. This is not the place for a detailed response, but clearly their analysis is not any challenge to our proposal concerning common ancestry.

Faure-Brac et al. (2024) and many others apparently see the multiple occurrences of whole-body endothermy in both lineages evolving de novo, without an evolutionary heritage. That is the long-held tenet. In sharp contrast, we take the more parsimonious view. We see its multiple occurrences in amniotes as homologous, with an ancient, shared heritage, just as is the case for the many other bodily functions that we share with today's birds.

6.4.2 Muscle NST and the Earliest Whole-body Endotherms

It seems that Rowland et al. (2015a) were the first to suggest that muscle NST was the earliest facultative thermogenic mechanism in vertebrates. Subsequently, Nowack et al. (2017) suggested that muscle NST could have a role not only in responding to cold (or, presumably, driving an arousal from hibernation) but also in "the static elevation of resting metabolic rate" (p. 9). They apparently had in mind that heat from muscle NST could contribute to the rise in Tb accompanying the evolution of endothermy according to the "aerobic capacity" evolutionary model put forward by Bennett and Ruben (1979). I think that specific possibility has not been discussed much, but I too have wondered about the possibility of a more general, longer-term use of muscle NST. Could it be, for example, a contributor to the increase in Tb accompanying fever in humans? Could it be instrumental in contributing to longer-term, facultative seasonal changes in BMR?

Were the earliest expressions of whole-body endothermy stimulated by warmth generated by muscle NST? Because of the apparent association between facultative NST and hibernation, evidence of hibernation in the fossil record could perhaps have provided some information about the age of NST. There is some osteohistological evidence of interruptions to growth in otherwise fast-growing individuals (lines of arrested growth [LAGs]), and Farlow et al. (1995) and Padian and Horner (2004) reported that many LAGs have been interpreted as evidence of torpor or hibernation. One striking example is an interpretation of growth lines in tusk dentine in a series of specimens of the Early Triassic endothermic dicynodont synapsid *Lystrosaurus* to show that seasonal torpor occurred in what was a polar climate whereas growth was uninterrupted in a warmer climate (Whitney and Sidor 2020). However, in our survey of the paleontological literature (Grigg et al. 2022) we found too few sources to provide anything concrete. The feasibility of using osteohistological information for diagnosing seasonal interruptions to growth in whole-body endotherms has been researched by Köhler et al. (2012), and such occurrences can result in normally homeothermic mammals from harsh seasons in mammals, so care needs to be exercised in making interpretations.

I doubt it is known whether some form of NST is obligatory for whole-body endothermy or not. Many adult humans have little or no BAT, giving rise to the idea that as adults we do not need facultative NST. However, that was before evidence for a form of NST in skeletal muscle emerged. It will be interesting to see if future research reveals muscle NST operational for long periods of time. Some form of facultative muscle NST could logically have gone hand in hand with selection for a higher metabolic intensity. Cherlin's (2023) colorful offering that skeletal muscle is essentially pre-adapted for "inevitably" being repurposed for "non-contractive thermogenesis" carries an implication that the implementation of that repurposing would be comparatively straightforward. Interestingly, this may be the case. In their comparison of heat production by isolated myofibrils of mouse and toad, Meizoso-Huesca et al. (2022) found that heat generation was critically dependent on the magnitude of the RyR1 Ca^{2+} leak from the sarcoplasmic reticulum, and reported that the leak from a mouse myofilament was about 10 times that from a toad; implementing the switch in function from contraction and relaxation to heat production may not require significant restructuring.

The case supporting a common ancestry (plesiomorphy) (Section 6.4.1) of whole-body endothermy and its characteristics as displayed by today's four extant examples is strong without making any assumption about the presence of facultative NST. However, because of their common ancestry and the association between facultative NST and torpor and hibernation in all four groups, the implication is strong that facultative NST contributed to the earliest occurrences of whole-body endothermy, strengthening the suggestion made originally by Rowland et al. (2015a, 2015b) and its endorsement by Nowack et al. (2017).

6.5 FUTURE RESEARCH

So many uncertainties arise from the comparatively recent recognition of what seems to be a ubiquity of non-BAT–based NST in whole-body endotherms that many of the long-standing tenets about thermoregulatory physiology may need to be revisited. Some things that suggested themselves to me about future research questions and opportunities now follow. To answer the skepticism, the reality of a source (or sources) of facultative NST from skeletal muscle needs to be unambiguously confirmed, or not, and the fine details of the mechanics of it (or them) worked out. Whereas injecting noradrenaline provides an assay for UCP1-modulated BAT, the lack of an effective assay for muscle NST is a severe drawback. Another area of uncertainty is about the quantum of heat that muscle NST can provide. That was questioned for SERCA-based NST by Campbell and Dicke (2018) and needs to be taken into consideration for any postulated source. Perhaps further miniaturization and automation of the approach identified by Watts et al. (2022) could be useful. Jastroch et al. (2021) pointed out the technical difficulties of "untangling the contribution of muscle shivering vs. non-shivering when both processes are located in the same cell" (p. 1092). That difficulty is even greater in eutherians, with at least two separate sources of NST as well as shivering. More research on marsupials could be particularly useful, taking advantage of their lack of BAT. Another significant issue, in humans particularly, and with reference to an animal model, concerns the questions that will arise in working out which of the two separate sources of NST in eutherians does what and when, and which muscles are operational and when, and how all that is coordinated. Not all whole-body endotherms show torpor or hibernation, but there are occurrences of both in all four groups. There is a vast amount of information about both these abandonments of homeothermy, but I was surprised how little is recorded about the relative durations of ST and (probable) NST (warming without any visible shivering, as in the echidna, Figure 6.5). Lømo et al. (2020) and Hohtola and González-Alonso (2020) have focused on the possible importance of tonic motor activity as a thermogenic source, implying something quite different from SERCA-based NST. Could tonic motor activity and/or the slow movement and occasional twitches we observed during what may have been an NST phase of the echidna's warm-up (Figure 6.5) be related? Finally, the single myofibril experiments mentioned above seem to provide a significant opportunity for making further comparison between muscle NST in mammals and birds. In the event that bird and eutherian muscle NST prove to have even more similarities than we found, it could open opportunities for research on a more easily managed species that lacks BAT.

ACKNOWLEDGMENTS

I am very grateful to Lyn Beard for her enormous contributions during many years of fieldwork on echidnas and many other animals, and for the determination and attention to detail she always showed. As an example, and in the context of this chapter, without her persistence and a strong reluctance to give up on what had already been a substantial search for an apparently lost animal we would not have the data shown in Figure 6.5. I am also very grateful to David Kirshner, coauthor with me of a large book about crocodylian biology, for redrawing Figure 6.5. Finally, I wish to express my sincere thanks to Naresh Bal for inviting me to contribute to this volume.

REFERENCES

Augee, M. L. (1978). Monotremes and the evolution of homeothermy. *Australian Zoologist* 20, 111–119.

Augee, M.L., and Gooden, B.A. (1992). Monotreme hibernation – some afterthoughts. In '*Platypus and Echidnas*. (Ed. M. L. Augee.) pp. 174–176. Royal Zoological Society of New South Wales, Mosman, NSW.

Bal, N. C., Maurya, S. K., Sopariwala, D. H., Sahoo, S. K., Gupta, S. C., Shaikh, S. A., et al. (2012). Sarcolipin is a newly identified regulator of muscle-based thermogenesis in mammals. *Nature Medicine* 18, 1575–1579. doi: 10.1038/nm.2897

Bal, N. C., Maurya, S. K., Singh, S., Wehrens, X. H., and Periasamy, M. (2016). Increased reliance on muscle-based thermogenesis upon acute minimization of brown adipose tissue function. *Journal of Biological Chemistry* 291, 17247–17257. doi: 10.1074/jbc.M116.728188

Bal, N. C., Sahoo, S. K., Maurya, S. K. and Periasamy, M. (2018). The role of sarcolipin in muscle non-shivering thermogenesis. *Frontiers in Physiology* 9,1217. https://doi.org/10.3389/fphys.2018.01217.

Beard, L. A., Grigg, G. C., and Augee, M. L. (1992). Reproduction by echidnas in a cold climsate. In In *Platypus and Echidnas* (ed. M. L. Augee), pp. 93–100. Royal Zoological Society of New South Wales, Mosman, NSW.

Bennett, A. F. and Ruben, J. A. (1979). Endothermy and activity in vertebrates. *Science* 206, 649–654.

Benson, R. B. J. (2012). Interrelationships of basal synapsids: cranial and postcranial morphological partitions suggest different topologies. *Journal of Systematic Palaeontology* 10, 601–624.

Benton, M. J. (2020). The origin of endothermy in synapsids and archosaurs and arms races in the Triassic. *Gondwana Research* 100, 261–289. https://doi.org/10.1016/j.gr. 2020.08.003.

Bernard, A., Lécuyer, C., Vincent, P., Amiot, R., Bardet, N., Buffetaut, E., Cuny, G., Fourel, F., Martineau, F., Mazin, J.-M. and Prieur, A. (2010). Regulation of body temperature by some Mesozoic marine reptiles. *Science* 328, 1379–1382.

Bicudo, J. E., Vianna, C. R. and Chaui-Berlinck, J. G. (2001). Thermogenesis in birds. *Bioscience Reports* 21, 181–188.

Bicudo, J. E., Bianco, A. C. and Vianna, C. R. (2002). Adaptive thermogenesis in hummingbirds. *Journal of Experimental Biology* 205, 2267–2273.

Blank, J. M., Morrissette, J. M., Farwell, C. J., Price, M., Schallert, R. J. and Block, B. A. (2007). Temperature effects on metabolic rate of juvenile Pacific bluefin tuna *Thunnus orientalis*. *Journal of Experimental Biology* 210, 4254–4261.

Block, B. A. (1994). Thermogenesis in muscle. *Annual Review of Physiology* 56, 535–577.

Block, B. and Finnerty, J. (1994). Endothermy in fishes: a phylogenetic analysis of constraints, predispositions, and selection pressures. *Environmental Biology of Fishes* P.40, 283–302.

Brice, P. H., Grigg, G. C., Beard, L. A., and Donovan, J. A. (2002*a*). Heat tolerance of short-beaked echidnas (*Tachyglossus aculeatus*) in the field. *Journal of Thermal Biology* **27**, 449–457. doi: 10.1016/S0306-4565(02)00015-3

Brice, P. H., Grigg, G. C., Beard, L. A. and Donovan, J. A. (2002*b*). Patterns of activity and inactivity in echidnas (*Tachyglossus aculeatus*) free ranging in a hot dry climate: correlates with ambient temperature, time of day and season. *Australian Journal of Zoology* **50**, 461–475. doi: 10.1071/ZO01080

Brice P. H. (2009). Thermoregulation in monotremes: Riddles in a mosaic. *Australian Journal of Zoology* 57, 255–263 (2009). doi:10.1071/ZO09039

Brice P.H. and Grigg G.C. (2023). Modelling gigantothermy endorses the whole-body tachymetabolic endothermy of ichthyosaurs, mosasaurs, and plesiosaurs (Sauropsida) pp.312–333. In "*Ruling Reptiles – Crocodylian Biology and Archosaur Paleobiology*" (Eds. Woodward, H.N. and J. O. Farlow) Indiana University Press.

Camens A. B. (2010). Were early Tertiary monotremes really all aquatic? Inferring paleobiology and phylogeny from a depauperate fossil record. *PNAS* 107 (4) E12. www.pnas.org/cgi/doi/10.1073/pnas.0912404107

Campbell, K. L. and Dicke, A. A. (2018). Sarcolipin makes heat, but is it adaptive thermogenesis? *Frontiers in Physiology* 9, 714. doi: 10.3389/fphys.2018.00714.

Cannon, B., Golozoubova, V., Matthias, A., Ohlsson, K. E., Jacobsson, A. and Nedergaard, J. (2000). Is there life in the cold without UCP1? Uncoupling proteins and thermoregulatory thermogenesis. In "*Life in the Cold*" (eds G. Heldmaier and M. Klingenspor). Springer, Berlin.

Cannon, B. and Nedergaard, J. (2004). Brown adipose tissue: function and physiological significance. *Physiological Reviews* 84, 277–359.

Carey, F.G., and Teal, J.M. (1966) Heat conservation in tuna fish muscle. *Proceedings of the National Academy of Science, USA* 56:1464–1469

Carey, F.G., and Teal, J.M. (1969) Mako and porbeagle: warm-bodied sharks. *Comparative Biochemistry and Physiology* 28:199–204. https://doi.org/10.1016/0010-406X(69)91335-8

Carey, F. (1982). A brain heater in the swordfish. *Science* 216:1327–1329. https://doi.org/10.1126science.7079766

Cherlin, V. (2023). Preadaptivity of non-contractive thermogenesis in the evolution of warm-bloodedness in vertebrates. *Biology Bulletin Reviews* 143(4):375–392. DOI:10.31857/S004213242304004X

Ciezarek, A. G., Osborne, O. G., Shipley, O. N., Brooks, E. J., Tracey, S. R., McAllister, J. D., Gardner, L. D., Sternberg, M. J. E., Block, B. & Savolainen, V. (2019). Phylotranscriptomic insights into the diversification of endothermic *Thunnus* tunas. *Molecular Biology and Evolution* 36, 84–96.

Clarke, A. and Rothery, P. (2008). Scaling of body temperatures in mammals and birds. *Functional Ecology* 2, 58–67.

Cubo, J., Sena, M. V. A., Aubier, P., Houee, G., Claisse, P., Faure-Brac, M. G., Allain, R., Andrade, R. C. L. P., Sayão, J., and Oliveira, G. R. (2020). Were Notosuchia (Pseudosuchia: Crocodylomorpha) warm-blooded? A palaeohistological analysis suggests ectothermy. *Biological Journal of the Linnean Society*, 131(1), 154–162.

Cubo, J., Aubier, P., Faure-Brac, M. G., Martet, G., Pellarin, R., Pelletan, I., and Sena, M. (2023). Paleohistological inferences of thermometabolic regimes in Notosuchia (Pseudosuchia: Crocodylomorpha) revisited. *Paleobiology* 49(2), 342–352.

de Araújo Sena M.V., F.C. Montefeltro, T.S. Marinho, et al., (2023). The cost of living in Notosuchia (Crocodyliformes, Mesoeucrocodylia), *Palaeogeography, Palaeoclimatology, Palaeoecology* 632, 111855 https://doi.org/10.1016/j.palaeo.2023.111855

de Meis, L. (2001). Uncoupled ATPase activity and heat production by the sarcoplasmic reticulum Ca^{2+}-ATPase. Regulation by ADP. *Journal of Biological Chemistry* 276, 25078–25087.

de Ricqlés, A., Padian, K. and Horner, J. R. (2003). On the bone histology of some Triassic pseudosuchian archosaurs and related taxa. *Annales de Paléontologie* 89, 67–101.

de Meis, L., Arruda, A. P. and Carvalho, D. P. (2005). Role of sarco/endoplasmic reticulum $Ca^{(2+)}$-ATPase in thermogenesis. *Bioscience Reports* 25, 181–190.

Eisentraut, M. (1960). Heat regulation in primitive mammals and tropical species. *Bulletin of the Museum of Comparative Zoology, Harvard* 124:31–43.

Farlow, J. O., Dodson, P. and Chinsamy, A. (1995). Dinosaur biology. *Annual Review of Ecology and Systematics* 26, 445–471.

Faure-Brac, M.G., Woodward, H.N., Aubier, P., and Cubo, J. (2024) On the origins of endothermy in amniotes. *iScience* 27, 109375. https://doi.org/10.1016/j.isci.2024.109375

Franck, J. P. C., Slight-Simcoe, E. and Wegner, N. C. (2019). Endothermy in the smalleye opah (*Lampris incognitus*): a potential role for the uncoupling protein sarcolipin. *Comparative Biochemistry and Physiology Part A Molecular Integrative Physiology* 233, 48–52.

Fritsches, K. A., Brill, R. W. and Warrant, E. J. (2005). Warm eyes provide superior vision in swordfishes. *Current Biology* 15, 55–58.

Gaudry, M.J. and Campbell, K.L. (2017). Evolution of UCP1 transcriptional regulatory elements across the mammalian phylogeny. *Frontiers of Physiology*. https://doi.org/10.3389/fphys.2017.00670

Geiser, F., and Seymour, R. S. (1989). Torpor in a pregnant echidna, *Tachyglossus aculeatus* (Monotremata: Tachyglossidae). *Australian Mammalogy* **12**, 81–82.

Geiser, F. and Broome L.S. (1991). Hibernation in the mountain pygmy possum *Burramys parvus* (Marsupialia) *Journal of Zoology, London. 223*, 593–602.

Geiser, F. and Martin, G.M. (2013). Torpor in the Patagonian opossum (*Lestodelphys halli*): implications for the evolution of daily torpor and hibernation. *Naturwissenschaften* 100:975–981

Geiser, F. (2021). *Ecological physiology of daily torpor and hibernation*. Springer Nature, Cham.

Geiser, F. and Cooper, C.E. (2023). Daily torpor, hibernation, and heterothermy in marsupials. In: Cáceres NC, Dickman CR (eds.) *American and Australasian marsupials: an evolutionary, biogeographical, and ecological approach*. Springer. https://doi.org/10.1007/978-3-030-88800-8_43-1

Grady, J.M., Enquist, B.J., Dettweiler-Robinson, E., Wright, N.A., Smith, F.A. (2014). Evidence for mesothermy in dinosaurs. *Science* 344, 1268–1272. (doi:10.1126/science.1253143)

Grant, T. R., and Dawson, T. J. (1978*a*). Temperature regulation in the platypus, *Ornithorhynchus anatinus*: production and loss of metabolic heat in air and water. *Physiological Zoology* **51**, 315–332.

Grant, T. R., and Dawson, T. J. (1978*b*). Temperature regulation in platypus, *Ornithorhynchus anatinus* – maintenance of body temperature in air and water. *Physiological Zoology* **51**, 1–6.

Grant, T. R. (1983). Body temperatures of free-ranging platypuses, *Ornithorhynchus anatinus* (Monotremata), with observations on their use of burrows. *Australian Journal of Zoology* **31**, 117–122. doi: 10.1071/ZO9830117

Grigg, G. C., Beard, L. A. and Augee, M. L. (1989). Hibernation in a monotreme, the echidna (*Tachyglossus aculeatus*). *Comparative Biochemistry and Physiology Part A Physiology* 92, 609–612.

Grigg, G. C., Beard, L., Grant, T. and Augee, M. (1992b). Body temperature and diurnal activity patterns in the platypus (*Ornithorhynchus anatinus*) during winter. *Australian Journal of Zoology* 40, 135–142.

Grigg, G.C., Augee, M.L. and Beard, L.A. (1992a). Thermal relations of free-living echidnas during activity and in hibernation in a cold climate. In *Platypus and Echidnas* (ed. M.L. Augee (160–173). Royal Zoological Society of New South Wales, Sydney.

Grigg, G. C., Seebacher, F., Beard, L. A. and Morris, D. (1998). Thermal relations of large crocodiles, *Crocodylus porosus*, free-ranging in a naturalistic situation. *Proceedings of the Royal Society B Biological Sciences* 265:1793–1799.

Grigg, G.C. and Seebacher, F. (1999). Field test of a paradigm: hysteresis of heart rate in thermoregulation by a free-ranging lizard, *Pogona barbata*. *Proceedings of the Royal Society, B*. 266:1–7

Grigg, G. C. and Beard, L. (2000). Hibernation by echidnas in mild climates: hints about the evolution of endothermy? In "*Life in the Cold*" 2000 (eds G. Heldmaier and M. Klingenspor), pp. 5–20. Springer-Verlag, Berlin.

Grigg, G.C., Beard, L.A., Barnes, J.A., Perry, L.I., Fry, G.J. and Hawkins, M. (2003). Body temperature in captive long-beaked echidnas (*Zaglossus bartoni*). *Comparative Biochemistry and Physiology Part A Physiology* 136, 911–916. doi: 10.1016/j.cbpb.2003.09.004

Grigg, G. C. (2004). An evolutionary framework for studies of hibernation and short-term torpor. In "*Life in the Cold*" 2004: Twelfth International Hibernation Symposium (LITC 2004) (eds B. M. Barnes and H. V. Carey), pp. 131–141. Institute of Arctic Biology, University of Alaska, Aboard the MS Veendam, from Vancouver to Seward, Alaska.

Grigg, G. C., Beard, L. A. and Augee, M. L. (2004). The evolution of endothermy and its diversity in mammals and birds. *Physiological and Biochemical Zoology* 77, 982–997.

Grigg, G.C. and Kirshner, D.S. (2015). "Biology and Evolution of Crocodylians" 649 Pp. CSIRO Publishing, Clayton, Victoria; Cornell University Press, New York. http://www.publish.csiro.au/book/7185/

Grigg, G.C., Nowack, J., Bicudo, J. E. P. W., Bal, N. C., Woodward, H. N. and Seymour, R. S. 2022. Whole-body endothermy: ancient, homologous and widespread among the ancestors of mammals, birds and crocodylians. *Biological Reviews* 97:766–801. doi: 10.1111/brv.12822

Gyhrs, C., Macedo, T., Bastos, B., Salgado-Irazabal, X., Hammadi, M., Bouarakia, O, and Boratyński, Z. (2022). High level of daily heterothermy in desert gerbils. *Journal of Tropical Ecology* 38, 451–453. https://doi.org/10.1017/S0266467422000360

Harlow, P. and Grigg, G. C. (1984). Shivering thermogenesis in the brooding Diamond Python, *Python spilotes spilotes*. *Copeia* 1984, 959–965.

Harrell, L., Perez-Huerta, A. and Suarez, C. (2016). Endothermic mosasaurs? Possible thermoregulation of Late Cretaceous mosasaurs (Reptilia, Squamata) indicated by stable oxygen isotopes in fossil bioapatite in comparison with coeval marine fish and pelagic seabirds. *Palaeontology* 59 351–363.

Hayward, J. S. and Lisson, P. A. (1992). Evolution of brown fat: its absence in marsupials and monotremes. *Canadian Journal of Zoology*, 70, 171–179.

Hervas, L.S., do Amaral-Silva, L., Sartori, M.R., Guadalupe-Silva, A., Gargaglioni, L.H., Lerchner, J., Oliveira, M.T., Bícego, K.C. (2024) Mitochondrial function in skeletal muscle contributes to reproductive endothermy in tegu lizards (Salvator merianae). *Acta Physiologica*. e14162.

Hillenius, W. J. (1992). The evolution of nasal turbinates and mammalian endothermy. *Paleobiology* 18, 17–29.

Hillenius, W. J. (1994). Turbinates in therapsids: evidence for Late Permian origins of mammalian endothermy. *Evolution* 48 207–229.

Hohtola E. (2004). Shivering thermogenesis in birds and mammals. In "*Life in the Cold: Evolution, Mechanisms, Adaptation, and Application*". (eds. Barnes, B.M., Carey, C.V.). Institute of Arctic Biology, University of Alaska, Number 27 Fairbanks, AK; 241–252

Hohtola, E. and González-Alonso, J. (2020). Motor unit function during cold induced thermogenesis in muscle—New perspectives on old concepts. *Acta Physiologica* 228: e13408. https://doi.org/10.1111/apha.13408

Hutchison, V. H., Dowling, H. D. and Vinegar, A. (1966). Thermoregulation in a brooding female Indian python, *Python molurus*. *Science* 151, 694–696.

Jastroch. M., Withers, K.W., Taudien, S., Frappell, P.B., Helwig, M., Fromme, T., Hirschberg, V., Heldmaier, G., McAllan, B.M., Firth, B.T., Burmester, T., Platzer, M. and Klingenspor, M. (2008). Marsupial uncoupling protein 1 sheds light on the evolution of mammalian nonshivering thermogenesis. *Physiological Genomics* 32:161–169. DOI: 10.1152/physiolgenomics.00183.2007

Jastroch, M., Withers, K.W., Stoehr, S. and Klingenspor, M. (2009). Mitochondrial proton conductance in skeletal muscle of a cold-exposed marsupial, *Antechinus flavipes*, is unlikely to be involved in adaptive nonshivering thermogenesis but displays increased sensitivity toward carbon-centered radicals. Physiological and Biochemical Zoology 82:447–454. https://doi.org/10.1086/603631

Jastroch, M., Polymeropolous, E. T. and Gaudry, M. J. (2021). Pros and cons for the evidence of adaptive nonshivering thermogenesis in marsupials. *Journal of Comparative Physiology B* 191, 1085–1095.

Kabat, A. P., Rose, R. W., Harris, J. and West, A. K. (2003a). Molecular identification of uncoupling proteins (UCP2 and UCP3) and absence of UCP1 in the marsupial Tasmanian bettong, *Bettongia gaimardi. Comparative Biochemistry and Physiology Part B Biochemistry and Molecular Biology* 134, 71–77.

Kabat, A. P., Rose, R. W. and West, A. K. (2003b). Non-shivering thermogenesis in a carnivorous marsupial, *Sarcophilus harrisii*, in the absence of UCP1. *Journal of Thermal Biology* 28, 413–420.

Knaus, P. L., Van Heteren, A. H., Lungmus, J. K. and Sander, P. M. (2021). Higher blood flow into the femur indicates elevated aerobic capacity in synapsids since the reptile-mammal split. *Frontiers in Ecology and Evolution* 2021, 751238.

Köhler, M., Marin-Moratalla, N., Jordana, X. and Aanes, R. (2012). Seasonal bone growth and physiology in endotherms shed light on dinosaur physiology. *Nature* 487, 358–361.

Körtner, G. and Geiser, F. (1998) Ecology of natural hibernation in the marsupial mountain pygmy possum (*Burramys parvus*). *Oecologia* 113:170–178

Kuchel, L. J. (2003). The energetics and patterns of torpor in free-ranging *Tachyglossus aculeatus* from a warm-temperate climate. Ph.D. Thesis, University of Queensland, Brisbane.

Lambertz, M., Grommes, K., Kohlsdorf, T. and Perry, S. F. (2015). Lungs of the first amniotes: why simple if they can be complex? *Biology Letters* 11, 20140848.

Lee, M. S. (1997). Pareiasaur phylogeny and the origin of turtles. *Zoological Journal of the Linnean Society* 120(3), 197–280.

Lee, M.S.Y. (2013). Turtle origins: insights from phylogenetic retrofitting and molecular scaffolds. *Journal of Evolutionary Biology* 26:2729–38

Legendre, L. J., Guenard, G., Botha-Brink, J. and Cubo, J. (2016). Palaeohistological evidence for ancestral high metabolic rate in archosaurs. *Systematic Biology* 65, 989–996.

Liu, M., Zhang, X.-Y., Wang, C.-Z., and Wang, D.-H. (2023). Recruitment of muscle genes as an effect of brown adipose tissue ablation in cold-acclimated Brandt's Voles (*Lasiopodomys brandtii*). *International Journal of Molecular Sciences* 24, 342. https://doi.org/10.3390/ijms24010342

Lømo, T., Eken, T., Bekkestad Rein, E. and Njå, A. (2020). Body temperature control in rats by muscle tone during rest or sleep. *Acta Physiolica* 228, e13348.

Loudon A, Rothwell N, Stock M (1985) Brown fat, thermogenesis and physiological birth in a marsupial. *Comparative Biochemistry and Physiology Part A Comparative Physiology* 81:815–819.

Lovegrove, B. G., Lawes, M. J. and Roxburgh, L. (1999). Confirmation of pleisiomorphic daily torpor in mammals: the round-eared elephant shrew *Macroscelides proboscideus* (Macroscelidea). *Journal of Comparative Physiology B* 169, 453–460.

Lovegrove, B. G. (2012). The evolution of endothermy in Cenozoic mammals: a plesiomorphic-apomorphic continuum. *Biological Reviews* 87, 128–162. doi: 10.1111/j.1469-185X.2011.00188.x

Lovegrove, B. G. (2019). *Fires of Life: Endothermy in Birds and Mammals*. Yale University Press.

Lyson, T.R. and Bever, G.R. (2020) Origin and evolution of the turtle body plan. *Annual Review of Ecology, Evolution and Systematics* 51, 143–66. https://doi.org/10.1146/annurev-ecolsys-110218-024746

Malan, A. (1996). The origins of hibernation: a reappraisal. In "Adaptations to the cold" 10th International Hibernation Symposium (eds. Geiser, F., Hulbert, A.J., Nicol, S.C.). University of New England Press, Armidale. p. 1–6.

Malan, A. (2014). The evolution of mammalian hibernation: lessons from comparative acid-base physiology. *Integrative and Comparative Biology* 54(3), 484–496. doi:10.1093/icb/icu002

Martin, C. J. (1903). Thermal adjustment and respiratory exchange in monotremes and marsupials – a study in the development of homeothermism. *Philosophical Transactions of the Royal Society of London. Series B*. 195, 1–37. doi: 10.1098/rstb.1903.0001

Martinez, Q., Okrouhlík, J., Šumbera, R. *et al.* (2023). Mammalian maxilloturbinal evolution does not reflect thermal biology. *Nature Communications* 14, 4425. doi: 10.1038/s41467-023-39994-1

Meizoso-Huesca, A., Pearce, L., Barclay, C. J. and Launikonis, B.S. (2022). Ca^{2+} leak through ryanodine receptor 1 regulates thermogenesis in resting skeletal muscle. *Proceedings of the National Academy of Science U.S.A.*. 119, e2119203119. DOI: 10.1073/pnas.2119203119

Mejías, C., Navedo, J.G., Sabat, P., Franco, L.M., Bozinovic, F. and Nespolo, R. F. (2022). Body Composition and Energy Savings by Hibernation: Lessons from the South American Marsupial *Dromiciops gliroides*. *Physiological and Biochemical Zoology* 95(3), 239–250. https://doi.org/10.1086/719932

Nicol, S.C. (1978). Non-shivering thermogenesis in the potoroo, *Potorous tridactylus* (Kerr). *Comparative Biochemistry and Physiology Part C Comparative Pharmacology* 59:33–37. https://doi.org/10.1016/0306-4492(78)90008-4

Nicol, S.C., Pavlides, D. and Andersen, N.A. (1997). Nonshivering thermogenesis in marsupials: absence of thermogenic response to β3-adrenergic agonists. *Comparative Biochemistry and Physiology Part A Physiology* 117:399–405. https://doi.org/10.1016/S0300-9629(96)00357-X

Nicol, S., and Andersen, N. A. (2002). The timing of hibernation in Tasmanian echidnas: why do they do it when they do? *Comparative Biochemistry and Physiology Part B Biochemistry and Molecular Biology* **131**, 603–611. doi: 10.1016/S1096-4959(02)00018-0

Nicol, S. C., Vedel-Smith, C. and Andersen, N. A. (2004). "Behavior, body temperature and hibernation in Tasmanian echidnas (*Tachyglossus aculeatus*). In "Life in the Cold: Evolution,Mechanisms, Adaptation, and Application". Twelfth International Hibernation Symposium, (eds Barnes B. M. and Carey H. V.) Institute of Arctic Biology, University of Alaska, Fairbanks, AK, 149–157.

Nicol, S. and Andersen, N. A. (2006). Body temperature as an indicator of egg-laying in the echidna, *Tachyglossus aculeatus*. *Journal of Thermal Biology* 31, 483–490. doi: 10.1016/j.jtherbio.2006.05.00

Nicol, S. C., Andersen, N. A., Arnold, W., and Ruf, T. (2009). Rewarming rates of two large hibernators: comparison of a monotreme and a eutherian. *Journal of Thermal Biology* 34, 155–159. doi: 10.1016/j.jtherbio.2009.01.003

Nicol, S. C. (2017). Energy homeostasis in monotremes. *Frontiers in Neuroscience* 11, 195. https://doi.org/10.3389/fnins.2017.00195.

Nicol, S.C., Morrow, G.E. and Harris, R.L. (2019). Energetics meets sexual conflict: the phenology of hibernation in Tasmanian echidnas. *Functional Ecology* 33, 2150–60.

Nowack, J., Giroud, S., Arnold, W. and Ruf, T. (2017). Muscle non-shivering thermogenesis and its role in the evolution of endothermy. *Frontiers in Physiology* 8, 889. https://doi.org/10.3389/fphys.2017.00889

Nowack, J., Vetter, S. G., Stalder, G., Painer, J., Kral, M., Smith, S., Le, M. H., Jurcevic, P., Bieber, C., Arnold, W. and Ruf, T. (2019). Muscle nonshivering thermogenesis in a feral mammal. *Science Reports* 9, 6378.

Ostrowski, S. and Williams, J.B. (2006). Heterothermy of free-living Arabian sand gazelles (*Gazella subgutturosa marica*) in a desert environment. *Journal of Experimental Biology* 209, 1421–1429. https://doi.org/10.1242/jeb.02151

Padian, K. and de Ricqlès, A.J. (2020). Inferring the physiological regimes of extinct vertebrates: methods, limits and framework. *Philosophical Transactions of the Royal Society, London B: Biological Sciences* 375, 20190147.

Padian, K. Horner, J. (2004). Dinosaur Physiology. In *The Dinosauria* (eds Weishampel, P. D. D. B., Dodson, P. and Osmólska, H.) pp. 660–671. University of California Press, Berkeley.

Pani, P. and Bal, N. C. (2022). Avian adjustments to cold and non-shivering thermogenesis: whats, wheres and hows. *Biological Reviews* 97, 2106–2126, DOI: https://doi.org/10.1111/brv.12885

Pani, P., Swalsingh, G., Pani, S., Senapati, U., Sahu, B., Pati, B., Rout, S. and Bal, N.C. (2023). Seasonal cold induces divergent structural/biochemical adaptations in different skeletal muscles of *Columba livia*: evidence for nonshivering thermogenesis in adult birds. *Biochemical Journal* 480, 1397–1409. Doi: 10.1042/BCJ20230245

Phillips, M.J., Bennett, T.H. and Lee, M.S. (2009). Molecules, morphology, and ecology indicate a recent, amphibious ancestry for echidnas. Proceedings of the National Adademy of Sciences, U.S.A. 106 (40), 17089–94.

Polymeropoulos, E.T., Jastroch, M. and Frappell, P.B. (2012). Absence of adaptive nonshivering thermogenesis in a marsupial, the fattailed dunnart (*Sminthopsis crassicaudata*). *Journal of Comparative Physiology B* 182:393–401. https://doi.org/10. 007/s00360-011-0623-x

Pontzer, H., Allen, V. and Hutchinson, J. R. (2009). Biomechanics of running indicates endothermy in bipedal dinosaurs. *PLoS One* 4, e7783.

Pough, W. H. (1980). The advantages of ectothermy for tetrapods. 115, 92–112.

Reisz, R.R. (1972) Pelycosaurian Reptiles from the Middle Pennsylvanian of North America. *Bulletin of the Museum of Comparative Zoology at Harvard*, 144:27–62.

Reynolds, W, and Hulbert, A.J. (1982). Cold acclimation in a small dasyurid marsupial: *Antechinus stuartii*. In *Carnivorous marsupials*. (ed. Michael Archer) pp. 279–283 Royal Zoological Society of New South Wales, Mosman, NSW.

Rismiller, P.D. and Mckelvey, M.W. (2000). Spontaneous arousal in reptiles? Body temperature ecology of Rosenberg's goanna, *Varanus rosenbergi*. In *Life in the cold* (eds. Heldmaier G, Klingenspor M). Springer:Berlin.

Robinson, S., Hechter, D., Almoumen, F. and Franck, J.P.C. (2024). Sarcolipin (sln) and sarcoplasmic reticulum calcium ATPase pump (serca1) expression increase in Japanese medaka (*Oryzias latipes*) skeletal muscle tissue following cold challenge, *Comparative Biochemistry and Physiology Part A Molecular and Integrative Physiology* 287, 111534. https://doi.org/10.1016/j.cbpa.2023.111534

Rose, R.W., West, A.K., Ye, J., McCormack, G.H. and Colquhoun, E.Q. (1999). Nonshivering thermogenesis in a marsupial (the Tasmanian Bettong *Bettongia gaimardi*) is not attributable to brown adipose tissue. *Physiological and Biochemical Zoology* 72:699–704. https://doi.org/10.1086/316709

Rowland, L. A., Bal, N. C. and Periasamy, M. (2015b). The role of skeletal-muscle based thermogenic mechanisms in vertebrate endothermy. *Biological Reviews* 90, 1279–1297. doi:10.1111/brv.12157

Rowland, L. A., Bal, N. C., Kozak, L. P. and Periasamy, M. (2015a). Uncoupling protein 1 and sarcolipin are required to maintain optimal thermogenesis, and loss of both systems compromises survival of mice under cold stress. *Journal of Biological Chemistry* 290, 12282–12289. doi: 10.1074/jbc.M115.637603.

Ruben, J. (1995). The evolution of endothermy in mammals and birds: from physiology to fossils. *Annual Review of Physiology* 57, 69–95.

Ruf, T. and Geiser, F. (2015). Daily torpor and hibernation in birds and mammals. *Biological Reviews* 90, 891–926.

Saito, S., Saito, C. T. and Shingai, R. (2008). Adaptive evolution of the uncoupling protein 1 gene contributed to the acquisition of novel nonshivering thermogenesis in ancestral eutherian mammals. *Gene* 408, 37–44.

Sanders, C.E., Tattersall, G.J., Reichert, M., Andrade, D.V., Abe, A.S. and W.K. Milsom (2015). Daily and annual cycles in thermoregulatory behaviour and cardio-respiratory physiology of black and white tegu lizards. *Journal of Comparative Physiology B* 185:905–915 DOI 10.1007/s00360-015-0928-2

Schucht, P.J., Klein, N. and Lambertz, M. (2021). What's my age again? On the ambiguity of histology-based skeletochronology. Proceedings of the Royal Society B 288, 20211166. https://doi.org/10.1098/rspb.2021.1166

Seebacher, F., Grigg, G.C. and Beard, L.A. (1999). Crocodiles as dinosaurs: behavioural thermoregulation in very large ectotherms leads to high and stable body temperatures. *Journal of Experimental Biology* 202, 77–86.

Seebacher, F. and Franklin, C. E. (2005). Physiological mechanisms of thermoregulation in reptiles: a review. *Journal of Comparative Physiology B* 175: 533–541. DOI 10.1007/s00360-005-0007-1

Seymour, R. S. (1976). Dinosaurs, endothermy and blood pressure. *Nature* 262, 207–208.

Seymour, R. S., Bennett-Stamper, C. L., Johnston, S. D., Carrier, D. R. and Grigg, G. C. (2004). Evidence for endothermic ancestors of crocodiles at the stem of archosaur evolution. *Physiological and Biochemical Zoology* 77, 1051–1067.

Seymour, R. S., Smith, S. L., White, C. R., Henderson, D. M. and Schwarz-Wings, D. (2012). Blood flow to long bones indicates activity metabolism in mammals, reptiles and dinosaurs. *Proceedings of the Royal Society of London. Series B, Biological Sciences* 279, 451–456.

Seymour, R. S. (2016). Cardiovascular physiology of dinosaurs. *Physiology* (Bethesda) 31, 430–441.

Seymour, R.S., Ezcurra, M., Henderson, D., Jones, M.E., Maidment, S.C., Miller, C.V., Nesbitt, S.J., Schwarz, D., Sullivan, C. & Wilberg, E. (2019). Large nutrient foramina in fossil femora indicate intense locomotor and metabolic activity in Triassic archosauromorphs and the pseudosuchian lineage. p. 190 in Society for Vertebrate Paleontology Meeting Abstracts. Society for Vertebrate Paleontology.

Shubin, N. (2008). "Your Inner Fish: A journey into the 3.5 billion year history of the human body" Pantheon Books ISBN 978-0-375-42447-2

Singh, D.P., Pearce, L., Choi, R.H., Meizoso-Huesca, A., Wette, S.G., Scott, J.W., Lamboley, C.R., Murphy, R.M. and Launikonis, B.S. (2023). Evolutionary isolation of ryanodine receptor isoform 1 for muscle-based thermogenesis in mammals. Proceedings of the National Academy of Sciences U.S.A. 120, e2117503120

Tattersall, G. J., Leite, C. A., Sanders, C. E., Cadena, V., Andrade, D. V., Abe, A. S. and Milsom, W. K. (2016). Seasonal reproductive endothermy in tegu lizards. *Science Advances* 2, e1500951.

Townsend, L.K., Wang, D., Wright, D.C. et al. (2023). Skeletal muscle, not adipose tissue, mediates cold-induced metabolic benefits. *Nature Metabolism* https://www.nature.com/articles/s42255-023-00837-4

van der Lans, A.A., Hoeks, J., Brans, B., Vijgen, G.H. et al. (2013). Cold acclimation recruits human brown fat and increases nonshivering thermogenesis. *Journal of Clinical Investigation* 123(8):3395–403. doi: 10.1172/JCI68993

Walter, I. and Seebacher, F. (2009). Endothermy in birds: underlying molecular mechanisms. *Journal of Experimental Biology* 212, 2328–2336.

Watts, C.A., Haupt, A., Smith, J., Welch, E., Malik, A., Giacomino, R., Walter, D., Mavundza, N., Shemery, A., Caldwell, H.K., Novak, C.M. (2022). Measuring skeletal muscle thermogenesis in mice and rats. *Journal of Visualized Experiments*,185, e64264, doi:10.3791/64264.

Wegner, N. C., Snodgrass, O. E., Dewar, H. and Hyde, J. R. (2015). *Animal physiology*. Whole-body endothermy in a mesopelagic fish, the opah, *Lampris guttatus*. Science 348, 786–789.

Whitney, M. R. and Sidor, C. A. (2020). Evidence of torpor in the tusks of *Lystrosaurus* from the Early Triassic of Antarctica. Communications Biology 3, 471.

Withers, P.C., Cooper, C.E., Maloney, S.K., Bozinovic, F. and Cruz-Neto, A.P. (2016). "Ecological and environmental physiology of mammals". Oxford University Press, Oxford

Woods, C.P. and Brigham, R.M. (2004). The avian enigma: "hibernation" by common poorwills (*Phalaenoptilus nuttallii*). In: Barnes BM, Carey HV (eds) "Life in the cold: evolution, mechanisms, adaptations and applications", vol 27, 12th edn. University of Alaska, Fairbanks, pp 231–240.

Woods, C.P., Czenze, Z.J. and Brigham, R.M. (2019). The avian "hibernation" enigma: thermoregulatory patterns and roost choice of the common poorwill. *Oecologia*. 189:47–53.

Wright, T., Davis, R.W., Pearson H.C., Murray M. and M.S. Sheffield-Moore (2021). Skeletal muscle thermogenesis enables aquatic life in the smallest marine mammal *Science* 373, 223–225

7 Non-Shivering Thermogenesis in Birds

A Role for Skeletal Muscle

Kênia Cardoso Bicego
São Paulo State University, Jaboticabal, Brazil

Punyadhara Pani
KIIT School of Biotechnology, Odisha, India

José Eduardo Pereira Wilken Bicudo
University of São Paulo, São Paulo, Brazil

7.1 NST IS A PLAUSIBLE PHYSIOLOGICAL TRAIT PRESENT IN BIRDS

Birds and mammals share a common reptilian-like ancestor and have evolved somewhat similar traits along their evolutionary history [1–3]. For example, both birds and mammals express high body temperatures, high resting, and high maximal metabolic rates [4, 5]. On the other hand, both terrestrial and aquatic birds possess feathers to insulate their bodies whereas most terrestrial mammals possess fur and most aquatic mammals possess blubber. The anatomies of their respiratory systems are radically different. Birds have a complex system of parabronchi, a series of parallel tubes supplied by a cross-current arrangement of blood capillaries, and a system of air sacs functioning as bellows [6, 7]. This anatomical arrangement permits a unidirectional flow of air and thus a more efficient extraction of oxygen into the blood stream, from the passing atmospheric air over the gas exchange surfaces. Mammals have alveoli—tiny air sacs—within their lungs, with atmospheric air flowing in and mixed pulmonary air flowing out by means of a tidal system, which renders them a less efficient oxygen extraction from incoming air into the blood capillaries surrounding the alveoli. One striking consequence of these anatomical and physiological differences is that birds can fly at very high altitudes while mammals are not capable to withstand low barometric pressures occurring at very high elevations. Although birds and mammals are widely distributed around the earth, birds have a wider mobility range than their mammal counterparts because of their flight abilities, thus allowing them to explore different ecological niches unavailable to mammals.

The range of body mass variation in mammals is much greater than in birds, between 2 grams in the Etruscan pigmy shrew to around 150 tons in the blue whale, and less than 1 gram in the bee hummingbird to about 100 kg in the ostrich. A large body mass, more than 20 kg, prevents large birds like ostriches, emus, and rheas from flying. Larger body sizes, on the other hand, can be advantageous to prevent the passive flow of thermal energy from the animal to a cooler surrounding environment because of surface-to-volume relationships. In this context, small birds and small mammals show very high resting metabolic rates per unit body mass when compared to larger birds and mammals, thus balancing out heat losses and allowing them to keep their body temperatures at normal

 DOI: 10.1201/9781003391470-9

elevated ranges (euthermia) [8]. On the subcellular level, studies have shown that mitochondrial coupling efficiency in skeletal muscles of small birds [9] and mammals [10] falls at low metabolic intensities, meaning that under such conditions small endotherms uncouple their skeletal muscle mitochondria in order to increase heat production. However, there are exceptions related to interspecific variations among small mammals and birds. For example, extremely small mammals, like the African pigmy mice, *Mus mattheyi* [11], and small birds, like cold acclimated black capped chickadees, *Poecile attricapillus* [12], show increased mitochondria coupling efficiency in their skeletal muscle and liver, respectively. In their specific cases it seems that other constraints, besides just body size, might be predominant, like frequent locomotion to forage in the former and shivering thermogenesis in the latter. Larger birds, like the king penguin, *Aptenodytes patagonicus* [13], under long-term fasting conditions, also show elevated mitochondrial coupling efficiency, which can impact positively their locomotor and foraging performances. As a general trend, however, and considering that skeletal muscle is an important thermogenic tissue [14–17] in endotherms, the similarities found in most small birds and mammals investigated so far might be thus envisaged as a lever to facilitate heat generation (i.e., energy dissipated as heat) under resting condition when heat losses have to be compensated [18, 19]. Moreover, uncoupling their skeletal muscle mitochondria (i.e., diminished efficiency) would allow them to further increase heat generation (i.e., energy dissipated as heat), thus an important advantage for them while at rest when activity is quite low [10].

Some small birds and mammals, whenever they cannot sustain such high rates of metabolism, usually because of environmental constraints, like scarce food supply or lower ambient temperatures, are able to decrease considerably their metabolic rates and body temperatures, thus guaranteeing their survival. This trait is commonly known as torpor, which has important thermoregulatory implications, particularly during the rewarming phase to reacquire euthermic body temperature values [20–24]. Many bird species facing extreme cold weather conditions, on the other hand, do not undergo torpor. They resort to continuous shivering thermogenesis. Many small-sized passerines such as the goldcrest (*Regulus regulus*) and titmice (*Paridae*), for example, which live in northern temperate zones, spend most of the winter at ambient temperatures below their thermoneutral zone [25]. As a result, they engage in shivering thermogenesis for long periods of time, in some cases throughout many days. This ability may be expressed through the partial recruitment of subsets of the combined mass of highly aerobic skeletal muscle fibers embedded in the pectoral muscle.

One of the features that starkly differentiate birds from mammals, in most cases, is the rather compact and fusiform body shape of birds. Of course, flying and diving can greatly benefit from such traits because of aero- and hydrodynamic considerations, depending on the case, of course, and by the support of their massive and powerful pectoral muscles compactly encased in their fusiform bodies to exert the required mechanical power. The pectoral muscles of birds comprise around 30% of their total body mass [26–28]. Flying depends mostly on aerobic metabolism, and the pectoral muscles are therefore extremely rich in mitochondria and blood capillaries [29–31]. Notoriously, migratory birds can fly long distances, like the Pacific golden plover, which can cover nearly 30,000 km from Alaska/Siberia to New Zealand/Australia. This feat requires not only very powerful pectoral muscles but also an efficient fuel to be burned, and fat is the most natural candidate.

All these things put together may lead us to build an argument that birds could potentially make use of their massive pectoral muscles as a non-shivering thermogenic organ. In fact, birds have a relatively significantly larger muscle mass when compared to mammals and ectothermic reptiles. Irrespective of their mode of locomotion—flying, diving, or running—all birds possess hyperplasic muscles. The evolution of birds has been the result of various selective pressures, some of them associated with the evolution of feathers and flight. Nevertheless, evidence suggests that the loss of a single gene codifying for the uncoupling protein 1 (UCP1) [32] in the reptilian-like bird ancestor [33] led to a presumed "crisis" in which the survival of the avian lineage depended on the striking expansion of the skeletal musculature as one of the various compensatory changes that occurred during its evolutionary history. It is quite remarkable that the evolutionary transitions leading to birds was characterized by the loss of various genes [34]. Thus, according to this view, the survival of the avian lineage depended, during this transitional critical period, on anatomical and physiological specializations that turned its skeletal musculature into a more thermogenic tissue than that of mammals.

When we look more closely at bird development, it is possible to establish an intrinsic connection between a large increase in muscle mass and the special features of the skeleton of modern birds, which also made it possible for them to fly. However, the muscular hyperplasia could have been a determinant factor for the survival of oviparous endotherms, thought to be the ancestors of birds in the Jurassic [2, 35, 36]. Thus, a rather significant muscle mass expressed in adults could have been a much more decisive trait with respect to thermal energy transfer during egg incubation than the thermal insulation provided by feathers.

7.2 SKELETAL MUSCLE AS A MAJOR SOURCE OF NST IN BIRDS

In many birds, skeletal muscle makes up more than half of the body mass. Moreover, in some of them, like the Japanese quail (*Coturnix japonica*), it can reach even up to 70% [37]. So, skeletal muscle can be a major site for NST in birds. Several birds including chickens, ducks, and penguins are known to develop regulatory NST upon prolonged cold exposure, increasing their thermogenic capacity and improvement of cold endurance [38–41]. Overwintering at cold latitudes leads to numerous biochemical changes in the muscles of birds, especially in the flight muscles, which comprise ~20% of their total body mass [26, 42]. These changes include improved lipid transport capacity and utilization. The experimental studies reported on Muscovy ducklings show that blood flow and oxygen supply to the skeletal muscles remain high during long-term cold exposure even in the absence of shivering [43–45]. These experiments were confirmed by simultaneous electromyogram recordings [45]. In chicken chicks, the acquisition of cold tolerance was associated with an increase in the redness of the skeletal muscle, concomitant with the increase in the expression of genes associated with slow-twitch fibers and the decrease in those associated with fast-twitch fibers [46, 47]. This indicates a prioritization of a muscle phenotype mainly composed of slow-twitch muscle fibers. In this case, the energy supply and heat production seem to be a result of enhanced expression of mitochondrial oxidative genes and of the chicken homologue of peroxisome proliferator-activated receptor-γ coactivator-1α (PGC1-α), a transcriptional co-activator able to upregulate mitochondrial biogenesis [48, 49]. PGC1-α is indeed known to be involved in adaptive thermogenesis in mammalian brown adipose tissue (BAT) and skeletal muscle tissue. Furthermore, rats also show a transition from glycolytic to oxidative muscle fiber when exposed to cold.

During bird´s ontogeny, it has been observed that the skeletal muscle development is associated with the attainment of endothermy, i.e., the capacity to respond to cold and other conditions increasing thermogenesis [50–53]. In precocial birds, such as ducks and chicken, the embryonic thigh muscles, rather than the breast, show increased mass and mitochondrial oxidative phosphorylation up to hatching, when they attain an endothermic phenotype [54]. Thigh muscles are indeed the major source of shivering thermogenesis and locomotor activity just after hatching in many precocial species. These muscles may also be a source of NST at this early stage when the increased mitochondrial biogenesis during development is accompanied by increases in membrane leakiness, although at a lower rate than oxidative phosphorylation. In contrast, for altricial species, which hatch naked, blind, and with limited locomotor capacity, thermoregulatory competence is attained many days after hatching, following an initial rapid growth phase. For instance, in the red-winged blackbird, this occurs 6 to 8 days after hatching, before they fledge the nest on days 11–12 [55]. In this case, the gradual increase in oxidative phosphorylation, leakiness (see discussion later in the chapter), and PGC1-α expression during the first week after hatching is predominant in breast rather than in thigh muscles. More bird species, with distinct patterns of development, must be studied to allow us to gain a better understanding of the differences in muscle development associated with the thermogenic capacity acquisition in birds, especially with respect to NST.

There have been several attempts to understand the molecular mechanisms behind cold-induced NST in avian skeletal muscle. Oxygen supply, free fatty acid load, and mitochondrial abundance all tend to increase in muscle after cold acclimation, consistent with the role for muscle NST in addition to that of shivering [56–61]. However, which of the proposed molecular mechanisms is the prime for NST in birds remains unresolved. Animal cells store metabolic energy as electrochemical gradients. Thus, the main cell sources of heat production in different conditions of an animal´s live comes from the transport of H^+ across the mitochondrial inner membrane, associated with oxidative phosphorylation and leakiness, and from the transport of Ca^{2+} across the endoplasmic reticulum membrane, associated with calcium storage for cell activity [62, 63]. Mitochondrial function will be approached first, simply because the mitochondrial machinery has evolved before the Ca^{2+} transport mechanisms during the evolution of animals.

7.3 MUSCLE NST VIA MITOCHONDRIAL LEAKINESS

The leakiness of any biological membrane can be regulated by the composition of membrane lipids and the modulation of ion channels [64]. These alterations may affect the amount of energy dissipated as heat and the cost of metabolic energy. One mechanism proposed to explain avian muscle NST arises through the inefficiency of mitochondrial metabolism. Any increase in the leakage of mitochondrial protons has been suggested to reduce mitochondrial efficiency and thus increase energy dissipation as heat. The presence of a proton gradient across the inner mitochondrial membrane, which is produced by the mitochondrial electron transport system and uses oxygen in the process, is necessary for the oxidative phosphorylation process that is required to produce ATP [13, 65, 66]. Leakage of protons across that membrane uncouples the link between oxygen consumption and ATP synthesis (OxPhos). This process leads to a reduction in efficiency and an increase in the dissipation of extra energy as heat. In this way, the animal's energy needs may affect how closely the electron transport chain and ATP production are coupled: during times of low substrate availability, proton leakage can be minimized to maximize ATP production. This has been studied in the mitochondria-rich BAT of eutherian mammals where such uncoupling is the basis for NST [67] (see Chapters 5 and 6). However, analogous processes have been reported for other tissues and organs, both in mammals and in organisms that lack BAT. It has been hypothesized that skeletal muscle mitochondrial leakiness can occur in several mammalian species [68–70], and there is scarce information on a similar process occurring in avian muscle (see discussion later in the chapter). Avian red blood cells (RBCs), which do not lose their nucleus and organelles during maturation, have also been found to have leaky mitochondria. In the RBCs of sympatric coal (*Periparusater*) and great tit (*Parus major*), both mitochondrial volume and leakiness are upregulated in the winter [71], but its possible role in avian NST remains to be determined.

7.3.1 UNCOUPLING VIA ADENINE NUCLEOTIDE TRANSLOCATOR (ANT)

Based on its correlation with a decreased mitochondrial membrane potential, ANT has been proposed as a component of the mechanism involved in mitochondrial uncoupling. ANT proteins are often the most abundant in the mitochondrial inner membrane, comprising ≈10% of all proteins. In avian skeletal muscle, ANT expression at the mRNA level has been seen to increase during cold acclimation. In gastrocnemius muscle of male Muscovy duckling, the ANT expression is 1.7-fold higher after prolonged exposure to cold [72]. Repeated exposure to cold water of juvenile king penguins (*Aptenodytes patagonicus*) leads to enhanced expression and activity of avANT [73, 74]. In the case of laying-type chickens, elevated expression of avANT mRNA occurs during an adaptive response to mild cold, with a concomitant increase in feeding efficiency [74]. Upregulation of ANT is also observed in the skeletal muscle of cold-acclimated mammals, suggesting that ANT is part of a cold-mediated remodeling of skeletal muscle [3, 75–77]. Furthermore, during ontogenesis of endothermy in chickens, not only the expression of ANT in skeletal muscle increases but also the proportion of ANT-induced mitochondrial uncoupling increases 8 days after hatching [78]. As a vital link between the mitochondrial and cytosolic compartments of cells, ANT's primary function is to catalyze the exchange of cytosolic ADP with mitochondrial ATP across the inner mitochondrial membrane; this protein is also called ATP/ADP carrier. Thus, ANT is one of the most ancient membrane proteins in the eukaryotic cell and appears to contribute to basal proton leak in a quantitative manner [79]. A parallel mechanism has been proposed for ANT to mediate proton leak in the skeletal muscle, like that of UCP1 in brown adipocytes. A recent study used a series of mitoplast-based electrophysiological experiments of different tissues, including skeletal muscle, from wild-type and ANT1-knockout mice. That study demonstrated the existence of fatty acid–activated proton leak dependent on the ANT1 and ANT2 isoforms in skeletal muscle. Both ANT and UCP1 require protonatable fatty acids (FA) for activation of proton transport, but FA appear to bind within the ANT translocation pathway as cofactors, while they are cotransported with H^+ through the UCP1 [80]. The action of ANT and UCP1 are negatively modulated by purine nucleotides, such as ADP for both ANT and UCP1, and GDP for UCP1. In the case of ANT, under situations of high ATP production, proton gradient dissipation by ATP synthase will be significant, and ANT activity will also be simultaneously amplified to export ATP, which, in turn, might be utilized by other mechanisms involved in heat production. Such a scenario of ANT uncoupling and direct effect on thermogenesis in avian skeletal muscle has not been carefully analyzed yet. However, it is conceivable to think that such scenario might also happen in birds, as recent evidence of the association between muscle thermogenesis and ANT uncoupling has been shown in a subtropical ectothermic reptile, the tegu lizard *Salvator merianae*, during spring reproduction when it presents facultative endothermy [81–83].

7.3.2 Action of Uncoupling Proteins (UCP)

The UCPs belong to a family of mitochondrial proteins, the same family of ANT, which may cause a leak of protons across the inner membrane, thus uncoupling the oxidation of reduced substrates from the phosphorylation of ADP to ATP. Although adaptive thermogenesis is the primary physiological function of the mammalian UCP1 (brown fat specific), the functions of the UCP2 (ubiquitous) and UCP3 (skeletal muscle and adipose tissue) still require careful consideration. There is some evidence that mitochondrial loose coupling may aid in the cold-induced heat production in skeletal muscle in the absence of both mammalian thermogenic BAT and UCP1 in birds. This mitochondrial mechanism had once gained support at the molecular level from the identification of a functional homologue of mammalian UCP3 in avian skeletal muscles, called avian UCP (avUCP). In cold-acclimated male Muscovy ducklings [84], chickens [78], and king penguins, higher levels of avUCP were found to be expressed in their skeletal muscles [73, 85]. In birds that had endured prolonged exposure to the cold, the mRNA of avUCP was upregulated, and in penguins that had become acclimated to cold water, its activity had increased [86]. Based on these findings, it was proposed that avUCP could form the basis of NST metabolic heat production in avian muscle analogous to the role of UCP1 in mammalian BAT. According to later studies, however, avUCP was not found to be involved in heat production and hence may not be the main mediator of muscle NST in birds. Moreover, fasting birds have increased levels of avUCP, which suggests that this protein may be crucial for free fatty acid mobilization into muscle cells [47, 87–89]. It also appears promising that avUCP may play a different, alternative role in controlling oxidative stress in skeletal muscle during cold acclimation. Besides avUCP, a HmUCP has been recognized in hummingbirds, specifically related to rewarming from torpor. The predicted amino acid sequence of HmUCP based on the cDNA fragment amplified has 72% identity with human UCP3 and 70% with rat UCP2 while only 55% identity with rat UCP1 (from BAT). The mRNA levels of HmUCP in the skeletal muscle (pectoral), as well as in the heart, were much higher during torpor than during euthermic state, indicating a possible role of the HmUCP in NST in the skeletal muscle during rewarming from torpor [31]. Therefore, further investigation is required to determine whether avUCP and HmUCP contribute to mandatory or regulatory thermogenesis.

The UCP1 is considered to be the exclusive UCP isoform of the mammalian BAT. For more than three decades, BAT has been deeply studied as the main site for NST in mammals [90]. In this context, because birds do not possess BAT, this has been an argument in the past to exclude birds from presenting NST. The BAT heat production is a result of a maximization of mitochondrial proton leak by the action of the uncoupling protein UCP1 as an inducible proton channel to release the proton motive force as heat [91]. Although the UCP1 orthologue exists in fish and amphibians, it looks like it has been lost in non-avian reptiles, birds, and many lineages of eutherian mammals. We now know that BAT and the importance of UCP1 for heat production seem to be an evolutionary acquisition restricted to eutherian mammals (absent in marsupials and monotremes). Even among eutherians, BAT does not seem to be a main thermogenic organ in adult larger mammals such as humans, ruminants, and rabbits. Moreover, skeletal muscle, but not BAT, is the tissue involved in NST in pigs [92, 93]. Thus, throughout the evolutionary history of vertebrates, before the acquisition of BAT in eutherian mammals, skeletal muscle has been a key site for NST, as demonstrated in fish, mammals, and birds. It is interesting to note, however, that BAT and skeletal muscle seem to be closely related tissues. These two cell types (brown adipocyte and skeletal fiber) originate from a common progenitor cell, the one that expresses myogenic factor 5 (Myf5), which was previously believed to be exclusive to skeletal muscle [94]. Furthermore, the brown adipocytes have origin and characteristics distinct from white adipocytes, which makes it curious that BAT has characteristics closer to muscle tissue than to white adipose tissue [95].

7.4 FUTILE CALCIUM CYCLING AND FREE FATTY ACIDS–MEDIATED NST IN BIRDS

Birds primarily depend on the ATP-dependent Ca^{2+} cycling via sarcoplasmic reticulum Ca^{2+} ATPase (SERCA), activated by Ca^{2+} leakage from the ryanodine receptor (RyR), which is mediated by fatty acids [96]. *Futile calcium cycling* refers to this Ca^{2+} cycling between the cytosol and sarcoplasmic reticulum (SR) through SERCA that is uncoupled from the action of contractile filaments. Studies on the skeletal muscle of birds gave rise to the hypothesis that SERCA can act as a heat pump [17]. In duckling skeletal muscle, cold acclimation was shown to upregulate Ca^{2+}-cycling proteins, such as SERCA and RyR, indicating a significant involvement in muscle thermogenesis [97–99]. Muscle homogenates and heavy SR microsomal fractions from cold-acclimated ducklings showed a 30–50% increase in Ca^{2+} transport activity [31, 98, 99]. Moreover, a recent study has reported the upregulation of SERCA2 and RYR1 in four different skeletal muscles (pectoralis, biceps, triceps, and iliotibialis) of tropical pigeons during winter season in comparison to that of summer [100].

According to some research, SERCA activity operating against the Ca^{2+} electrochemical gradient may experience uncoupling, or the separation of Ca^{2+} uptake from ATP hydrolysis, which would result in higher ATP utilization and heat production [101–103]. The extent to which ionic concentration is responsible, whether alone or with other factors having a role in controlling SERCA uncoupling, is still a mystery and needs careful consideration in future research. Sarcolipin (SLN), a SERCA uncoupler, has been identified as a potential regulating mechanism [104]. The role of avian SLN had not been fully investigated for years, even though SLN function in muscle NST has been characterized in mammals. SLN is a single transmembrane protein that interacts with SERCA in the intramembranous area, according to in vitro and structural investigations. To increase ATP utilization and heat production, SLN can bind to the Ca^{2+}-bound form of SERCA and enable its ATP hydrolysis activity while obstructing Ca^{2+} transport across the SR membrane. It has been demonstrated that cold acclimation raises SLN expression in skeletal muscle in mammals [105–108]. Surprisingly, SLN expression was found to be downregulated after cold acclimation in a study on adult dark-eyed juncos [109]. However, this investigation involved just exposure to extreme cold and was conducted via transcriptome analysis without mRNA level validation or SERCA/SLN ratio assessment. In contrast to previous findings, a recent study has shown that along with SERCA2 and

RYR1, SLN is also highly upregulated in free-ranging pigeon skeletal muscle NST during tropical cold [100]. This study is the first to document the role of SERCA-SLN mechanism in avian skeletal muscle. SLN is upregulated 25-fold during the period of migratory preparation, according to a recent study on migrating white-throated sparrows (*Zonotrichia albicollis*), which may indicate a role in avian flight. SLN may be involved in fattening for migration, as suggested for grey catbirds (*Dumetella carolinensis*) submitted to photoperiod manipulation to simulate changes in season [110]. In this case, in response to long-day photo stimulation (preparing for migration), the increase in the SLN transcription was found to be strongly correlated with fat mass accumulation but not with SERCA activity, SR membrane Ca^{2+} leak, Ca^{2+} uptake rates, pumping efficiency, or mitochondrial biogenesis. Recent molecular-level research on the relationship between SLN and SERCA, however, revealed that SLN increases SERCA-based heat production, at least in mammalian muscle. Moreover, the outcomes of the recent study on free-ranging adult pigeons mentioned above will ignite research to delineate SERCA-SLN–based NST experiments in other avian systems [100]. The experiments in this study were conducted in a tropical climate where birds were divided into two groups as per seasons, i.e., winter and summer birds. This study is the first of its kind to report unique thermogenic recruitment of different muscles in adult pigeon in the face of seasonal cold (Figure 7.1). In contrast to the common belief, this study highlights that pectoralis is not representative of all the avian skeletal muscles, as other muscles show very distinct winter-induced adaptations in free-ranging adult pigeons. However, the role of SLN and the extent of SERCA-mediated NST in different avian species might be different and need further research.

Although it appears that the core of NST is the activation of SERCA-mediated Ca^{2+} cycling in muscle in response to increased cytosolic Ca^{2+} levels, it is not entirely clear how the cytosolic Ca^{2+} concentration can be increased without causing contraction. The role of other putative mechanisms, such as Ca^{2+} entry through the sarcolemma via store-operated calcium entry (SOCE), has not yet been explored. RyR1-mediated SR Ca^{2+} release has been hypothesized as a mechanism [111]. The stromal interaction molecule (STIM) and Orai proteins, which are situated on the SR and sarcolemma, respectively, promote SOCE by physically interacting to transport Ca^{2+} from the extracellular fluid into the sarcoplasm. Earlier research suggested that SOCE is only recruited when the SR store has been exhausted, but later investigations showed that SOCE-mediated Ca^{2+} entry can operate as a spatially localized signal to start particular Ca^{2+}-signaling processes [112–114]. It still must be clearly demonstrated whether SOCE can be controlled by an unidentified upstream mechanism during exposure to cold that can activate SERCA and establish futile Ca^{2+} cycling without muscular contraction or shivering.

Free fatty acids (FFAs), particularly palmitoyl carnitine, may have a role in the release of SR Ca^{2+}. Palmitoyl carnitine increased ryanodine binding in isolated SR and promoted Ca^{2+} release from RyR1 in cold-acclimated Muscovy duckling muscles [115, 116]. Recent research has demonstrated that various pathophysiological circumstances in animals, in which muscle FFA levels may be high, include RyR1-mediated Ca^{2+} release [117, 118]. Some birds and most mammals are known to have higher levels of FFAs following cold adaptation [119, 120]. In cold-acclimated ducklings and pigeons, increased FFA absorption in the skeletal muscle was found along with a corresponding increase in mitochondrial respiration [121]. A higher level of ATP generation in skeletal muscle may facilitate ATP utilization through the Ca^{2+}-cycling mechanism, according to studies of mitochondrial membrane potential. It appears likely that FFAs trigger RyR1-mediated Ca^{2+} leakage in addition to activating mitochondrial ATP production, resulting in fruitless Ca^{2+} cycling in the absence of muscle contraction. The mechanism triggering futile Ca^{2+} cycling in avian skeletal muscle is still unclear, and endogenous agonists including thyroid hormones, glucagon, and adrenaline may play a part [122, 123]. Interestingly, as mentioned earlier, ANT is involved in the uncoupling effect of FFAs on mitochondria [79]. Moreover, the upregulation of ANT expression in skeletal muscle upon cold acclimation in rodents is highly correlated with the expressions of SLN and SERCA [77, 124]. Thus, there is a possibility of a synergistic action of SERCA and ANT for uncoupling sarcoplasmic Ca^{+2} transport and mitochondrial OxPhos, respectively, with consequent further heat production.

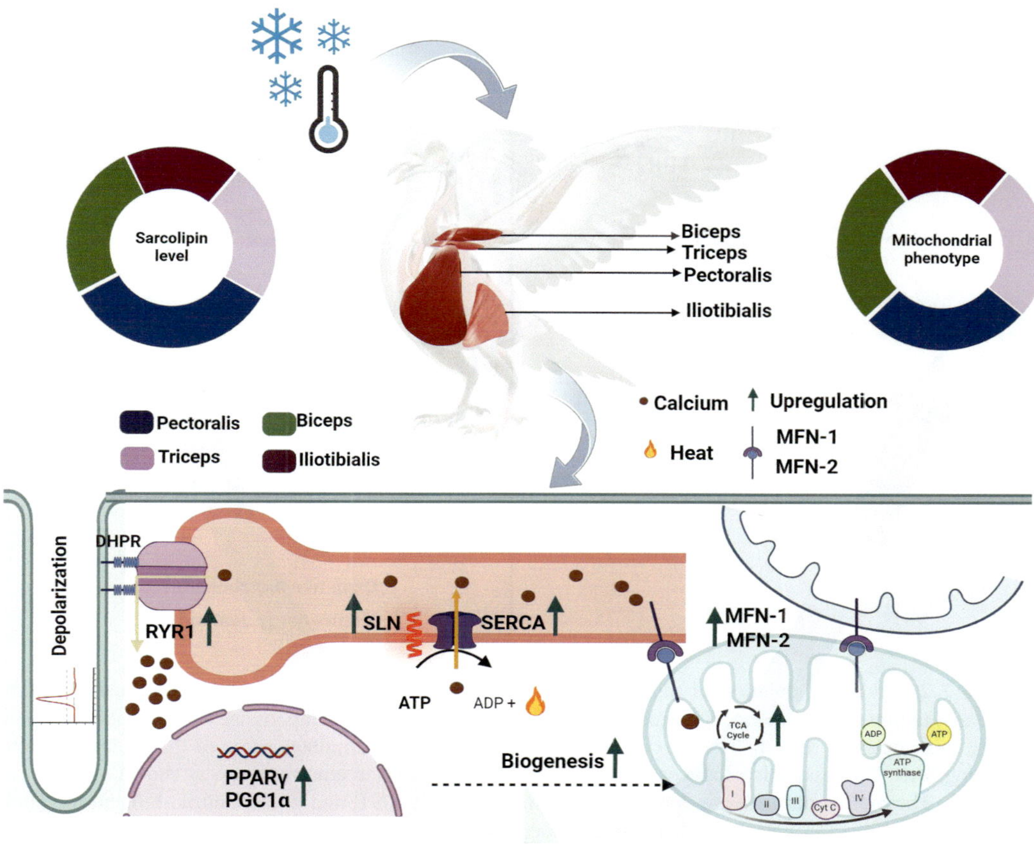

FIGURE 7.1 Schematic representation of the primary mechanism of skeletal muscle non-shivering thermogenesis (NST) in cold-adapted tropical pigeons. This scheme is intended to illustrate how the interaction between sarcolipin (SLN) and sarcoendoplasmic-reticulum calcium ATPase (SERCA) impacts muscle metabolism. SERCA actively transports Ca^{2+} from the cytosol into the sarcoplasmic reticulum lumen by hydrolyzing adenosine triphosphate (ATP). During Ca^{2+} transport, SLN and Ca^{2+} compete for binding to SERCA. The binding of SLN to SERCA does not interfere with ATP hydrolysis, but rather prevents Ca^{2+} transport through a mechanism known as uncoupling, in which Ca^{2+} slips back into the cytosol. SERCA uncoupling leads to futile cycling of the SERCA pump, which further results in increased ATP hydrolysis/heat production, creating energy demand. When SERCA is uncoupled, there is an abrupt increase in cytosolic Ca^{2+}. This promotes Ca^{2+} entry into the mitochondrial matrix, activating ATP synthesis and oxidative metabolism. The left pie chart represents the change in SLN level in various cold-adapted skeletal muscles (the percentage is represented by extracting the data from a recently published research [100]) and the right pie chart represents the modulation of mitochondrial phenotype, i.e., the altered biogenesis of cold-adapted mitochondria. Some of the transcription factors, including PGC-1α and PPARϒ, upregulate mitochondrial biogenesis and induce mitochondria to undergo more fusion than fission, increasing their size and interconnectivity across the skeletal muscle. Ca^{2+} entry into the mitochondria can signal swift enhancement of ATP output during muscle NST activation. SR–mitochondria proximity is modulated by proteins like mitofusins (MFN-1 and MFN-2) that are upregulated during cold adaptation. *ADP*, adenosine diphosphate; *DHPR*, dihydropyridine receptor; *ETC*, electron transport chain; *MFN1*, mitofusin 1; *MFN2*, mitofusin 2; *PGC1α*, peroxisome proliferator-activated receptor-gamma coactivator α; *Pi*, inorganic phosphate; *PPARγ*, Peroxisome proliferators–activated receptor γ; *RyR1*, ryanodine receptor 1.

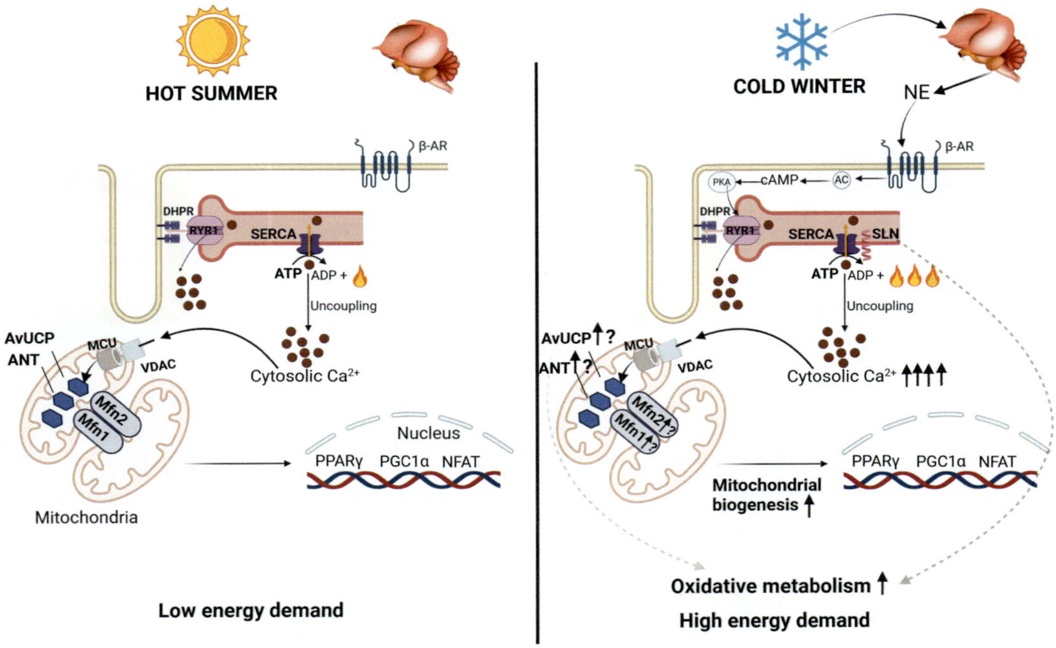

FIGURE 7.2 Avian modulation of muscle NST mechanisms. Some of the avian skeletal muscles exhibit unique adaptations to cold-induced thermogenesis. During summer, when temperature is high, energy demand of most avian muscles remains low. In contrast, during winter, cold stimulates some of the skeletal muscles expressing βAR. This activation induces long-chain reaction of downstream processes, as shown on the right-side figure. The most prominent are upregulation of SLN, avUCP, ANT, and mitochondrial biogenesis. These processes are involved in muscle NST, but how these pathways cross-talk to each other is currently not fully understood. *AC*, adenylate cyclase; *βAR*, beta adrenergic receptor; *DHPR*, dihydropyridine receptor; *MCU*, mitochondrial calcium uniporter; *Mfn1*, mitofusin 1; *Mfn2*, mitofusin 2; *NE*, norepinephrine; *NFAT*, nuclear factor of activated T-cells; *PGC1α*, peroxisome proliferator-activated receptor-gamma coactivator α; *PKA*, protein kinase A; *PPARγ*, peroxisome proliferators-activated receptor γ; *VDAC*, voltage-dependent anion channel.

Higher levels of ANT during cold acclimation would also supply the cytoplasm with extra ATP, facilitating SERCA heat production, because of the ANT primary action as a transporter of ATP out of mitochondria (Figure 7.2).

7.5 CONCLUSION AND PERSPECTIVES

Based on the discussion of all the mechanisms in this chapter, it can be concluded that there is solid evidence to date for SERCA-SLN–mediated NST to be the major player in cold-acclimated birds.

Avian NST might be differentially recruited depending on the severity of cold challenges, allowing improved survival while minimizing energy cost. Another potential mechanism of muscle NST in birds is ANT-based futile mitochondrial energy dissipation, which has gained the attention of researchers in recent years; whether this mechanism is true for birds is still a matter of debate. Moreover, it is becoming evident that avUCP functions to prevent mitochondrial ROS production and oxidative stress in avian skeletal muscle under situations of intense metabolic activity, hence indirectly serving as a facilitator of NST. Facts that are known are based on the hormones and related parameters of cold-acclimated adult birds, but molecular aspects have not been thoroughly investigated. More studies of structural and biochemical adjustments in the skeletal muscle after cold acclimation in adult birds are needed to illuminate the still obscure parts of bird NST.

REFERENCES

1. Bennett, A.F., J.W. Hicks, and A.J. Cullum, *An experimental test of the thermoregulatory hypothesis for the evolution of endothermy*. Evolution, 2000. **54**(5): p. 1768–1773.
2. Grigg, G.C., L.A. Beard, and M.L. Augee, *The evolution of endothermy and its diversity in mammals and birds*. Physiological and Biochemical Zoology, 2004. **77**(6): p. 982–997.
3. Grigg, G., et al., *Whole-body endothermy: ancient, homologous and widespread among the ancestors of mammals, birds and crocodylians*. Biological Reviews, 2022. **97**(2): p. 766–801.
4. Bicego, K.C., R.C. Barros, and L.G. Branco, *Physiology of temperature regulation: comparative aspects*. Comparative Biochemistry and Physiology Part A: Molecular & Integrative Physiology, 2007. **147**(3): p. 616–639.
5. Clarke, A. and H.O. Pörtner, *Temperature, metabolic power and the evolution of endothermy*. Biological Reviews, 2010. **85**(4): p. 703–727.
6. Whittow, G.C., *Sturkie's avian physiology*. 1999: Elsevier.
7. Maina, J., *The lung-air sac system of birds: development, structure, and function*. 2005: Springer Science & Business Media.
8. Amaral-Silva, L.D., et al., *Regulated hypothermia in response to endotoxin in birds*. The Journal of Physiology, 2021. **599**(11): p. 2969–2986.
9. Barbe, J., J. Watson, and D. Roussel, *The allometry of mitochondrial efficiency is tissue dependent: a comparison between skeletal and cardiac muscles of birds*. 2023. **226**(23).
10. Mélanie, B., et al., *Allometry of mitochondrial efficiency is set by metabolic intensity*. Proceedings of the Royal Society B, 2019. **286**(1911): p. 20191693.
11. Boël, M., et al., *Improved mitochondrial coupling as a response to high mass-specific metabolic rate in extremely small mammals*. Journal of Experimental Biology, 2020. **223**(5): p. jeb215558.
12. Milbergue, M.S., et al., *How does mitochondrial function relate to thermogenic capacity and basal metabolic rate in small birds?* Journal of Experimental Biology, 2022. **225**(12): p. jeb242612.
13. Bourguignon, A., et al., *Increased mitochondrial energy efficiency in skeletal muscle after long-term fasting: its relevance to animal performance*. Journal of Experimental Biology, 2017. **220**(13): p. 2445–2451.
14. Janský, L., *Non-shivering thermogenesis and its thermoregulatory significance*. Biological Reviews of the Cambridge Philosophical Society, 1973. **48**(1): p. 85–132.
15. Rowland, L.A., et al., *Uncoupling Protein 1 and Sarcolipin Are Required to Maintain Optimal Thermogenesis, and Loss of Both Systems Compromises Survival of Mice under Cold Stress*. The Journal of Biological Chemistry, 2015. **290**(19): p. 12282–12289.
16. Bennett, A.F. and J.A. Ruben, *Endothermy and activity in vertebrates*. Science, 1979. **206**(4419): p. 649–654.
17. Block, B.A., *Thermogenesis in muscle*. Annual Review of Physiology, 1994. **56**(1): p. 535–577.
18. Nespolo, R.F., et al., *Using new tools to solve an old problem: the evolution of endothermy in vertebrates*. Trends in Ecology & Evolution, 2011. **26**(8): p. 414–423.
19. Barbe, J., D. Roussel, and Y. Voituron, *Effect of physiological hyperthermia on mitochondrial fuel selection in skeletal muscle of birds and mammals*. Journal of Thermal Biology, 2023. **117**: p. 103719.
20. Geiser, F., *Seasonal Expression of Avian and Mammalian Daily Torpor and Hibernation: Not a Simple Summer-Winter Affair(†)*. Frontiers in Physiology, 2020. **11**: p. 436.
21. Körtner, G., R.M. Brigham, and F. Geiser, *Winter torpor in a large bird*. Nature, 2000. **407**(6802): p. 318.
22. Körtner, G., R.M. Brigham, and F. Geiser, *Torpor in free-ranging tawny frogmouths (Podargus strigoides)*. Physiological and Biochemical Zoology, 2001. **74**(6): p. 789–797.
23. Doucette, L.I., et al., *Roost type influences torpor use by Australian owlet-nightjars*. Naturwissenschaften, 2011. **98**(10): p. 845–854.
24. Nogueira-de-Sá, P.G., J.E.P.W. Bicudo, and J.G. Chaui-Berlinck, *Energy and time optimization during exit from torpor in vertebrate endotherms*. Journal of Comparative Physiology B, 2023. **193**(4): p. 461–475.
25. Bicudo, J.E.P.W., et al., *Adaptations: Living in Specific Environments*, in *Ecological and Environmental Physiology of Birds*. 2010, Oxford University Press. Chapter 5.
26. Dawson, W.R. and C. Carey, *Seasonal acclimatization to temperature in cardueline finches*. Journal of Comparative Physiology, 1976. **112**(3): p. 317–333.
27. Dawson, W.R. and R.L. Marsh, *Metabolic acclimatization to cold and season in birds*, in *Physiology of cold adaptation in birds*. 1989, Springer. p. 83–94.

28. Bicudo, J.E.P.W. and J.G. Chaui-Berlinck, *Locomotion and thermogenesis in hummingbirds*. Comparative Biochemistry and Physiology Part B: Biochemistry and Molecular Biology, 1998. **120**(1): p. 27–33.
29. Mathieu-Costello, O., et al., *Fiber capillarization and ultrastructure of pigeon pectoralis muscle after cold acclimation*. Journal of Experimental Biology, 1998. **201**(23): p. 3211–3220.
30. Petit, M. and F. Vézina, *Phenotype manipulations confirm the role of pectoral muscles and haematocrit in avian maximal thermogenic capacity*. Journal of Experimental Biology, 2014. **217**(6): p. 824–830.
31. Bicudo, J.E.P., A.C. Bianco, and C.R. Vianna, *Adaptive thermogenesis in hummingbirds*. Journal of Experimental Biology, 2002. **205**(15): p. 2267–2273.
32. Palou, A., et al., *The uncoupling protein, thermogenin*. The International Journal of Biochemistry & Cell Biology, 1998. **30**(1): p. 7–11.
33. Schwartz, T.S., S. Murray, and F. Seebacher, *Novel reptilian uncoupling proteins: molecular evolution and gene expression during cold acclimation*. Proceedings of the Biological Sciences, 2008. **275**(1637): p. 979–985.
34. Newman, S.A., N.V. Mezentseva, and A.V. Badyaev, *Gene loss, thermogenesis, and the origin of birds*. Annals of the New York Academy of Sciences, 2013. **1289**: p. 36–47.
35. Ruben, J., *The evolution of endothermy in mammals and birds: from physiology to fossils*. Annual Review of Physiology, 1995. **57**: p. 69–95.
36. Newman, S.A., *Thermogenesis, muscle hyperplasia, and the origin of birds*. Bioessays, 2011. **33**(9): p. 653–656.
37. Fowler, S.P., et al., *An analysis of skeletal muscle response to selection for rapid growth in Japanese quail (Coturnix coturnix Japonica)*. Growth, 1980. **44**(3): p. 235–252.
38. El Halawani, M.E.-S., W.O. Wilson, and R.E. Burger, *Cold-acclimation and the role of catecholamines in body temperature regulation in male Leghorns*. Poultry Science, 1970. **49**(3): p. 621–632.
39. Duchamp, C., et al., *Nonshivering thermogenesis and adaptation to fasting in king penguin chicks*. American Journal of Physiology. Regulatory, Integrative and Comparative Physiology, 1989. **257**(4): p. R744–R751.
40. Barré, H., *Metabolic and insulative changes in winter-and summer-acclimatized king penguin chicks*. Journal of Comparative Physiology B, 1984. **154**: p. 317–324.
41. Duchamp, C. and H. Barre, *Skeletal muscle as the major site of nonshivering thermogenesis in cold-acclimated ducklings*. American Journal of Physiology. Regulatory, Integrative and Comparative Physiology, 1993. **265**(5): p. R1076–R1083.
42. Dawson, W.R., et al., *Seasonal and geographic variation of cold resistance in house finches Carpodacus mexicanus*. Physiological Zoology, 1983. **56**(3): p. 353–369.
43. Barre, H., et al., *Potentiated muscular thermogenesis in cold-acclimated muscovy duckling*. American Journal of Physiology. Regulatory, Integrative and Comparative Physiology, 1985. **249**(5): p. R533–R538.
44. Teulier, L., et al., *Ontogeny of non-shivering thermogenesis in Muscovy ducklings (Cairina moschata)*. Comparative Biochemistry and Physiology Part A: Molecular & Integrative Physiology, 2014. **175**: p. 82–89.
45. Duchamp, C., et al., *Increased role of skeletal muscle in the calorigenic response to glucagon of cold-acclimated ducklings*. American Journal of Physiology. Regulatory, Integrative and Comparative Physiology, 1993. **265**(5): p. R1084–R1091.
46. Sweeney, L.J., P.D. Brodfuehrer, and B.L. Raughley, *An introductory biology lab that uses enzyme histochemistry to teach students about skeletal muscle fiber types*. Advances in Physiology Education, 2004. **28**(1): p. 23–28.
47. Collin, A., et al., *Cold-induced enhancement of avian uncoupling protein expression, heat production, and triiodothyronine concentrations in broiler chicks*. General and Comparative Endocrinology, 2003. **130**(1): p. 70–77.
48. Ueda, M., et al., *Possible role for avPGC-1α in the control of expression of fiber type, along with avUCP and avANT mRNAs in the skeletal muscles of cold-exposed chickens*. FEBS Letters, 2005. **579**(1): p. 11–17.
49. Ventura-Clapier, R., A. Garnier, and V. Veksler, *Transcriptional control of mitochondrial biogenesis: the central role of PGC-1α*. Cardiovascular Research, 2008. **79**(2): p. 208–217.
50. Stevens, E., *The evolution of endothermy*. Journal of Theoretical Biology, 1973. **38**(3): p. 597–611.
51. Walter, I. and F. Seebacher, *Endothermy in birds: underlying molecular mechanisms*. Journal of Experimental Biology, 2009. **212**(15): p. 2328–2336.
52. Rowland, L.A., N.C. Bal, and M. Periasamy, *The role of skeletal-muscle-based thermogenic mechanisms in vertebrate endothermy*. Biological Reviews, 2015. **90**(4): p. 1279–1297.

53. Price, E.R. and E.M. Dzialowski, *Development of endothermy in birds: patterns and mechanisms*. Journal of Comparative Physiology B, 2018. **188**(3): p. 373–391.
54. Sirsat, S.K., et al., *Development of endothermy and concomitant increases in cardiac and skeletal muscle mitochondrial respiration in the precocial Pekin duck (Anas platyrhynchos domestica)*. Journal of Experimental Biology, 2016. **219**(8): p. 1214–1223.
55. Sirsat, S.K., et al., *The 12-day thermoregulatory metamorphosis of Red-winged Blackbirds (Agelaius phoeniceus)*. Journal of Comparative Physiology. B, 2016. **186**(5): p. 651–663.
56. Stager, M., D.L. Swanson, and Z.A. Cheviron, *Regulatory mechanisms of metabolic flexibility in the dark-eyed junco (Junco hyemalis)*. The Journal of Experimental Biology, 2015. **218**(5): p. 767–777.
57. Smith, R.L., et al., *Metabolic flexibility as an adaptation to energy resources and requirements in health and disease*. Endocrine Reviews, 2018. **39**(4): p. 489–517.
58. Harlow, P. and G. Grigg, *Shivering thermogenesis in a brooding diamond python, Python spilotes spilotes*. Copeia, 1984. 1984: p. 959–965.
59. Hutchison, V.H., H.G. Dowling, and A. Vinegar, *Thermoregulation in a brooding female Indian python, Python molurus bivittatus*. Science, 1966. **151**(3711): p. 694–696.
60. Berner, N.J., *Oxygen consumption by mitochondria from an endotherm and an ectotherm*. Comparative Biochemistry and Physiology Part B: Biochemistry and Molecular Biology, 1999. **124**(1): p. 25–31.
61. Mak, I., E. Shrago, and C. Elson, *Modification of liver mitochondrial lipids and of adenine nucleotide translocase and oxidative phosphorylation by cold adaptation*. Biochimica et Biophysica Acta (BBA)-Bioenergetics, 1983. **722**(2): p. 302–309.
62. Vigelsø, A., N.B. Andersen, and F. Dela, *The relationship between skeletal muscle mitochondrial citrate synthase activity and whole body oxygen uptake adaptations in response to exercise training*. International Journal of Physiology, Pathophysiology and Pharmacology, 2014. **6**(2): p. 84.
63. Ruas, J.S., et al., *Underestimation of the maximal capacity of the mitochondrial electron transport system in oligomycin-treated cells*. PLoS One, 2016. **11**(3): p. e0150967.
64. Else, P.L., N. Turner, and A. Hulbert, *The evolution of endothermy: role for membranes and molecular activity*. Physiological and Biochemical Zoology, 2004. **77**(6): p. 950–958.
65. Salin, K., et al., *Variation in the link between oxygen consumption and ATP production, and its relevance for animal performance*. Proceedings of the Royal Society B: Biological Sciences, 2015. **282**(1812): p. 20151028.
66. Roussel, D., et al., *Skeletal muscle metabolism in sea-acclimatized king penguins. I. Thermogenic mechanisms*. Journal of Experimental Biology, 2020. **223**(21): p. jeb233668.
67. Guderley, H., et al., *Why are some mitochondria more powerful than others: insights from comparisons of muscle mitochondria from three terrestrial vertebrates*. Comparative Biochemistry and Physiology Part B: Biochemistry and Molecular Biology, 2005. **142**(2): p. 172–180.
68. Arruda, A.P., et al., *Cold tolerance in hypothyroid rabbits: role of skeletal muscle mitochondria and sarcoplasmic reticulum Ca2+ ATPase isoform 1 heat production*. Endocrinology, 2008. **149**(12): p. 6262–6271.
69. Bakken, G.S. and D.M. Gates, *Heat-transfer analysis of animals: some implications for field ecology, physiology, and evolution*, in *Perspectives of biophysical ecology*. 1975, Springer. p. 255–290.
70. Beeckmans, S., *Some structural and regulatory aspects of citrate synthase*. International Journal of Biochemistry, 1984. **16**(4): p. 341–351.
71. Nord, A., et al., *Avian red blood cell mitochondria produce more heat in winter than in autumn*. The FASEB Journal, 2021. **35**(5): p. e21490.
72. Roussel, D., et al., *Increase in the adenine nucleotide translocase content of duckling subsarcolemmal mitochondria during cold acclimation*. FEBS Letters, 2000. **477**(1–2): p. 141–144.
73. Talbot, D.A., et al., *Uncoupling protein and ATP/ADP carrier increase mitochondrial proton conductance after cold adaptation of king penguins*. The Journal of Physiology, 2004. **558**(1): p. 123–135.
74. Mujahid, A., et al., *Expression of avian UCP and ANT in skeletal muscle of cold-exposed laying and meat-type chickens*. The Journal of Poultry Science, 2005. **42**(3): p. 230–237.
75. Kunji, E.R., et al., *The transport mechanism of the mitochondrial ADP/ATP carrier*. Biochimica et Biophysica Acta (BBA)-Molecular Cell Research, 2016. **1863**(10): p. 2379–2393.
76. Klingenberg, M., *The ADP and ATP transport in mitochondria and its carrier*. Biochimica et Biophysica Acta (BBA) - Biomembranes, 2008. **1778**(10): p. 1978–2021.
77. Bal, N.C., et al., *Increased reliance on muscle-based thermogenesis upon acute minimization of brown adipose tissue function*. Journal of Biological Chemistry, 2016. **291**(33): p. 17247–17257.

78. Toyomizu, M., et al., *Cold-induced mitochondrial uncoupling and expression of chicken UCP and ANT mRNA in chicken skeletal muscle*. FEBS Letters, 2002. **529**(2–3): p. 313–318.
79. Andreyev, A.Y., et al., *The ATP/ADP-antiporter is involved in the uncoupling effect of fatty acids on mitochondria*. European Journal of Biochemistry, 1989. **182**(3): p. 585–592.
80. Bertholet, A.M., et al., *H+ transport is an integral function of the mitochondrial ADP/ATP carrier*. Nature, 2019. **571**(7766): p. 515–520.
81. Hervas, L.S., et al., *Mitochondrial function in skeletal muscle contributes to reproductive endothermy in tegu lizards (Salvator merianae)*. Acta Physiologica (Oxford, England), . 2024: p. e14162. https://doi.org/10.1111/apha.14162
82. Tattersall, G.J., C.A. Leite, and C.E. Sanders, *Seasonal reproductive endothermy in tegu lizards*. Science Advances, 2016. **2**(1): p. e1500951.
83. Sanders, C.E., et al., *Daily and annual cycles in thermoregulatory behaviour and cardio-respiratory physiology of black and white tegu lizards*. Journal of Comparative Physiology. B, 2015. **185**(8): p. 905–915.
84. Rey, B., et al., *Up-regulation of avian uncoupling protein in cold-acclimated and hyperthyroid ducklings prevents reactive oxygen species production by skeletal muscle mitochondria*. BMC Physiology, 2010. **10**: p. 1–12.
85. Talbot, D.A., et al., *Superoxide activates a GDP-sensitive proton conductance in skeletal muscle mitochondria from king penguin (Aptenodytes patagonicus)*. Biochemical and Biophysical Research Communications, 2003. **312**(4): p. 983–988.
86. Teulier, L., et al., *Cold-acclimation-induced non-shivering thermogenesis in birds is associated with upregulation of avian UCP but not with innate uncoupling or altered ATP efficiency*. Journal of Experimental Biology, 2010. **213**(14): p. 2476–2482.
87. Evock-Clover, C.M., et al., *Expression of an uncoupling protein gene homolog in chickens*. Comparative Biochemistry and Physiology Part A: Molecular & Integrative Physiology, 2002. **133**(2): p. 345–358.
88. Abe, T., et al., *Possible role of avian uncoupling protein in down-regulating mitochondrial superoxide production in skeletal muscle of fasted chickens*. FEBS Letters, 2006. **580**(20): p. 4815–4822.
89. Toyomizu, M., et al., *Progressive alteration of UCP and ANT in skeletal muscle of fasted chickens*. The Journal of Poultry Science, 2006. **43**(2): p. 167–172.
90. Jastroch, M. and F. Seebacher, *Importance of adipocyte browning in the evolution of endothermy*. Philosophical Transactions of the Royal Society B, 2020. **375**(1793): p. 20190134.
91. Cannon, B. and J. Nedergaard, *Brown adipose tissue: function and physiological significance*. Physiological Reviews, 2004. https://doi.org/10.1152/physrev.00015.2003
92. Trayhurn, P., N.J. Temple, and J. Van Aerde, *Evidence from immunoblotting studies on uncoupling protein that brown adipose tissue is not present in the domestic pig*. Canadian Journal of Physiology and Pharmacology, 1989. **67**(12): p. 1480–1485.
93. Herpin, P., M. Damon, and J. Le Dividich, *Development of thermoregulation and neonatal survival in pigs*. Livestock Production Science, 2002. **78**(1): p. 25–45.
94. Seale, P., et al., *PRDM16 controls a brown fat/skeletal muscle switch*. Nature, 2008. **454**(7207): p. 961–967.
95. Jung, S.M., J. Sanchez-Gurmaches, and D.A. Guertin, *Brown Adipose Tissue Development and Metabolism*. Handbook of Experimental Pharmacology, 2019. **251**: p. 3–36.
96. Pani, P. and N.C. Bal, *Avian adjustments to cold and non-shivering thermogenesis: whats, wheres and hows*. Biological Reviews, 2022. **97**(6): p. 2106–2126.
97. Meissner, G. and A. El-Hashem, *Ryanodine as a functional probe of the skeletal muscle sarcoplasmic reticulum Ca 2+ release channel*. Molecular and Cellular Biochemistry, 1992. **114**: p. 119–123.
98. Dumonteil, E., H. Barre, and G. Meissner, *Sarcoplasmic reticulum Ca (2+)-ATPase and ryanodine receptor in cold-acclimated ducklings and thermogenesis*. American Journal of Physiology-Cell Physiology, 1993. **265**(2): p. C507–C513.
99. Dumonteil, E., H. Barre, and G. Meissner, *Expression of sarcoplasmic reticulum Ca2+ transport proteins in cold-acclimating ducklings*. American Journal of Physiology-Cell Physiology, 1995. **269**(4): p. C955–C960.
100. Pani, P., et al., *Seasonal cold induces divergent structural/biochemical adaptations in different skeletal muscles of Columba livia: evidence for nonshivering thermogenesis in adult birds*. Biochemical Journal, 2023. **480**(17): p. 1397–1409.
101. de Meis, L., A.P. Arruda, and D.P. Carvalho, *Role of sarco/endoplasmic reticulum Ca2+-ATPase in thermogenesis*. Bioscience Reports, 2005. **25**(3–4): p. 181–190.

102. Kjelstrup, S., et al., *Is the Ca 2+-ATPase from sarcoplasmic reticulum also a heat pump?* European Biophysics Journal, 2008. **38**: p. 59–67.
103. Marie, V. and J.E. Silva, *Calcium pool size modulates the sensitivity of the ryanodine receptor channel and calcium-dependent ATPase of heavy sarcoplasmic reticulum to extravesicular free calcium concentration*. Journal of Cellular Physiology, 1998. **175**(3): p. 283–294.
104. Bal, N.C., et al., *Sarcolipin is a newly identified regulator of muscle-based thermogenesis in mammals*. Nature Medicine, 2012. **18**(10): p. 1575–1579.
105. Pant, M., N.C. Bal, and M. Periasamy, *Sarcolipin: a key thermogenic and metabolic regulator in skeletal muscle*. Trends in Endocrinology and Metabolism, 2016. **27**(12): p. 881–892.
106. Pant, M., N.C. Bal, and M. Periasamy, *Cold adaptation overrides developmental regulation of sarcolipin expression in mice skeletal muscle: SOS for muscle-based thermogenesis?* The Journal of Experimental Biology, 2015. **218**(Pt 15): p. 2321–2325.
107. Maurya, S.K., et al., *Sarcolipin Is a Key Determinant of the Basal Metabolic Rate, and Its Overexpression Enhances Energy Expenditure and Resistance against Diet-induced Obesity*. The Journal of Biological Chemistry, 2015. **290**(17): p. 10840–10849.
108. Campbell, K.L. and A.A. Dicke, *Sarcolipin makes heat, but is it adaptive thermogenesis?* Frontiers in Physiology, 2018. **9**: p. 368603.
109. Stager, M. and Z.A. Cheviron, *Is there a role for sarcolipin in avian facultative thermogenesis in extreme cold?* Biology Letters, 2020. **16**(6): p. 20200078.
110. Elowe, C.R., M. Stager, and A.R. Gerson, *Sarcolipin relates to fattening, but not sarco/endoplasmic reticulum Ca2+-ATPase uncoupling, in captive migratory gray catbirds*. Journal of Experimental Biology, 2024. **227**(1): p. jeb246897.
111. Sztretye, M., et al., *SOCE is important for maintaining sarcoplasmic calcium content and release in skeletal muscle fibers*. Biophysical Journal, 2017. **113**(11): p. 2496–2507.
112. Michelucci, A., et al., *Role of STIM1/ORAI1-mediated store-operated Ca2+ entry in skeletal muscle physiology and disease*. Cell Calcium, 2018. **76**: p. 101–115.
113. Pan, Z., M. Brotto, and J. Ma, *Store-operated Ca2+ entry in muscle physiology and diseases*. BMB Reports, 2014. **47**(2): p. 69.
114. Edwards, J.N., et al., *Ultra-rapid activation and deactivation of store-operated Ca2+ entry in skeletal muscle*. Cell Calcium, 2010. **47**(5): p. 458–467.
115. Dumonteil, E., H. Barré, and G. Meissner, *Effects of palmitoyl carnitine and related metabolites on the avian Ca (2+)-ATPase and Ca2+ release channel*. The Journal of Physiology, 1994. **479**(1): p. 29–39.
116. Meissner, G., *Ryanodine receptor/Ca2+ release channels and their regulation by endogenous effectors*. Annual Review of Physiology, 1994. **56**(1): p. 485–508.
117. Muslikhov, E., I. Sukhanova, and P. Avdonin, *Arachidonic acid activates release of calcium ions from reticulum via ryanodine receptor channels in C2C12 skeletal myotubes*. Biochemistry (Moscow), 2014. **79**: p. 435–439.
118. Lamboley, C., et al., *Ca2+ leakage out of the sarcoplasmic reticulum is increased in type I skeletal muscle fibres in aged humans*. The Journal of Physiology, 2016. **594**(2): p. 469–481.
119. Barré, H., J. Nedergaard, and B. Cannon, *Increased respiration in skeletal muscle mitochondria from cold-acclimated ducklings: uncoupling effects of free fatty acids*. Comparative Biochemistry and physiology B, Comparative Biochemistry, 1986. **85**(2): p. 343–348.
120. Staples, J.F. and J.C. Brown, *Mitochondrial metabolism in hibernation and daily torpor: a review*. Journal of Comparative Physiology B, 2008. **178**: p. 811–827.
121. Parker, G., *Changes in muscle, liver, and plasma free fatty acid levels in the pigeon on acute exposure to cold*. Archives Internationales de Physiologie et de Biochimie, 1978. **86**(4): p. 771–777.
122. Little, A.G. and F. Seebacher, *The evolution of endothermy is explained by thyroid hormone-mediated responses to cold in early vertebrates*. Journal of Experimental Biology, 2014. **217**(10): p. 1642–1648.
123. Roussel, D., et al., *Effects of cold acclimation and palmitate on energy coupling in duckling skeletal muscle mitochondria*. FEBS Letters, 1998. **439**(3): p. 258–262.
124. Shabalina, I.G., et al., *Cold tolerance of UCP1-ablated mice: a skeletal muscle mitochondria switch toward lipid oxidation with marked UCP3 up-regulation not associated with increased basal, fatty acid- or ROS-induced uncoupling or enhanced GDP effects*. Biochimica et Biophysica Acta (BBA)-Bioenergetics, 2010. **1797**(6–7): p. 968–980.

8 Ryanodine Receptor 1 Calcium Leak in Non-Shivering Thermogenesis

Aldo Meizoso-Huesca and Bradley S. Launikonis
The University of Queensland, Brisbane, Australia

8.1 INTRODUCTION

In skeletal muscle, the generation of force is crucial for facilitating coordinated movements and maintaining posture, thereby enabling animals to engage in a wide range of physical activities. However, this tissue serves another physiological role: thermogenesis. Remarkably, the skeletal muscle utilizes some of the intracellular players involved in force production for the regulated generation of heat. The sarcoplasmic reticulum (SR) calcium release channel, ryanodine receptor 1 (RyR1), is not solely essential for force production required for locomotion. The RyR1 plays an important role in skeletal muscle–mediated thermogenesis even when (endothermic) animals are not engaging in physical activity (non-shivering thermogenesis). This aspect holds crucial importance in the physiology of endotherms, allowing them to uphold their core temperature without relying on (a) external heat sources and (b) heat produced by the activation of the contractile filaments. Pivotal to this, certain organisms capitalize on the gating and regulatory properties of their RyRs, setting the basal RyR1 Ca^{2+} leakage rates to fulfill the heat requirements. In what follows, we will delve into this topic, taking a particular emphasis on the role of the RyR1 in skeletal muscle thermogenesis and its regulation.

Force production is the outcome of a complex intracellular mechanism known as excitation-contraction (EC) coupling. EC coupling involves the coordinated functioning of two membranous systems. First, the transverse tubules are invaginations of the plasma membrane (known as plasmalemma) that are distributed regularly along the muscle fiber (Franzini-Armstrong and Porter, 1964). These tubules play a crucial role in transmitting electrical signals from the surface into the muscle fiber. Second, the sarcoplasmic reticulum (SR), a specialized type of endoplasmic reticulum, consists of a network of tubules and cisternae. The SR plays a vital role in the regulation of intracellular calcium acting as the primary storage organelle for calcium ions (Ca^{2+}) (Constantin and Podolsky, 1965). Within the muscle fiber, the transverse tubules and the SR establish physical interactions, forming a highly integrated Ca^{2+} signaling hub that is critical for efficient EC coupling. The intricate interplay between these membranous systems ensures the precise control of calcium release and contraction in skeletal muscle.

Notably, the SR can be distinguished into two distinct structural domains, each serving different functional roles in calcium handling (Ogata and Yamasaki, 1997). The longitudinal SR, also known as the free SR, runs parallel to the long axis of the muscle fiber along the A and I bands of the sarcomeres and wraps around the contractile filaments of the skeletal muscle. This domain constitutes most of the SR and plays a crucial role in calcium uptake, where approximately 80% of its protein content is made up of the sarcoplasmic/endoplasmic reticulum Ca^{2+} ATPase (SERCA), which is responsible for actively transporting calcium ions into the SR (Scales and Inesi, 1976).

 DOI: 10.1201/9781003391470-10

In contrast, the junctional SR, also known as the SR terminal cisternae, is situated at the intersection of the A–I bands. This region is specialized in calcium release, primarily mediated by the RyR1. The junctional SR acts as a focal point for calcium release during EC coupling, facilitating the coordinated contraction of muscle fibers (Adams et al., 1990).

During EC coupling, an action potential triggered by the release of acetylcholine from the axon terminal into the motor end plate propagates into the fiber through the t-tubules via voltage-gated Na^+ channels (Luff and Atwood, 1972; Hodgkin and Horowicz, 1960). Within the fiber, this depolarization is sensed by the voltage-gated Ca^{2+} channels (Cav1.1), also known as dihydropyridine receptors (DHPRs). In skeletal muscle, the DHPR is embedded in the t-tubule membrane and establishes a physical interaction with the RyR1, located in the opposed SR terminal cisternae membrane. At rest, the DHPR exerts an inhibitory effect on the RyR1. However, t-tubule membrane depolarization induces conformational changes on the DHPR, releasing this inhibition and ultimately activating the RyR1 channels. This depolarization-induced conformational change in the DHPR plays a crucial role by triggering the release of calcium ions from the sarcoplasmic reticulum into the cytosol, following their concentration gradient. This release increases the free cytoplasmic calcium concentration ($[Ca^{2+}]_{cyto}$) approximately 10-fold (from ~100 nM to ~1 mM) and leads to the binding of Ca^{2+} ions to troponin C at the myofilaments. This interaction displaces troponin C from the contractile filaments, allowing for the interaction between actin and myosin filaments. Finally, SERCA pumps Ca^{2+} back into the SR, promoting muscle relaxation by lowering the $_{free}[Ca^{2+}]_{cyto}$ back to its resting levels.

Hence, the correct functioning of skeletal muscle during both rest and active periods is heavily dependent on the precise control of intracellular calcium levels. In addition to the mobilization of intracellular calcium, maintaining a well-regulated baseline level of intracellular calcium is essential for muscle function, metabolism, and thermogenesis. An integral player in this process is the RyR1.

Being the largest known ion channel, the RyR1 is a homotetramer with a molecular weight of 2.2 mega Daltons. It is composed of four protomers, each consisting of 5,037 residues (Lanner et al., 2010). Within the muscle fiber, the RyR1 is situated in the membrane of the SR terminal cisternae, facing the cytosolic space between the SR membrane and the transverse tubule (Franzini-Armstrong, 1970). This arrangement of membranes is consistent throughout the muscle fiber and forms a repeating structure called the *triad*. The triad is comprised of a transverse tubule surrounded by two SR terminal cisternae, which are spaced approximately 10–15 nanometers apart (Takekura et al., 2001).

Pioneer visualizations of the RyR1 were made just 5 years after the term *triad* was coined in the late 1950s (Porter and Palade, 1957). The earliest reports pointing to the existence of RyR1 date back to 1962, when an electron microscopy study revealed the presence of 100 angstroms-wide "bridges" or projections between the sarcoplasmic reticulum (SR) membrane and the t-tubule in bat muscle (Revel, 1962). However, the exact nature of these projections remained undefined.

Advancements in electron microscopy enabled further characterization of these structures, which were later identified as SR "feet" (Franzini-Armstrong, 1970). Subsequent analysis of isolated SR vesicles provided additional insights, revealing that these feet-like structures existed as tetramers. Analyzing these structures, it became clear that a small portion of them was embedded in the SR membrane, while the majority faced the cytosol (Ferguson et al., 1984).

A key breakthrough in understanding the molecular identity of the feet-like structures came with the discovery of the capacity of ryanodine, a plant alkaloid, to act as a high-affinity ligand for a calcium channel in the junctional SR (Fleischer et al., 1985). This significant advancement paved the way for purifying the ryanodine-binding molecule, which was subsequently recognized as the RyR. Notably, the identified RyR shared physical properties with the previously observed SR feet, leading to the conclusion that these receptors were indeed the SR feet—a calcium release channel located in the SR terminal cisternae (Imagawa et al., 1987; Inui et al., 1987).

Recent advancements in cryoelectron microscopy have not only validated the initial observations regarding the RyR1 but also provided valuable structural insights into the channel at atomic resolution, enhancing our understanding of its structural-functional relationship. The RyR1 structure exhibits a mushroom-like architecture, comprising a cytosolic cap that constitutes approximately 80% of the channel, a transmembrane domain housing the channel pore, and an intraluminal domain. Of particular importance, the large cytosolic domain is connected to the transmembrane domain through electrodense domains, which play a crucial role in the allosteric coupling between the two domains. In addition, the cytosolic cap exhibits a myriad of globular domains that serve as interaction domains with regulatory ligands (Van Petegem, 2012).

The functional regulation of RyR1 encompasses interactions with a diverse array of ligands, such as proteins, small molecules, and ions. Additionally, post-translational modifications play a pivotal role in modulating the activity of this channel. These regulatory mechanisms will be further examined and discussed in the following section.

8.2 REGULATION OF RYANODINE RECEPTOR 1

8.2.1 Endogenous Effectors

8.2.1.1 Ions and Small Molecules

Calcium exerts a biphasic concentration-dependent modulation of RyR1 activity, where low micromolar calcium concentrations increase channel's activity whereas millimolar calcium concentrations inhibit it. This hormetic effect is explained by the presence of two different calcium-binding sites: (1) a high-affinity stimulatory binding site and (2) a low-affinity inhibitory binding site (Smith et al., 1986).

Magnesium ions inhibit RyR1 via two different mechanisms. First, they compete with calcium for high-affinity Ca^{2+} stimulatory binding sites (Laver et al., 1997). Of note, in resting conditions, the free magnesium concentration in the cytosol is ~1 mM, whereas the calcium concentration is ~100 nM, thus at rest, Mg^{2+} outcompetes Ca^{2+} ions for the high-affinity Ca^{2+}-binding sites, thus inhibiting the channel. Additionally, Mg^{2+} ions also bind to low-affinity Ca^{2+}-binding sites, resulting in RyR1 inhibition (Meissner et al., 1986).

ATP stimulates RyR1 activity in the millimolar range by binding to interdomains within the carboxy-terminal domain of the channel. It is worth noting that other adenine nucleotides, including ADP, cAMP, and adenosine, can also stimulate RyR1 activity, although with lower efficacy compared to ATP (Meissner, 1984).

8.2.2 Modulatory Proteins

8.2.2.1 Dihydropyridine Receptor (Ca_v1.1)

The dihydropyridine receptor is the voltage-dependent Ca^{2+} channel (Ca_v1.1) that is located in the t-tubule membrane and mediates action potential–induced SR Ca^{2+} release via RyR1 activation during EC coupling. The DHPR is formed by five different subunits: a1, a2, b, g, and d, where the a1 constitutes the channel structure as well as contains the voltage-sensor domain of the complex (Catterall, 2011). The activation of RyR1 by the DHPR occurs through physical interactions between the two proteins, mediated by the loops II and III of the a1 subunit. In addition to its role in modulating voltage-induced SR Ca^{2+} release, interaction of the DHPR with the RyR1 in resting muscle inhibits the leak-associated conformations of the RyR1s, thus reducing RyR1 leak (Eltit et al., 2011).

8.2.2.2 Calmodulin

Calmodulin is a cytosolic 17 kDa Ca^{2+}-binding protein that interacts with the RyR1 in both a Ca^{2+}-free and Ca^{2+}-bound form (Moore et al., 1999). This interaction occurs with a stoichiometry of four calmodulin molecules per RyR1 tetramer, and it modulates RyR1 through binding to the calmodulin-binding region located in residues 3614–3643 (McCarthy et al., 2020). Calmodulin exerts a [Ca^{2+}]-dependent modulation on the channel, with nanomolar concentrations of calcium activating the channel, whereas micromolar calcium concentrations inhibit it (Rodney et al., 2000; Tripathy et al., 1995).

8.2.2.3 FK506-Binding Protein 12 (FKBP12)

FK506-binding protein 12 (FKBP12), also known as Calstabin1 (calcium channel-stabilizing binding protein), is part of the immunophilin proteins that interact with immunosuppressive drugs such as cyclosporin and rapamycin. FKBP12 binds to the RyR1 with a stoichiometry of four FKBP12s per RyR1 homotetramer. By interacting with the RyR1 channel, FKBP12 stabilizes it, limiting Ca^{2+} leak (Gonano and Jones, 2017). Moreover, functional evidence points to FKBP12 playing a role in modulating the physical coupling between the DHPR and RyR1 (Tang et al., 2004).

8.2.2.4 Calsequestrin

Calsequestrin is the main calcium buffer inside the SR (Royer and Rios, 2009). Fast-twitch skeletal muscles express calsequestrin 1, whereas calsequestrin 2 is majorly found in cardiac muscle and in slow-twitch skeletal muscle. In addition to buffering Ca^{2+} within the SR, calsequestrin 1 oligomers establish interactions with the RyR1s, regulating its activity. This interaction is mediated by proteins anchored to the SR membrane, including triadin and junction, reducing the open probability of the channel (Beard et al., 2009).

8.2.3 Post-Translational Modifications

8.2.3.1 Phosphorylation

RyR1 is phosphorylated by various kinases, such as protein kinase A (PKA) and Ca^{2+}/calmodulin-dependent protein kinase II (CamKII), forming complexes with these enzymes (Reiken et al., 2003). On the other hand, RyR1 interacts with phosphatases, specifically protein phosphatases 1 and 2A, which counteract the modulatory effects of the kinases. Phosphorylation of RyR1 is generally linked to an increase in calcium leak through the channel. Thus far, the best-characterized phosphorylation site in the RyR1 is the Ser2843, which is associated with FKBP12 dissociation (Andersson et al., 2011)

8.2.3.2 Oxidation and S-nytrosylation

Oxidation and S-nitrosylation play key regulatory roles in the function of RyR1. The RyR1 protein contains ~100 cysteine residues per monomer, and modifications in their redox state (primarily those in the cytosolic face of the channel and in the transmembrane domain) can occur through oxidizing agents like H_2O_2 or nitrosylating agents such as nitric oxide, inducing the formation of disulfide bonds between neighboring cysteines. These modifications induce the formation of disulfide bonds between adjacent cysteines, leading to structural changes and reversible modulation of the channel's function. For instance, S-nitrosylation of cysteine3635 disrupts the inhibitory effect of calmodulin on RyR1, resulting in channel activation (Moore et al., 1999). Additionally, RyR1 oxidation and nitrosylation have been associated with decreased binding affinity of RyR1 to FKBP12. On the other hand, evidence for inhibitory effects of s-nitrosylation on RyR1 has been reported, suggesting that the effect of oxidation and s-nitrosylation is dependent on the specific residue modified (Dulhunty et al., 2000).

8.2.3.3 Fragmentation

Although less explored, RyR1 fragmentation has been proposed as a regulatory mechanism of RyR1 leak. RyR1 fragmentation has been shown to be oxidation dependent and to take place 6–24 hours after an exercise intervention. This modification of the channel has been proposed to increase RyR1 leakage, contributing to the mechanisms of adaptation to exercise (Place et al., 2015; Tripp et al., 2022).

Recent studies propose RyR1 fragmentation as a potential regulatory mechanism for RyR1 leak. This process, which occurs in an oxidation-dependent manner, has been observed to occur within 6–24 hours following an exercise intervention. The fragmentation of RyR1 is believed to enhance

RyR1 leakage, playing a role in the adaptive mechanisms associated with exercise (Place et al., 2015; Tripp et al., 2022). Although the precise mechanisms and functional implications of RyR1 fragmentation are still being explored, these findings suggest its involvement in modulating RyR1 activity during exercise-induced muscle adaptations.

8.2.4 Exogenous Ligands

Through the decades of research on the RyR1, several exogenous ligands of this channel have been discovered. Ryanodine was found to display a high binding affinity for the RyR1 channel, inducing an increase in the open probability of the channel at low nanomolar concentrations while inhibiting at micromolar concentrations. Caffeine is widely used as a tool to induce SR Ca^{2+} release by activating the RyR1 at low millimolar concentrations. Mechanistically, binding of caffeine to the RyR1 enhances the affinity of RyR1 activating Ca^{2+}-binding sites while decreasing the affinity of RyR1 inhibitory Mg^{2+}-binding sites. In addition, caffeine works in a cooperative fashion with ATP, increasing the sensitivity of RyR1 for this endogenous modulator (Charisani et al., 2021). Dantrolene was initially developed as a muscle relaxant and was later discovered to inhibit RyR1 in a way dependent of calmodulin and Mg^{2+} (Snyder et al., 1967). Of note, since its implementation, the use of this drug has significantly reduced the rate of mortality associated with malignant hyperthermia from ~80% to less than 2%. Rycals (Ryanodine receptor calcium release channel stabilizers) constitute a novel family of drugs that bind to the RyR1 and promote the interaction between RyR1 and FKBP12 (Melville et al., 2022).

8.3 RYR1 CA^{2+} LEAK: IMPLICATIONS IN MUSCLE PHYSIOLOGY

Under physiological conditions, approximately 85% of the calcium in a healthy, unstressed muscle fiber is stored within the SR. This corresponds to approximately 1.3 mM Ca^{2+} per unit fiber volume. The remaining 15% of calcium is distributed throughout the cytosol, mitochondria, and other organelles such as the nucleus. In these locations, calcium plays crucial regulatory roles in signaling, metabolism, and gene expression regulation, respectively.

As noted in the previous section, the binding partners of the RyR1 are essential for regulating its gating properties, enabling precise control over the rate of RyR1-mediated Ca^{2+} leakage from the SR into the cytosol. This mechanism is utilized by organisms to integrate muscle physiology with varying energetic demands in different contexts such as cold exposure and obesity. It is worth noting that endotherms have developed the ability to exploit the plasticity of RyR1 Ca^{2+} leak without compromising muscle function. This is illustrated in a comparison of RyR1 leak rates between fast-twitch fibers from wild-type and heterozygous mouse carrying a malignant hyperthermia (MH)-causative mutation. MH is a detrimental reaction triggered by volatile anesthetics such as halothane or isofluorane. Under this condition, the skeletal muscle undergoes calcium dyshomeostasis, displaying uncontrolled release of Ca^{2+} ions from the SR into the cytosol. This leads to muscle rigidity and a hypermetabolic response that ultimately causes a potentially lethal increase in whole-body temperature. Despite this, MH-susceptible individuals do not display a pathological phenotype on a daily basis unless their muscles are compromised with stressors (e.g., anesthetics, heat stress). Over the years, murine models of MH have been instrumental in understanding the plasticity of calcium handling in skeletal muscle, particularly in response to changes in RyR1 Ca^{2+} leak in resting muscle. Modeling-derived data show that RyR1 leak rate in wild-type mouse model is 17.1 $\mu M\ s^{-1}$, whereas in the MH mouse model this value increases to 185.6 $\mu M\ s^{-1}$. Despite this over 10-fold difference in RyR1 leak rate, the MH mouse model exhibits a life span similar to wild-type mice, only compromised by external stressors. More significantly, functional assessments of their muscles reveal levels of force production comparable to wild-type mice *ex vivo*, indicating a functional adaptation of the muscle to the 10-fold increase in RyR1 leak rate (Lamboley et al., 2021). This adaptation is achieved through subsequent modulation of intracellular mechanisms. However, before exploring these mechanisms, let us address the question: What determines SR calcium content?

Calcium content in the SR is determined by the balance between basal Ca^{2+} leakage and uptake at equilibrium (often referred as calcium cycling), mediated by the RyR1 channels and SERCAs, respectively. It is important to note that maintaining SR calcium content represents a significant energetic demand for resting muscles. This demand is estimated to contribute between 20% and 50% of the tissue's basal energy expenditure. In this sense, the energy required to maintain SR calcium content is influenced by two factors: (1) the rate of RyR1 Ca^{2+} leak and (2) the efficiency of SR Ca^{2+} uptake, which is determined by the coupling between Ca^{2+} translocation and ATP hydrolysis by SERCAs. The latter component, like the RyR1 leak rate, can be subject to regulation, and more detailed information can be found elsewhere (Gamu et al., 2020; Xu and Remmen, 2021).

In skeletal muscle fibers, RyR1 Ca^{2+} leak has a direct influence on the activity of SERCA, amplifying its rate of ATP hydrolysis by defining the $[Ca^{2+}]_{cyto}$ at the cytosolic face of the pump (Meizoso-Huesca et al., 2022). Of note, the nonlinear relationship between SERCA ATPase activity and cytosolic calcium concentration ($[Ca^{2+}]_{cyto}$), combined with the endogenous resting $[Ca^{2+}]_{cyto}$ at the nanomolar levels, creates an optimal range of $[Ca^{2+}]_{cyto}$ for the regulation of SERCA activity in relation to ATP expenditure and thermogenesis in resting skeletal muscle (Figure 8.1). Importantly, this range is limited by the threshold of $[Ca^{2+}]_{cyto}$ required for muscle contraction, which is approximately 400 nM. Before reaching this threshold, SERCA activity can be tuned to the resting $[Ca^{2+}]_{cyto}$ levels to modulate ATP utilization for thermogenesis without triggering muscle contraction, effectively coupling calcium handling with energy expenditure during resting conditions.

The nonlinear relationship of SERCA activity across increasing $[Ca^{2+}]_{cyto}$ has been explored by different research groups and implies that small changes in $[Ca^{2+}]_{cyto}$ within this range can have significant effects on SERCA activity and subsequent ATP hydrolysis. For instance, considering the Kd of SERCA to calcium, we have estimated that an increase of 100 to 125 nM $[Ca^{2+}]_{cyto}$ drive a 63% increase in SERCA activity, from a consumption of 31 μM (ATP) s^{-1} to 50 μM s^{-1}.

Nevertheless, *in vivo*, the increase in SERCA activity resulting from RyR1 leak is not solely driven by SR calcium. It is also influenced by an increase in the bulk $[Ca^{2+}]_{cyto}$, which is mediated by an elevated basal calcium permeability of the t-tubule membrane. This increase in calcium permeability involves both store-dependent channels, which are activated by calcium depletion in the SR, as well as transient receptor potential (TRP) channels. These store-dependent and TRP channels

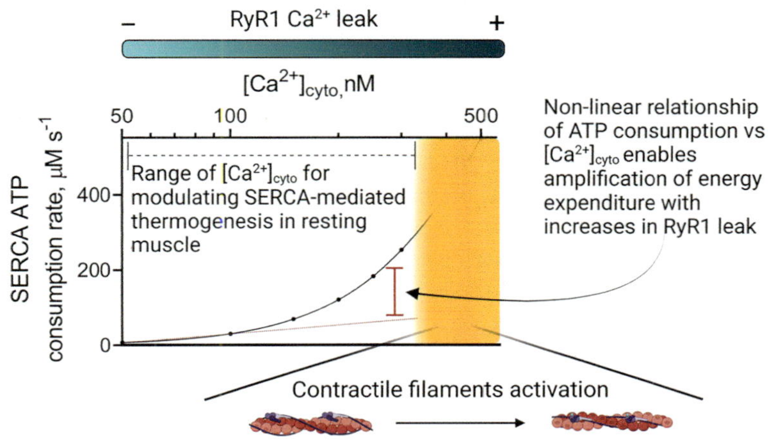

FIGURE 8.1 Relationship between SERCA ATP expenditure and free cytosolic calcium concentration ($[Ca^{2+}]_{cyto}$). Model of the SERCA ATP consumption rate across a range of 500 to 50 nM $[Ca^{2+}]_{cyto}$ obtained from Barclay and Launikonis (2021).

located in the t-tubule membrane allow a basal calcium entry into the muscle cell, and their activity is proportional to the degree of RyR1 leak (Lopez et al., 2021). As a result, these calcium entry channels contribute to the elevated resting $[Ca^{2+}]_{cyto}$ observed in situations characterized by leaky RyR1 channels. For example, Lopez et al. reported an increase from 120 nM $[Ca^{2+}]_{cyto}$ in wild-type mice to 156 nM in the MH mouse model described above.

Along with the 10-fold increase in RyR1 leak rate and the increase in free $[Ca^{2+}]_{cyto}$, we have found that SR calcium content decreases from 1.31 mmols per volume of fiber to 1.15. This decrease represents a change in the percentage of SR Ca^{2+} content relative to the total compartmentalized Ca^{2+} content in the fiber, from 84% (WT) to 76% (MH). Accordingly, mathematical modeling predicts a 15% decrease in calcium release in response to single stimulus compared to wild-type. In this context, two questions arise: (1) How is force production maintained despite the reduced calcium release? (2) Where is the missing calcium that is not stored in the SR but still compartmentalized within in the fiber? The answers to both questions are related.

First, in the context of a maintained capacity of force production despite a reduced SR calcium release, the excess of resting cytosolic calcium plays a key role in compensating for the decreased Ca^{2+} released from the SR. Mathematical modeling indicates that it achieves this by pre-saturating cytosolic binding sites that, in a wild-type context, would be occupied by the released SR calcium. In this manner, the elevated $[Ca^{2+}]_{cyto}$ helps alleviate the amount of Ca^{2+} release necessary for force generation during EC coupling.

A second implication of the increased $[Ca^{2+}]_{cyto}$ is related to how the muscle adapts to this change to meet the associated SERCA-dependent energetic demand. It is important to recognize that any equilibrium point represents a dynamic state where the opposing calcium fluxes mediated by RyR1s and SERCAs reach a steady state, implying a basal SERCA-mediated ATP consumption. In this scenario, the activity of SERCA, as acknowledged above, will be increased by the $[Ca^{2+}]_{cyto}$ (i.e., increased calcium translocation at higher $[Ca^{2+}]_{cyto}$). Consequently, the basal rate of ATP hydrolysis by SERCA is increased in resting muscle in conditions of leaky RyR1s. Yet, it is worth noting that in the context of leaky RyR1 channels and the resulting increase in $[Ca^{2+}]_{cyto}$ the cytosolic concentration of ATP remains relatively constant. This suggests that the supply of ATP available for SERCA utilization is augmented to meet the increased ATP demands in this context.

To cope with this, mitochondria have developed systems that adjust ATP production according to the energetic demands related to calcium handling. Although a potential physiological role of mitochondria as a calcium-buffering system for the cell is contentious (Reggiani and Marucci, 2022), it is widely accepted that mitochondrial calcium handling plays an important role in the function of this organelle. Calcium enters the mitochondria from the cytosol and moves into the intermembrane space through the voltage-dependent anion channel 1 (VDAC1). VDAC1 is a large pore-like structure located in the mitochondrial outer membrane that facilitates the translocation of various ions and metabolites into the intermembrane space. It is generally accepted that due to the large pore size of VDAC1 (18–20 A) (Choundhary et al., 2014) coupled with its large density in the outer mitochondrial membrane, mitochondrial $[Ca^{2+}]$ in the intermembrane space is comparable to the $[Ca^{2+}]_{cyto}$. Some evidence challenges this view, however, proposing that VDAC1 gating properties can be modulated through different mechanisms, altering its Ca^{2+} conductance (Sander et al., 2021). In the inner mitochondrial membrane, Ca^{2+} is translocated into the mitochondrial matrix through the mitochondrial calcium uniporter (MCU) complex.

The resting free calcium concentration in the mitochondrial matrix ($[Ca^{2+}]_{mito}$) has been the subject of investigation by various research groups and is generally acknowledged to fall within the nanomolar (nM) to micromolar (μM) range. Calcium plays a pivotal role within the mitochondria, as it can influence the rate of several mechanisms involved in ATP synthesis through allosteric modulation of key components.

One notable effect of calcium within the mitochondria is its ability to enhance the activity of dehydrogenases involved in the Krebs cycle (Denton, 2009; Williams et al., 2015). An example of this is the upregulation of pyruvate dehydrogenase, the first identified Ca^{2+}-sensitive Krebs cycle enzyme, with an apparent K_D of approximately 10 μM. Furthermore, calcium increases the affinity of isocitrate dehydrogenase for its ligand, in a manner dependent on the concentration of ADP. Similarly, the affinity of α-ketoglutarate dehydrogenase for α-ketoglutarate is enhanced in a calcium-dependent fashion.

In addition to its effects on the production of electron donors, calcium also exerts regulatory influence on the electron transport chain itself, increasing electron flow through complexes I, III, and IV (Glancy et al., 2013), thereby enhancing the overall activity of the electron transport chain.

Furthermore, evidence from cardiac muscle suggests that calcium could enhance ATP synthesis by regulating the ATP synthase activity. Calcium has been shown to bind to the ATP synthase inhibitory factor 1, leading to its dissociation from the ATP synthase and subsequently increasing the activity of the enzyme (Harris and Das, 1991). This regulatory mechanism provides an additional means by which calcium can modulate ATP synthesis within the mitochondria.

Nowadays, it is well acknowledged that conditions of leaky RyR1s are associated with increased $[Ca^{2+}]_{mito}$ as is the case with muscles after exercise interventions, where RyR1 leak plays an important role in muscle adaptation to exercise, as well as in congenital conditions such as MH. Together, the diverse effects of mitochondrial calcium in respiration and ATP synthesis underscore its vital role in regulating mitochondrial bioenergetics. By modulating the activity of various enzymes involved in ATP synthesis and influencing the electron transport chain, calcium within this organelle tunes the production of ATP to meet the dynamic demands associated with changes in intracellular calcium levels.

In this way, a redistribution of intracellular calcium between the SR, cytosol, and mitochondria, which occurs as a result of RyR1 leak and is mediated by chronic RyR1 leak-dependent extracellular Ca^{2+} entry, effectively contributes to the preservation of muscle function in relation to both force production and thermogenesis.

8.4 FUNCTIONAL FEATURES OF RYR1 ISOFORM FOR MUSCLE-BASED THERMOGENESIS, ADVANTAGES OVER α/β RYANODINE RECEPTORS

In contrast to lower vertebrates, mammals utilize their resting skeletal muscles for thermoregulation. This suggests that there are inherent differences in the muscle characteristics between these two groups that allow mammals to effectively generate and regulate heat. From a different perspective, it can be inferred that certain characteristics of lower vertebrate muscles may impose limitations on their capacity to serve as a regulated heat source.

Notably, there are two key differences in the RyR-mediated calcium leak between muscles of lower vertebrates and mammals. First, when comparing the leak rate of RyR in an ectothermic model such as the amphibian *Rhinella marina* to that observed in mammals, a significant disparity is evident. The leak rate in the ectotherm is approximately 2–3 μM per second, whereas in mammals it is 14–30 μM per second.

A second clear distinction between these organisms lies in the presence of two isoforms of RyRs in lower vertebrates: RyRα and RyRβ, as opposed to the RyR1 isoform found in mammals. Exploring the activation mechanisms of these different RyR isoforms can help us understand how they may influence the capacity of skeletal muscle for regulated heat generation between lower vertebrates and mammals.

RyRα and RyRβ are functionally analogous to RyR1 and RyR2, respectively. In lower vertebrates, the activation of RyRα is mediated by the DHPR, which is activated by voltage, in a similar manner to RyR1 in mammals. On the other hand, the subpopulation of RyRs corresponding to RyRβ is activated by increases in cytosolic calcium at the junctional space between the t-tubule membrane and the SR terminal cisternae. Once triggered, the activation of RyRβ leads to a regenerative SR calcium release mechanism known as Ca^{2+}-induced Ca^{2+} release (CICR). This mechanism resembles the SR calcium release mechanism mediated by RyR2 in the heart.

Why is the Ca^{2+} insensitivity of RyR1 relevant for thermogenesis in resting muscle? We have recently proposed that the cytosolic Ca^{2+} insensitivity of RyR1 plays a crucial role in allowing the muscle to regulate the rate of RyR1 Ca^{2+} leakage under physiological conditions when there is a need for increased energy expenditure and thermogenesis (Singh et al., 2023). Unlike in lower vertebrates, where RyRβ isoforms would respond to increased $[Ca^{2+}]_{cyto}$ with abrupt activation, RyR1 in mammals remains insensitive to these fluctuations. This is important because, to utilize RyR1 as a heat source without compromising muscle function, it is crucial that increases in resting $[Ca^{2+}]_{cyto}$, especially in the junctional space facing RyR1, do not trigger channel activation. If RyR1 were sensitive to these increases, it would result in uncontrolled release of calcium, which is observed in pathological conditions such as malignant hyperthermia. By maintaining the Ca^{2+} insensitivity of RyR1, the muscle can ensure controlled and regulated heat production without detrimental consequences.

On the other hand, the sensitivity of βRyRs to cytosolic calcium can be seen as a distinct feature observed in lower vertebrates that contributes to their lower basal metabolic rate compared to endotherms. By being responsive to even small changes in cytosolic calcium levels, βRyRs restrict the loading of calcium into the SR. The controlled leak rate of RyRs in lower vertebrates, mediated by the sensitivity of βRyRs to cytosolic calcium, has important implications for their energy utilization. By limiting this feature, these organisms conserve energy by minimizing the rate of ATP hydrolysis mediated by SERCA. As a result, the demand for both food intake and energy associated with foraging is reduced, optimizing their energy utilization and minimizing energy expenditure.

8.5 PHYSIOLOGICAL CONDITIONS ASSOCIATED WITH CHANGES IN RYANODINE RECEPTOR 1 LEAK

8.5.1 Cold Acclimation

The role of skeletal muscle, particularly RyR1, in the thermogenic mechanisms activated during cold acclimation has been investigated in studies using murine models. These studies provide evidence supporting the involvement of RyR1 in cold-induced thermogenesis. For example, when wild-type mice are treated with the RyR1 leak inhibitor dantrolene and exposed to a temperature of 4°C, they exhibit a decrease in their core body temperature compared to control mice (Bal et al., 2012). This suggests that RyR1 plays a role in the regulation of body temperature and contributes to the adaptive response to cold conditions.

In murine models lacking brown adipose tissue (BAT), the reliance on skeletal muscle for thermogenesis, specifically non-shivering thermogenesis, becomes more prominent. This emphasizes the capacity of skeletal muscle to be recruited for heat production in the absence of BAT. Mechanistically, cold acclimation drives an increase in *RYR1* expression. Additionally, it increases RyR1 phosphorylation, as well as the dissociation of calstabin 1 from the channel, destabilizing it to make it more leaky (Aydin et al., 2008; Bal et al., 2016). These molecular changes are accompanied by an elevation in oxygen consumption, indicative of increased energy demands, and alterations in mitochondrial dynamics that are associated with adaptations to meet these heightened energy requirements.

8.5.2 Obesity

While much of the research on modifications in energy expenditure related to Ca^{2+} handling in obesity has focused on the modulation of SERCA function, there is evidence suggesting that obesity may also have a modulatory effect on RyR1s. Jacque-Fernandez et al. (2020) discovered an elevated basal RyR1 activity in muscle fibers from mice fed with a high-fat, high-sucrose diet compared to those fed with a standard diet. Importantly, despite this increase in activity, no significant disruptions in calcium handling during EC coupling were observed. Mechanistically, this increase in RyR1 activity takes place along an increase in RyR1 S-nitrosylation, which is induced by an elevation in reactive oxygen species (ROS) derived from the mitochondria, as described by Jain et al. (2014). These findings highlight the potential impact of obesity on RyR1 function and suggest a link between altered calcium handling and metabolic disturbances associated with obesity.

8.5.3 Exercise

Exercise interventions have been shown to elicit changes in the function of RyR1, which play a crucial role in the adaptive response to training and the modulation of metabolism. Studies conducted in mice have revealed that even low-intensity exercise performed over a period of three weeks can induce notable alterations in RyR1 leak. During this exercise regime, an increase in RyR1 leak occurs, associated with the dissociation of calstabin 1 from the channel. This dissociation leads to an increase in resting $[Ca^{2+}]_{cyto}$, triggering a cascade of molecular events that contribute to training adaptation. One important consequence of increased RyR1 leak and elevated $[Ca^{2+}]_{cyto}$ is the induction of mitochondrial biogenesis, which is supported by the observed upregulation of key markers associated with mitochondrial biogenesis, including PGC1-α, MEF2c, and TFAM.

A notable finding from a study conducted by Ivarsson et al. (2019) indicates that inducing RyR1 leak alone, using pharmacological agents that destabilize RyR1, can trigger the same mitochondrial adaptations observed with exercise. This discovery provides strong evidence for the crucial role of exercise-induced RyR1 leak in driving mitochondrial adaptations.

Recent research has shed further light on the connection between RyR1 Ca^{2+} leak and exercise-induced mitochondrial adaptations. Specifically, in response to sprint interval training, RyR1 Ca^{2+} leak is also increased, leading to a subsequent rise in mitochondrial Ca^{2+} content. Intriguingly, when mitochondrial calcium uptake is inhibited, the exercise-induced adaptations are blunted, underscoring the significance of this process in the beneficial effects of exercise (Zanou et al., 2021).

The combined evidence from these studies emphasizes the critical involvement of RyR1 and its Ca^{2+} leak in the complex interplay of exercise-induced mitochondrial adaptations. Whether induced by exercise or through pharmacological means, RyR1 leak appears to play a central role in triggering mitochondrial biogenesis and other molecular events that enhance muscle function and modulate metabolism.

8.6 RYANODINE RECEPTOR 1 IN DISEASE

Mutations in the *RYR1* gene can disrupt the normal function of the RyR1 channel, leading to various pathologies. Dominant *RYR1* variants have been associated with mutations in three specific regions of the RyR1 channel: the N-terminal region (amino acids 35–614), the central region (amino acids 2163–2458), and the C-terminal region (amino acids 4550–4940) (Lawal, Todd, and Meilleur, 2018; Zhou et al., 2007). On the other hand, mutations linked to recessive *RYR1* variants have not been associated with any specific region of the protein.

Most of the pathologies associated with *RyR1* mutations primarily affect skeletal muscle (Lawal, Todd, and Meilleur, 2018). These conditions encompass a group of myopathies characterized by muscle hypotonia and progressive muscle weakness, with varying degrees of severity among the different diseases. Some examples of these myopathies include central core disease, centronuclear myopathy, multiminicore disease, core-rod myopathy, and fiber-type disproportion (Clarke et al., 2010; McCarthy et al., 2000). In addition to these dystrophy-like phenotypes, other pathological conditions associated with dysregulation of RyR1 have been identified, such as atypical periodic paralysis, rhabdomyolysis, and malignant hyperthermia (Dlamini et al., 2013; McCarthy et al., 2000; Rosenberg et al., 2015).

Furthermore, excessive Ca^{2+} leak due to RyR1 dysregulation has been implicated in the underlying mechanisms of various muscle disorders, making it a promising therapeutic target for multiple conditions. In the MDX mouse model of Duchenne muscular dystrophy, targeting of RyR1 with the molecule s107, which enhances the affinity of calstabin 1 to nitrosylated RyR1, resulted in reduced markers of muscle damage along with an improved muscle function (Bellinger et al., 2009). Additionally, RyR1 leak has been associated with skeletal muscle dysfunction observed in conditions such as aging-related muscle weakness (Andersson et al., 2011) and cancer-associated muscle weakness (Waning et al., 2015). More importantly, evidence shows that targeting this leak prevents the muscle impairment characteristic of these conditions. This further highlights the significance of RyR1 leak as a potential therapeutic target for different myopathies that appear to involve oxidative stress as a common factor.

Overall, dysregulation of RyR1 disrupts calcium homeostasis within the muscle fiber, often leading to excessive calcium leakage that creates persistent cellular stress. This stress initiates a feedback loop, causing oxidative stress that further impairs RyR1 function and potentially contributes to the progression of dystrophies and muscle weakness–related disorders. However, in certain conditions, such as malignant hyperthermia, where the individuals experience a relatively normal life, the extent of RyR1 leak present permits the muscle to adapt. This adaptation entails alterations in calcium handling, calcium compartmentalization, and calcium handling–associated metabolism, thereby maintaining a stable state and averting evident pathological phenotypes under basal circumstances. Nevertheless, exposure to factors like volatile anesthetics, heat stress, or exercise can push the system to its limits, posing life-threatening risks. Alternatively, different environmental scenarios or physiological conditions can result in alterations of the RyR1 gating properties, modifying its leakage rate and setting the basal ATP consumption rate by SERCA. The heat involved in this reaction along with the heat generated throughout the metabolic reactions necessary to meet the ATP demands allow endothermic organisms to use their skeletal muscles as an important thermogenic source.

REFERENCES

Adams B. A., Tanabe T., Mikami A., Numa S., Beam K. G. Intramembrane charge movement restored in dysgenic skeletal muscle by injection of dihydropyridine receptor cDNAs. *Nature* 346(6284) 569–72 (1990).

Andersson D. C., Betzenhauser M. J., Reiken S., Meli A. C., Umanskaya A., Xie W., Shiomi T., Zalk R., Lacampagne A., Marks A. R. Ryanodine receptor oxidation causes intracellular calcium leak and muscle weakness in aging. *Cell Metab* 14(2) 196–207 (2011).

Aydin J., Shabalina I. G., Place N., Reiken S., Zhang S, Bellinger A. M., Nedergaard J., Cannon B., Marks A. R., Bruton J. D., Westerblad H. Nonshivering thermogenesis protects against defective calcium handling in muscle. *FASEB J* 22(11) 3919–24 (2008).

Bal N. C., Maurya S. K., Singh S., Wehrens X., Periasamy M. Increased reliance on muscle-based thermogenesis upon acute minimization of brown adipose tissue function. *J Biol Chem* 291(33) 17247–57 (2016).

Bal N. C., Maurya S.K., Sopariwala D. H., Sahoo S. K., Gupta S. C., Shaikh S. A., Pant M., Rowland L. A., Bombardier E., , Tupling A. R., Molkentin J. D., Periasamy M. Sarcolipin is a newly identified regulator of muscle-based thermogenesis in mammals. *Nat Med* 18(10) 1575–1579 (2012).

Barclay C. J., Launikonis B.S. Components of activation heat in skeletal muscle. *J Muscle Res Cell Motil* 42(1) 1–16 (2021).

Beard N. A., Wei L., Dulhunty A. F. Ca^{2+} signaling in striated muscle: the elusive roles of triadin, junctin, and calsequestrin. *Eur Biophys J* 39 27–36 (2009).

Bellinger A. M., Reiken S., Carlson C., Mongillo M., Liu X., Rothman L., Matecki S., Lacampagne A., Marks A.R. Hypernitrosylated ryanodine receptor/calcium release channels are leaky in dystrophic muscle. *Nat Med* 15(3) 325–30 (2009).

Catterall W. A. Voltage-Gated Calcium Channels. *Cold Spring Harb Perspect Biol* 3(8) a003947 (2011).

Charisani V. R., Pasek D. A., Meissner G. Structural and functional interactions between the Ca^{2+}-, ATP-, and caffeine-binding sites of skeletal muscle ryanodine receptor (RyR1). *J Biol Chem* 297(3) 101040 (2021).

Choundhary O. P., Paz A., Adelman J. L., Colletier J., Abramson J., Grabe M. Structure-guided simulations illuminate the mechanism of ATP transport through VDAC1. *Nat Struct Mol Biol* 21(7) 626–32 (2014).

Clarke N. F., Waddell L. B., Cooper S. T., Perry M., Smith R. L., Kornberg A. J., Muntoni F., Lillis S., Straub V., Bushby K., Guglieri M., King M. D., Farrell M., Marty I., Lunardi J., Monnier N., North K. Recessive mutations in RYR1 are a common cause of congenital fiber type disproportion. *Hum Mutat* 31(7) E1544–50 (2010).

Constantin, L. L., Podolsky R. J. Calcium localization and the activation of striated muscle fibers. *J Physiol* 24(5) 1141–5 (1965).

Denton R. M. Regulation of mitochondrial dehydrogenases by calcium ions. *Biochem Biophys Acta* 1787(11) 1309–1316 (2009).

Dlamini N., Voermans N. C., Lillis S., Stewart K, Kamsteeg E-J., Drost G., Quinlivan R., Snoeck M., Norwood F., Radunovic A., Straub V., Roberts M., Vrancken A. F., van der Pol W. L., de Coo R. I., Manzur A. Y., Yau S., Abbs S., King A., Lammens M., Hopkins P. M., Mohammed S., Treves S., Muntoni F., Wraige E., Davis M. R., van Engelen B., Jungbluth H. Mutations in RYR1 are a common cause of exertional myalgia and rhabdomyolysis. *Neuromuscul Disord* 23(7) 540–8 (2013).

Dulhunty A., Haarman C., Green D., Hart J. W. How Many Cysteine Residues Regulate Ryanodine Receptor Channel Activity? *Antioxid Redox Signal* 2(1) 27–34 (2000).

Eltit J. M., Li H., Ward C. W., Molinski T., Pessah I. N., Allen P. D., Lopez J. R. Orthograde dihydropyridine receptor signal regulates ryanodine receptor passive leak. *Proc Natl Acad Sci* 108(17) 7046–51 (2011).

Ferguson, D. G., Schwartz, H. W., Franzini-Armstrong, C. Subunit structure of junctional feet in triads of skeletal muscle: A freeze-drying, rotary-shadowing study. *J Cell Biol*, 99(5) 1735–42 (1984).

Fleischer S., Ogunbunmi E. M., Dixon M. C., Fleer E. A. Localization of Ca^{2+} release channels with ryanodine in junctional terminal cisternae of sarcoplasmic reticulum of fast skeletal muscle. *Proc Natl Acad Sci* 82(21) 7256–9 (1985).

Franzini-Armstrong C. STUDIES OF THE TRIAD: I. Structure of the Junction in Frog Twitch Fibers. *J Cell Biol* 47(2) 488–99 (1970).

Franzini-Armstrong C., Porter K. R. Sarcolemmal invaginations constituting the t system in fish muscle fibers. *J Cell Biol* 22(3) 675–96 (1964).

Gamu D., Juracic E. S., Hall K. J., Tupling R. A. The sarcoplasmic reticulum and SERCA: a nexus for muscular adaptive thermogenesis. *Appl Physiol Nutr Metab* 45(1) 1–10 (2020).

Glancy B., Willis W. T., Chess D. J., Balaban R. S. Effect of Calcium on the Oxidative Phosphorylation Cascade in Skeletal Muscle Mitochondria. *Biochem* 52 (16) 2793–809 (2013).

Gonano L. A., Jones P. P. FK506-binding proteins 12 and 12.6 (FKBPs) as regulators of cardiac Ryanodine Receptors: Insights from new functional and structural knowledge. *Channels* 11(5) 415–425 (2017).

Harris D. A., Das A. M. Control of mitochondrial ATP synthesis in the heart. *Biochem J* 280(3) 561–573 (1991).

Hodgkin, A.L., Horowicz P. Potassium contractures in single muscle fibres. *J Physiol* 153(2) 386–403 (1960).

Imagawa T., Smith J. S., Coronado R., Campbell K. P. Purified ryanodine receptor from skeletal muscle sarcoplasmic reticulum is the Ca^{2+}-permeable pore of the calcium release channel. *J Biol Chem* 262(34) 16636–43 (1987).

Inui M., Saito A., Fleischer S. Purification of the ryanodine receptor and identity with feet structures of junctional terminal cisternae of sarcoplasmic reticulum from fast skeletal muscle. *J Biol Chem* 262(4) 1740–7 (1987).

Ivarsson N., Mattsson C. M., Cheng A. J., Bruton J. D., Ekblom B., Lanner J. T., Westerblad H. SR Ca^{2+} leak in skeletal muscle fibers acts as an intracellular signal to increase fatigue resistance. *J Gen Physiol* 151(4) 567–577 (2019).

Jain S. S., Paglialunga S., Vigna C, Ludzki A., Herbst E. A., Lally J. S., Schrauwen P., Hoeks J., Tupling A. R., Bonen A., Holloway G. P. High-fat diet-induced mitochondrial biogenesis is regulated by mitochondrial-derived reactive oxygen species activation of CaMKII. *Diabetes* 63(6) 1907–13 (2014).

Jaque-Fernandez F., Beaulant A., Berthier C., Monteiro L., Allard B., Casas M., Rieusset J., Jacquemond V. Preserved Ca^{2+} handling and excitation-contraction coupling in muscle fibres from diet-induced obese mice. *Diabetologia* 63(11) 2471–2481 (2020).

Lamboley C.R., Pearce L., Seng C., Meizoso-Huesca A., Singh D. P., Frankish B. P., Kaura V., Lo H. P., Ferguson C., Allen P. D., Hopkins P. M., Parton R. G., Murphy R. M., Van der Poel C., Barclay C. J., Launikonis B. S. Ryanodine receptor leak triggers fiber Ca^{2+} redistribution to preserve force and elevate basal metabolism in skeletal muscle. *Sci Adv* 7(44) eabi7166 (2021).

Lanner J. T., Georgiu D. K., Joshi A. D., Hamilton S. L. Ryanodine receptors: structure, expression, molecular details, and function in calcium release. *Cold Spring Harb Perspect Biol* 2(11) a003996 (2010).

Laver D. R., Baynes T. M., Dulhunty A. F.. Magnesium inhibition of ryanodine-receptor calcium channels: evidence for two independent mechanisms. *J Membr Biol* 156(3) 213–229 (1997).

Lawal T. A., Todd J. J., Meilleur K. G. Ryanodine receptor 1-related myopathies: Diagnostic and therapeutic approaches. *Neurother* 15(4) 885–899 (2018).

Lopez J. R., Uryash A., Adams J., Hopkins P. M., Allen P. D. Molecular Modification of Transient Receptor Potential Canonical 6 Channels Modulates Calcium Dyshomeostasis in a Mouse Model Relevant to Malignant Hyperthermia. *Anesthesiology* 134(2) 234–247 (2021).

Luff A., Atwood H. Membrane properties and contraction of single muscle fibers in the mouse. *Am J Phys* 222(6) 1435–40 (1972).

McCarthy T. V., Quane K. A., Lynch P. J. Ryanodine receptor mutations in malignant hyperthermia and central core disease. *Hum Mutat* 15(5) 410–7 (2000).

McCarthy M. R., Savich Y., Cornea R.L., Thomas D. D. Resolved Structural States of Calmodulin in Regulation of Skeletal Muscle Calcium Release. *Biophys J* 118(5) 1090–1100 (2020).

Meissner, G. Adenine nucleotide stimulation of Ca^{2+}-induced Ca^{2+} release in sarcoplasmic reticulum. J Biol Chem 259(4) 2365–74 (1984).

Meissner, G., Darling, E., Eveleth, J. Kinetics of Rapid Ca^{2+} Release by Sarcoplasmic Reticulum. Effects of Ca^{2+}, Mg^{2+}, and Adenine Nucleotides. *Biochem*, 25(1) 236–44 (1986).

Meizoso-Huesca A., Pearce L., Barclay C. J., Launikonis B. S. Ca^{2+} leak through ryanodine receptor 1 regulates thermogenesis in resting skeletal muscle. *Proc Natl Acad Sci* 119(4) e2119203119 (2022).

Melville Z., Dridi H., Yuan Q., Reiken S., Wronska A., Liu Y., Clarke O.B., Marks A. R. A drug and ATP binding site in type 1 ryanodine receptor. *Structure* 30(7) 1025–1034.e4 (2022).

Moore C.P., Rodney G., Zhang J. Z., Santacruz-Toloza L., Strasburg G., Hamilton S. L. Apocalmodulin and Ca^{2+} calmodulin bind to the same region on the skeletal muscle Ca^{2+} release channel. *Biochem* 38(26) 8532–7 (1999).

Moore C.P., Zhang J. Z., Hamilton S. L. A role for cysteine 3635 of RyR1 in redox modulation and calmodulin binding. *J Biol Chem* 274(52) 36831–36834 (1999).

Ogata, T., Yamasaki Y. Ultra-High-Resolution Scanning Electron Microscopy of Mitochondria and Sarcoplasmic Reticulum Arrangement in Human Red, White, and Intermediate Muscle Fibers. *Myology* 248(2) 214–23 (1997).

Place N., Ivarsson N., Venckunas T., Neyroud D., Brazaitis M., Cheng A. J., Ochala J., Kamandulis S., Girard S., Volungevičius G., Paužas H., Mekideche A., Kayser B., Martinez-Redondo V., Ruas J. L., Bruton J., Truffert A., Lanner J. T., Skurvydas A., Westerblad H. Ryanodine receptor fragmentation and sarcoplasmic reticulum Ca^{2+} leak after one session of high-intensity interval exercise. *Proc Natl Acad Sci* 112(50) 15492–7 (2015).

Porter K., Palade G. E. Studies on the Endoplasmic Reticulum III. Its form and distribution in striated muscle cells. *J Biophisic Biochem Cytol* 3(2) 269–300 (1957).

Reggiani C., Marucci L. A controversial issue: Can mitochondria modulate cytosolic calcium and contraction of skeletal muscle fibers? *J Gen Physiol* 154(9) e202213167 (2022).

Reiken S., Lacampagne A., Zhou H., Kherani A., Lehnart S. E., Ward C., Huang F., Gaburjakova M., Gaburjakova J., Rosemblit N., Warren M. S., He K., Yi G., Wang J., Burkhoff D., Vassort G., Marks A. R. PKA phosphorylation activates the calcium release channel (ryanodine receptor) in skeletal muscle: defective regulation in heart failure. *J Cell Biol* 160(6) 919–28 (2003).

Revel J. P. The sarcoplasmic reticulum of the bat cricothyroid muscle. *J Cell Biol* 12(3) 571–588 (1962).

Rodney G. G., Williams B. Y., Strasburg G. M., Beckingham K., Hamilton S.L. Regulation of RYR1 activity by Ca(2+) and calmodulin. *Biochem* 39(26) 7807–12 (2000).

Rosenberg H., Pollock N., Schiemann A., Bulger T., Stowellcorresponding K. Malignant hyperthermia: a review. *Orphanet J Rare Dis* 10(93) (2015).

Royer L., Rios E. Deconstructing calsequestrin. Complex buffering in the calcium store of skeletal muscle. *J Physiol* 587(13) 3101–11 (2009).

Sander P., Gudermann T., Schredelseker J. A calcium guard in the outer membrane: Is VDAC a regulated gatekeeper of mitochondrial calcium uptake? *Int J Mol Sci* 22(2) 946 (2021).

Singh, D. P., Pearce L., Choi R. H., Meizoso-Huesca A., Wette S. G., Scott J. W., Lamboley C. R., Murphy R. M., Launikonis B. S. Evolutionary isolation of ryanodine receptor isoform 1 for muscle-based thermogenesis in mammals. *Proc Natl Acad Sci* 120(4) e2117503120 (2023).

Smith, J. S., Coronado, R. and Meissner, G. Single Channel Measurements of the Calcium Release Channel from Skeletal Muscle Sarcoplasmic Reticulum. *J Gen Physiol* 88(5) 573–88 (1986).

Snyder H. R. Jr, Davis C. S., Bickerton R. K., Halliday R.P. 1-[(5-arylfurfurylidene) amino]-hydantoins. A new class of muscle relaxants. *J Med Chem* 10(5) 807–10 (1967).

Takekura, H., Flucher, B. E. and Franzini-Armstrong, C. Sequential docking, molecular differentiation, and positioning of T-tubule/SR junctions in developing mouse skeletal muscle. *Dev Biol* 239(2) 204–14 (2001).

Tang W., Ingalls C. P., Durham W. J., Snider J., Reid M. B., Wu G., Matzuk M.M., Hamilton S.L. Altered excitation-contraction coupling with skeletal muscle specific FKBP12 deficiency. *FASEB J* 18(13) 1597–9 (2004).

Tripathy A., Xu L., Mann G., Meissner G. Calmodulin activation and inhibition of skeletal muscle Ca^{2+} release channel (ryanodine receptor). *Biophys J* 69(1) 106–19 (1995).

Tripp T. R., Frankish B. P., Lun V., Wiley J. P. ., Shearer J., Murphy R. M., MacInnis M. J. Time course and fibre type-dependent nature of calcium-handling protein responses to sprint interval exercise in human skeletal muscle. *J Physiol* 600(12) 2897–2917 (2022).

Van Petegem. Ryanodine Receptors: Structure and Function. *J Biol Chem* 287(38) 31624–31632 (2012).

Waning D. L., Mohammad K. S., Reiken S., Xie W., Andersson D. C., John S., Chiechi A., Wright L. E., Umanskaya A., Niewolna A., Trivedi T., Charkhzarrin S., Khatiwada P., Wronska A., Haynes A., Benassi M. S., Witzmann F. A., Zhen G., Wang X., Cao X., Roodman G. D., Marks A., Guise T. A. Excess TGF-β mediates muscle weakness associated with bone metastases in mice. *Nat Med* 21(11) 1262–1271 (2015).

Williams G. S. B., Boyman L., Lederer W. J. Mitochondrial calcium and the regulation of metabolism in the heart. *J Mol Cell Cardiol* 78 35–45 (2015).

Xu H., Remmen H. V. The SarcoEndoplasmic Reticulum Calcium ATPase (SERCA) pump: a potential target for intervention in aging and skeletal muscle pathologies. *Skeletal Muscle* 11(1) 25 (2021).

Zanou N., Dridi H., Reiken S., Imamura de Lima T., Donnelly C., De Marchi U., Ferrini M., Vidal J., Sittenfeld L., Feige J. N., Garcia-Roves P. M., Lopez-Mejia I. C., Marks A. R., Auwerx J., Kayser B., Place N. Acute RyR1 Ca^{2+} leak enhances NADH-linked mitochondrial respiratory capacity. *Nat Commun* 12(1) 7219 (2021).

Zhou H., Jungbluth H. A. Sewry C. A., Feng L., Bertini E., Bushby K., Straub V., Roper H., Rose M. R., Brockington M., Kinali M., Manzur A., Robb S., Appleton R., Messina S., D'Amico A., Quinlivan R., Swash M., Müller C. R., Brown S., Treves S., Muntoni F. Molecular mechanisms and phenotypic variation in RYR1-related congenital myopathies. *Brain* 130(8) 2024–36 (2007).

9 Futile SERCA-Mediated Calcium Cycling Activity as the Basis for Non-Shivering Thermogenesis in the Skeletal Muscle

Naresh Chandra Bal
KIIT University, Bhubaneswar, India

Muthu Periasamy
University of Central Florida, Orlando, Florida, United States

9.1 INTRODUCTION

Vertebrates that are endothermically homeotherms (like most mammals and birds) produce heat inside their body when exposed to cold ambient temperature. The first line of defense against cold challenge is shivering, which is uncoordinated contraction-relaxation of skeletal muscle that generates significant heat without work. However, it is observed that after a short period, shivering subsides but the animal is able to sustain constant body temperature indicating the existence of non-shivering thermogenesis (NST). Since the early days of research scientists have been trying to identify the sites for NST. Discoveries in the 1960s provided first evidences that brown adipose tissue (BAT) forms the major site of NST in some of the mammals, including humans [1]. The role of BAT in thermogenesis and its various aspects of body temperature maintenance has been covered in other chapters in Unit 1. Soon it was discovered that several animals do not have functional BAT but are able to maintain their body temperature quite efficiently, suggesting possible existence of other sites and/or mechanisms of NST. Some researchers have proposed quite early on that the skeletal muscle harbors mechanisms of NST independent of shivering [2–5]. Kayser et al. in the 1920s showed that spinal cord–denervated pigeons (*Columba livia*) exposed to cold can maintain body temperature suggested the existence of NST [6]. Since then researchers have been trying to define the mechanistic details of muscle NST, and some of the studies as early as the 1950s have reported on developments in this direction. The first conclusive demonstration of existence of NST in the skeletal muscle came from cold exposure studies on dogs by Davis in 1967 [3]. However, it was the next logical question to differentiate the heat-generating mechanism of muscle NST from shivering, which in fact still remains debated. Some of the pioneering studies in the 1980s provided evidence that skeletal muscle NST is regulated by thyroid and catecholamines [7–11]. During this time, the strongest support for muscle NST hypothesis came from the discovery of "heater organs" in endothermic fish. These heater organs located in the ocular muscle (forebrain) dedicated to produce heat enable them to maintain a higher cranial temperature than that of the ambient environment [12].

DOI: 10.1201/9781003391470-11

Ultrastructural and biochemical studies showed that the heater organ is actually a modified skeletal muscle [13–15]. This topic is covered in detail in Chapter 5. The fish heater organ utilizes calcium (Ca^{2+}) cycling through the sarcoplasmic reticulum (SR)-like structures and is highly enriched with mitochondria to produce heat [16, 17]. Thus, SR Ca^{2+}cycling has been tapped for NST long before the advent of mammals during the early steps of evolution of vertebrates; this is discussed in detail in Chapter 2.

The observation that SR-like structures can utilize Ca^{2+}cycling for thermogenesis inspired researchers to explore such a mechanism in the skeletal muscle NST. However, research in muscle NST during the 1990s and 2000s was pushed to the background by primary focus on NST in the brown adipose tissue (BAT). Pioneering molecular research in the BAT field during this period led to the establishment of the dogma that in eutherian mammals BAT is the only site of NST (discussed in Chapter 2). Further, the belief has been established that other homeothermic clades such as birds, monotremes, and marsupials lacking BAT possess shivering as the only mechanism of facultative thermogenesis. This idea is still held by some researchers completely refuting the possibility of muscle NST. Utilizing neonatal birds as a model, researchers acquired evidence that establish the existence of muscle NST. In 1987, Barre et al. showed that duckling skeletal muscles undergo an increase in oxidative capacity upon cold adaptation indicating the role of mitochondrial metabolism in muscle NST [18]. Another study showed that cold adaptation in rat increases Ca^{2+}-loading capacity of skeletal muscle mitochondria [19]. A major breakthrough was achieved when some studies reported upregulation of the SR Ca^{2+} release channel, namely ryanodine receptor (RyR) 1 along with SERCA in the skeletal muscle upon cold adaptation in ducklings [20–22]. It was proposed that the Ca^{2+}leak from RyR1 might activate SERCA kinetics leading to futile ATP hydrolysis and heat production. Further, recent investigations show that such a release and reuptake of Ca^{2+} also work in resting mammalian muscles (discussed in Chapter 4), providing additional evidence for the original idea. The potential role of SERCA in muscle NST in mammals was proposed by the studies of De Meis and colleagues [23–29]. In addition to SERCA, some other mechanisms such as Na^+-K^+ ATPase [30–32] and mitochondrial proton leak [33–35] have also been proposed as the basis for muscle NST, although the molecular details are still missing.

The mechanistic basis of muscle NST has been fragmented for several decades, and systematic research has been slow. However, with the discovery of sarcolipin (SLN), a small regulator of SERCA as the missinglink in muscle NST in 2012, catalyzed much needed investigation in this area [36]. Many studies have been designed to gain insight about the role of futile SERCA activity in the skeletal muscle and its regulation for controlled activation/deactivation [37–39]. Recent data suggest that the muscle NST can be triggered by nervous system and is discussed in the next chapter of the current unit. In this chapter, we will cover the various aspects of SERCA-based thermogenesis in skeletal muscle. Especially we will discuss how the coupling between the two primary functions of SERCA, i.e., Ca^{2+} transport into the SR lumen and ATP hydrolysis is modulated to alter energy efficiency of SERCA generating heat. In addition, we will discuss the role of various factors in modulating SERCA efficiency including cations and micropeptides like SLN. Another aspect of discussion is SLN-mediated regulation of SERCA in production of local Ca^{2+} signal that serves as an integrator between mitochondrial metabolism and thermoregulation thereby regulating whole-body energy expenditure.

9.2 SERCA AS THE HEAT PUMP

SERCA is ubiquitously expressed in the endoplasmic reticulum (ER) or SR (the major intracellular Ca^{2+} store) of every cell types. Among mammals, more than 10 different isoforms of SERCA have been reported [40, 41]. SERCA is responsible to maintain a low cytosolic Ca^{2+} concentration by transporting Ca^{2+} into ER/SR against aconcentration gradient by use of energy from ATP hydrolysis. In skeletal muscles, two SERCA isoforms are expressed: while SERCA1 exhibits high abundance in the fast-twitch muscles, medium abundance in intermediate muscles, and lowabundance in pure slow-twitch muscles, SERCA2 expression exhibits the reverse trend with high abundance in slow-twitch muscles that are associated with posture-bearing activities, i.e., sustaining low-level contractionover a long period of time [42]. The primary function of SERCA is to serve as the determinant of cytosolic Ca^{2+} that governs relaxation of the skeletal muscle. Increases in cytosolic Ca^{2+} concentration activate SERCA, and it is believed that various isoforms of SERCA have different activation thresholds. In the skeletal muscle, SERCA is activated at very low Ca^{2+}, and its Ca^{2+} removal is the principal cause of muscle relaxation. SERCA being an ATP-driven pump, researchers have been trying to determine the amount of Ca^{2+} transported (i.e., work done by the pump) per the amount of ATP consumed (i.e., fuel utilized), and this is known as the efficiency of SERCA. The amount of energy generated from ATP hydrolysis by SERCA that is not coupled to Ca^{2+} uptake is liberated as heat, and this heat-split ratio might vary depending on local milieu. The relationships between different factors that can influence SERCA-Ca^{2+} transport efficiency have been recently tested [43, 44].

Comparative physico-chemical analyses of various SERCA isoforms indicate that the energy efficiency of SERCA (i.e., the coupling between ATP hydrolysis and Ca^{2+} uptake) is highly variable. Biophysical and mathematical studies shows that a major portion (18–55%) of the energy from ATP hydrolyzed by SERCA is not trapped into work and is dissipated as heat; therefore, these studies termed it the *heat pump* [5, 14, 28]. Moreover, recent *in vitro* data show that ATP hydrolysis by SERCA is activated even at a very low Ca^{2+} concentration. Based on this it has been suggested that SERCA activity in resting muscle can be recruited by minor fluctuations in cytosolic Ca^{2+}, thereby contributing to more than 30% of basal metabolic rate and is the best candidate for muscle NST. The SR in the skeletal muscle stores a high concentration of Ca^{2+} and is released to the cytosol only through a tightly regulated channel, ryanodine receptor (RyR). The amplitude and duration of Ca^{2+} coming out via RyR is variable based on different signals and are given different terms such as Ca^{2+} release and Ca^{2+} leak. The number of SERCA recruited and the rate at which they function are mainly regulated by this Ca^{2+} flow from the SR to the cytosol, thereby regulating SERCA-based thermogenesis. It has been suggested that the difference between cytosolic and SR luminal Ca^{2+} concentration or the Ca^{2+} electrochemical gradient against which SERCA could greatly influence its coupling efficiency impacting heat generation [25], [28], Marie & Silva, 1998). The other factor that regulates heat production from SERCA is the relative expression of its regulatory microproteins. Two better-known SERCA regulators are SLN and phospholamban (PLB), while few othershave been recently identified; myoregulin (MLN) [45], dwarf open reading frame (DWORF) [46], and ankyrin 1 [47]. Another factor proposed to affect SERCA efficiency is the SR membrane-lipid composition [48], by modulating the rate of conformational, transition thereby impacting heat liberation ratio. Concentrations of some cations such as Mg^{2+}, H^+, K^+ are shown to influence SERCA kinetics in isolated molecular experiments, biological relevance needs to be understood in details. All these factors affecting SERCA function might influence heat production from SERCA, but it can be expected that modulation by regulatory proteins can be more efficient.

9.2.1 SLN is the Major Regulators of SERCA Heat Production

As mentioned earlier, the SERCA function is regulated by micropeptides such as SLN and PLB. The mechanistic details of PLB (52 amino acids) and SLN (31 amino acids) regulating SERCA activity have been wellstudied. Both the microproteins are single transmembrane peptides having only few residues on the SR lumen side. PLB having larger N-terminus bears a cytosolic α-helix with higher structured N-terminus, whereas SLN has a small unstructured N-terminus [49]. Interestingly, these regulators show very distinct tissue level distribution: PLB is highly abundant in ventricular (heart) tissue, while SLN is only expressed in the skeletal muscles and atria [49]. Molecular-level studies show that PLB modulates the Ca^{2+} affinity of SERCA pump by binding to the Ca^{2+}-free E2 confirmation of SERCA and gets dislodged from the interaction groove when Ca^{2+} binds to SERCA, therefore affecting the SERCA activation kinetics. Contrastingly, SLN can bind to Ca^{2+}-bound SERCA and remains bound through the conformational transitions during Ca^{2+} transport cycle due to N-terminal flexibility. The detailed mechanistic interaction of SLN with SERCA was studied by Professor Anthony Lee in 2002 [50]. They showed that SLN hinders SERCA-mediated Ca^{2+} transport (aborts prematurely Ca^{2+} release inside the SR) but allows ATP hydrolysis leading to liberation of heat; they termed this process *uncoupling*. Their studies also showed that SLN promotes uncoupling and heat generation in a concentration (SLN ratio to SERCA)-dependent manner, indicating its role in thermogenesis, which stimulated us to study this phenomena using mouse models. Our investigation showed that SLN-knockout (KO) mice are unable to maintain body temperature (Tc) when exposed to cold, but the reintroduction of SLN expression in the SLN-KO mice rescues the cold-sensitive phenotype. In contrast, PLB-KO showed no impairment in Tc maintenance, indicating PLB is not involved in thermogenesis. Our biochemical (protein cross-linking) investigation defined the SLN-binding transmembrane groove on SERCA and found that SLN can also bind the Ca^{2+}-bound SERCA confirmation [51, 52]. Our observations were further confirmed by x-ray crystallography of SERCA/SLN co-crystal structures solved by independent research groups [53–56]. It was found that SLN can reduce the total Ca^{2+} uptake, namely Vmax of SERCA, but does not affect ATP utilization [57]. Due to this, for similar amount of Ca^{2+}uptake, higher ATP utilization occurs, the energy of which is released as heat. Our experiments also showed that SLN-mediated uncoupling of SERCA is not absolute, and even at high Ca^{2+} levels we estimated that only ~30% of the pumps are uncoupled, thereby not affecting muscle function. Most of the published studies on SLN so far have utilized mammalian system, while SLN gene is observed in all vertebrates, and the role of SLN in other organisms has not been carefully studied (discussed later in the chapter).

9.3 SERCA (SLN) ACTIVITY IN WHOLE-BODY ENERGY HOMEOSTASIS

Thermogenic processes utilize energy to produce heat; that is why they modulate energy usage pattern of different organs, thereby impacting whole-body energy homeostasis. Therefore, energy (ATP) utilization by SERCA regardless of ST or NST significantly affect muscle metabolism. Recent data indicate that futile Ca^{2+} cycling by SERCA does not impair muscle contractility while increasing energy expenditure [58, 59]. Meizoso-Huesca et al. showed that futile Ca^{2+} cycling operates in resting skeletal muscle of mice but not toads [60]. This suggests that mammals have evolved a mechanism to cause minimal Ca^{2+} leak from RyR1 to recruit SERCA-based ATP hydrolysis without activating myosin ATPase and cross-bridge cycling [60]. This SERCA-based NST can be further amplified by SLN-mediated uncoupling providing a further layer of sophistication for muscle NST. Data from several labs suggest that in addition to thermogenesis, futile SERCA activity and, along with it, SLN action have a major impact on whole-body energy homeostasis [61, 62]. Lowering of the rate of SERCA-mediated Ca^{2+}uptake serves to generate a local Ca^{2+} signal, which has two parallel cellular effects. First, this local Ca^{2+} signal enters mitochondria via channels such as VDAC and MCU, boosting ATP production. Second, it can be relayed into the nucleus through cascade of protein activation, upregulating genes related to mitochondrial biogenesis and oxidative metabolism and leading to enrichment of the muscle with hardware of higher metabolism. SERCA is one of the most abundant proteins in skeletal muscle, and in SR preparation it is highly enriched. Therefore, a fraction of SERCA can be uncoupled by agents such as SLN without significant impact on the rate of Ca^{2+} uptake under low to moderate physical activity conditions. In fact, it has been shown that SERCA overexpression in striated muscles improve their substrate utilization, leading to reduction in insulin resistance [63, 64].

9.4 SLN EXPRESSION IS DYNAMICALLY REGULATED

The expression of SLN both at protein and transcript levels is highly responsive to external stimuli and quickly regulated by various physiological as well as pathological conditions. The dynamic regulation of SLN expression is exemplified by its very pattern observed during development of rodents [65]. SLN expression start to appear around the 15th day post coitum in most of the skeletal muscles examined in mouse and rat [66, 67]. Its expression gradually peaks around the day of birth and stays high in the first week of postnatal life [67]. Then it is downregulated slowly in fast-twitch fibers by weaning age and becomes restricted to slow-twitch muscles such as soleus [67]. SLN protein is abundant in most of the human skeletal muscle tissues analyzed. Further, dramatic upregulation of SLN has been reported in the skeletal muscles in many types of myopathies. Although the exact reason for SLN upregulation in diseases is not known, it further gives evidence that SLN expression is very dynamically regulated.

9.4.1 Cold and Exercise

The first evidence for the role of SLN in cold adaptation came from our studies that SLN-KO mice cannot withstand acute cold, while muscle-specific SLN overexpression rescues the phenotype [38]. Thereafter it was found that cold adaptation of neonate mice reverses the developmental program and retains the SLN expression in many of the muscles, whereas abundance is significantly downregulated when reared at normal room temperature. The next evidence for cold as a stimulus regulating SLN expression in the skeletal muscle was provided by cold adaptation studies of interscapular BAT (iBAT)-ablated mice [68]. The BAT-based NST is compromised in the iBAT-ablated mice, and upon adaptation to severe cold (4°C), their muscles exhibit upregulation of SLN [68]. Further, UCP1-KO that also has impaired BAT-based NST also showed SLN upregulation in some of its skeletal muscle upon cold adaptation. A recent study showed that mice with impaired BAT or under hypothyroidism recruit muscle NST via upregulation of SLN gene expression [69]. Another study showed that UCP1 deletion in rat induces SLN expression in skeletal muscle upon cold adaptation [70, 71]. Cold adaptation has been shown to stimulate SLN expression in skeletal muscles of several other animals including wild boars, plateau pika, and even free-ranging pigeon [72–74]. However, a transcriptomics study reported that during acclimatization to extreme cold (–20°C) of adult dark-eyed juncos, SLN expression is downregulated, indicating a different species or degree of cold challenge might have a differential effect on muscle SLN abundance [75].

The other biological stimuli shown to affect SLN expression in the skeletal muscle is exercise (Figure 9.1). Our initial studies showed that skeletal muscle–specific expression of SLN increases exercise ability in mice [67, 76]. It was also reported by others that the knockdown of SLN in human muscle cells impairs metabolic and substrate uptake capacity [77]. Interestingly, however, different types of exercise were found to exert distinct effects on SLN expression in humans and animal models. Morales-Alamo et al. reported that sprinting training reduced SLN level only in vastus lateralis, a muscle where SLN is highly abundant in untrained participants, indicating that the lowered uncoupling of SERCA by SLN facilitates muscle performance [78]. A more recent study showed that maternal exercise in mice enhances muscle-based thermogenesis partly by upregulating SLN expression [79]. Another study also reported upregulation of SLN mRNA in mice skeletal muscle upon exercise training [80]. In birds, stimuli that mimic migration possibly induce adaptations of exercise-upregulated SLN expression—for example, ~30 times in white-throated sparrows (*Zonotrichia albicollis*) [81]. Thus, cold seems to be a univocal inducer of SLN expression, while species-dependent effects might be there for exercise as a stimulus.

9.4.2 Species and Different Muscle Groups

Another interesting but poorly studied aspect is how SLN expression differs between species and across muscle groups. Babu et al. were the first to show that SLN is differentially expressed between a large (i.e., dog) and a small (i.e., rat) mammalian model [66]. Their study also showed that different muscles of rodents follow a distinct expressional pattern during their development. Further studies from various labs made it clear that SLN expression varies across mammals and muscle groups, including humans. However, in birds, the expression pattern of SLN remained uncertain mainly because all commercially available SLN antibodies do not react to avian SLN. We have recently

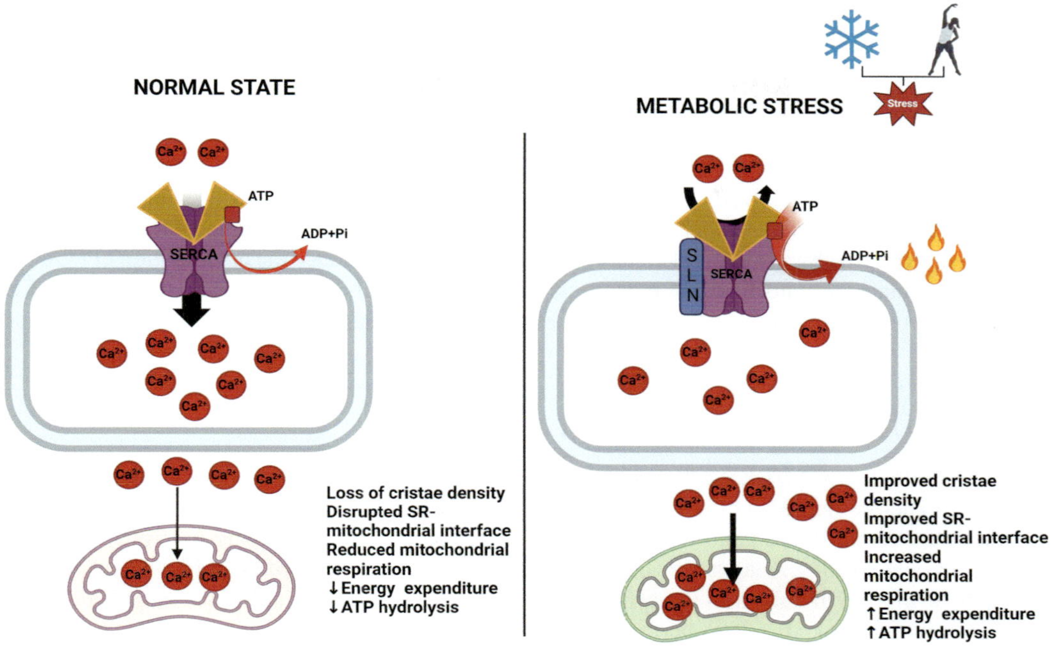

FIGURE 9.1 Role of sarcolipin (SLN) in enhancing the heat resulted from SERCA activity. Most skeletal muscles can modulate the SLN expression in response to signals. There are diverse physiological and pathological conditions that upregulate SLN expression in skeletal muscle. Thus, these signals can be taken together as metabolic stresses. SLN acts as a dual catalyst. On one hand, it increases mitochondrial Ca^{2+} intake, boosting energy production, while on the other hand, it enhances SERCA inefficiency leading to higher ATP usage and increased local thermogenesis.

shown that tropical seasonal cold adaptation can bring about upregulation of SLN expression in skeletal muscles in free-ranging adult Indian pigeons (*Columba livia*). It was also found that different muscles of pigeon possess differing levels of SLN abundance and also relative to SERCA expression [74, 82]. We have data to show that different muscles in goat express varying levels of SLN protein. Further, during prenatal and postnatal development of goats, different muscles exhibit differential SLN expression pattern that overlaps with programmed changes in SERCA1a expression. These data indicate that the ratio of SLN to SERCA is uniquely adapted for each muscle, and this stoichiometry undergoes changes under various physio-pathological stimuli.

9.4.3 SLN Regulates Muscle Mitochondrial Phenotype

Research from various labs indicates that SLN function is closely associated with enhanced mitochondrial capacity. Most of the physiological conditions associated with SLN upregulation also induce an increased mitochondrial phenotype. Studies revealed that SLN increased the local Ca^{2+} signal by uncoupling the SERCA Ca^{2+} uptake, which is relayed to mitochondria by mediators of Ca^{2+} entry such as MCU/VDAC. We have observed that the mitochondrial Ca^{2+} uptake is further facilitated by mitochondrial fusion and enhanced SR-mitochondrial proximity during the conditions when SLN is upregulated, such as cold adaptation [74]. The iBAT-ablated mice adapted to severe cold upregulate SLN-based muscle NST to a great level and simultaneously exhibit dramatic remodeling of the skeletal muscles both biochemically and structurally. Data showed that all the mitochondrial fusion proteins (Mfn1, Mfn2, and Opa1) are upregulated, especially MFN2, which is superinduced [68]. Our transmission electron microscopic (TEM) studies showed that intramyofibrilar mitochondria are placed close to SR, indicating an enlarged mitochondria-associated SR membrane (MAM) that might be correlated with SLN-mediated NST [83]. In addition, SLN function is intertwined with mitochondrial phenotype in an indirect fashion. The local Ca^{2+} signal generated by SLN at SERCA is decoded by Ca^{2+}-sensitive transcription factors and regulators such as calmodulin (CaM)-dependent kinase (CAMK) and calcineurin (CAN). While CaM in muscle cytosol activate CaM-dependent protein kinase (CaMK), especially CaMKII, CAN boosts function of the transcription factor called *nuclear factor of activated T cells* (NFAT). These transcription factors relay the signal to the nucleus for upregulating the genes involved in mitochondrial biogenesis depending on the intensity of the signal. Both CamKII and NFAT pathways in the skeletal muscle are proposed to induce the upregulation of PGC1α, the master regulator of mitochondrial biogenesis [84, 85]. However, studies indicate that the CamKII pathway plays a major role in the long-term effects of SLN signaling–mediated skeletal muscle remodeling.

9.5 EVOLUTIONARY RELEVANCE OF SLN

Skeletal muscle–based NST has a huge significance for evolution of endothermy (Chapter 2) as well as shifting to heterothermy in primarily endothermic vertebrates (Chapter 1). Intriguingly, SLN-like sequences are found in the genomic database of all vertebrate taxa, including heterothermic fish, amphibians, and reptiles. Analysis of SLN sequences suggests that this protein might have appeared during the evolution of fish from the cyclostomes, which lack SLN-like sequence in their genome. If the only function of SLN is to perform thermogenesis, then its appearance in fish is difficult to envision. Recently, however, there was a study focusing on understanding the role of SLN in an endothermic mesopelagic fish called opah (*Lamprisincognitus*) [86]. Opah does not possess a heater organ but maintains a Tc ~10°C above the ambient temperature through mechanisms yet to be determined. It was shown that the deep red pectoral muscles (more amenable to heat production) of opah express an abundant amount of SLN mRNA, and the SLN-SERCA transcript ratio is much higher than in white muscles, indicating potential for uncoupling and NST via SLN. Although the role of SLN in other ectotherms has not yet been studied, it might be feasible that SLN facilitates production of some basal level of local heat sufficient to warm up the muscles for locomotion in cold temperatures. In fact, cold adaptation in fish enhances mitochondrial biogenesis and oxidative capacity in skeletal muscle similar to that in mammals. An alternative proposition is that the initial function of SLN might be to generate local Ca^{2+} signal at the SERCA, but thermogenic action was introduced into the molecule later during the evolution of endothermy.

SLN sequence is found in all genomes from reptiles that gave rise to the endothermic taxa, mammals and birds. However, the relevance of reptilian SLN is not known currently, and the expression of protein by the reptilian SLN gene or its expression in muscle has not been studied to date. Some researchers have proposed that reptilian stock that gave rise to mammals and birds had already evolved mechanisms to sustain a primitive level of endothermy. It has been documented that tachymetabolism is more widespread than believed among ancestors of endotherms. A recent study on a seasonally endothermic reptile called "Tegu lizard" has shown it to maintain body temperature (Tc) ~10°C higher than its ambient environment without involving increased physical activity and/or shivering, suggesting muscle NST [87]. Further, brooding pythons that generate heat are suggested to have long gaps between episodes of shivering, supporting the possibility of existence of recruitable muscle NST [87–90]. Fascinatingly, lizards and snakes possess SLN genes that have conserved the transmembrane helix part, indicating it can interact with reptilian SERCA [91]. Any additional spotlight on the function of reptilian SLN will be of high significance to gain insight into evolution of endothermy.

9.6 FUTURE DIRECTIONS

Based on the presently available data it is established that the SERCA pump is a thermogenesis mechanism that can be recruited by both ST and NST. Much of the current work on SLN has been carried out in rodents using transgenic mouse models, which demonstrate its role in NST and metabolism. Unlike in rodents, SLN is much more abundant in larger mammals and shows a higher SLN-to-SERCA ratio. Studies in larger mammals including dogs, goats, and humans suggest that muscle NST may play an important role because these animals wean off their BAT activity as they mature and often do not contain functional BAT in adult life. There is an urgent need for understanding SLN function in humans, and future studies should involve large mammals including dogs, monkeys, and humans. SLN is bound to SERCA within the SR membrane and can dynamically regulate the SERCA pump activity. Although SLN binding to SERCA has been defined by co-crystals and x-ray diffractions, a better knowledge of SLN interaction with the SERCA pump at the molecular level is highly desirable for understanding its role in muscle physiology as well as designing novel drugs that mimic SLN function and increase energy metabolism. The relationship between SERCA and SLN during vertebrate evolution will add not only fundamental knowledge but also a better understanding of SLN's role in comparative muscle physiology. The relevance of the SLN:SERCA ratio in different muscle diseases is just beginning to be explored and needs further studies using both animal and human experimentation. Thus, SLN is not just a peptide but a physiological modulator of thermogenesis and energy metabolism.

REFERENCES

1. Smith, R.E. and B.A. Horwitz, *Brown fat and thermogenesis*. Physiol Rev, 1969. **49**(2): p. 330–425.
2. Grav, H.I. and A.S. Blix, *A source of nonshivering thermogenesis in fur seal skeletal muscle*. Science, 1979. **204**(4388): p. 87–9.
3. Davis, T.R., *Contribution of skeletal muscle to nonshivering thermogenesis in the dog*. Am J Phys, 1967. **213**(6): p. 1423–6.
4. Duchamp, C., et al., *Nonshivering thermogenesis in king penguin chicks. II. Effect of fasting*. Am J Phys, 1991. **261**(6 Pt 2): p. R1446–54.
5. Duchamp, C. and Barré, H., *Skeletal muscle as the major site of nonshivering thermogenesis in cold-acclimated ducklings*. Am J Phys, 1993. **265**(5 Pt 2): p. R1076–83.
6. Kayser, C., *Contribution à l'étude du mécanisme nerveux de la régulation thermique: Régulation chimique et métabolisme de l'eau*. 1928: Doin.
7. Clausen, T. and J.A. Flatman, *The effect of catecholamines on Na-K transport and membrane potential in rat soleus muscle*. J Physiol, 1977. **270**(2): p. 383–414.
8. Mejsnar, J. and L. Jansky, *Nonshivering thermogenesis and calorigenic action of catecholamines in the white mouse*. Physiol Bohemoslov, 1971. 20(2): p. 157–162.
9. Clausen, T. and J.A. Flatman, *The effect of catecholamines on Na—K transport and membrane potential in rat soleus muscle*. J Physiol, 1977. **270**(2): p. 383–414.
10. Biron, R., et al., *Thyroid hormones and the energetics of active sodium-potassium transport in mammalian skeletal muscles*. J Physiol, 1979. **297**(0): p. 47–60.
11. Lin, M. and Akera, T., *Increased (Na+, K+)-ATPase concentrations in various tissues of rats caused by thyroid hormone treatment*. J Biol Chem, 1978. **253**(3): p. 723–726.
12. Carey, F.G., *A brain heater in the swordfish*. Science, 1982. **216**(4552): p. 1327–9.
13. Block, B.A., *Structure of the brain and eye heater tissue in marlins, sailfish, and spearfishes*. J Morphol, 1986. **190**(2): p. 169–89.
14. Block, B.A., *Thermogenesis in muscle*. Annu Rev Physiol, 1994. **56**: p. 535–77.
15. Dickson, K.A. and Graham, J.B., *Evolution and consequences of endothermy in fishes*. Physiol Biochem Zool, 2004. **77**(6): p. 998–1018.
16. Block, B. and Kim, S., *Localization of junctional Sr proteins in the thermogenic heater cells of billfish*. Faseb J. 1992, 6. 20814–3998.
17. Block, B.A., J. O'Brien, and G. Meissner, *Characterization of the sarcoplasmic reticulum proteins in the thermogenic muscles of fish*. J Cell Biol, 1994. **127**(5): p. 1275–87.
18. Barré, H., L. Bailly, and J.L. Rouanet, *Increased oxidative capacity in skeletal muscles from cold-acclimated ducklings: a comparison with rats*. Comp Biochem Physiol B, 1987. **88**(2): p. 519–22.
19. Kavanagh, N.I., E.K. Ainscow, and M.D. Brand, *Calcium regulation of oxidative phosphorylation in rat skeletal muscle mitochondria*. Biochim Biophys Acta, 2000. **1457**(1–2): p. 57–70.

20. Dumonteil, E., H. Barré, and G. Meissner, *Sarcoplasmic reticulum Ca(2+)-ATPase and ryanodine receptor in cold-acclimated ducklings and thermogenesis*. Am J Phys, 1993. **265**(2 Pt 1): p. C507–13.
21. Dumonteil, E., H. Barré, and G. Meissner, *Expression of sarcoplasmic reticulum Ca2+ transport proteins in cold-acclimating ducklings*. Am J Phys, 1995. **269**(4 Pt 1): p. C955–60.
22. Meissner, G., *Ryanodine receptor/Ca2+ release channels and their regulation by endogenous effectors*. Annu Rev Physiol, 1994. **56**: p. 485–508.
23. Arruda, A.P., et al., *Cold tolerance in hypothyroid rabbits: role of skeletal muscle mitochondria and sarcoplasmic reticulum Ca2+ ATPase isoform 1 heat production*. Endocrinology, 2008. **149**(12): p. 6262–71.
24. Arruda, A.P., et al., *Thermogenic activity of Ca2+-ATPase from skeletal muscle heavy sarcoplasmic reticulum: the role of ryanodine Ca2+ channel*. Biochim Biophys Acta, 2007. **1768**(6): p. 1498–505.
25. de Meis, L., *Uncoupled ATPase activity and heat production by the sarcoplasmic reticulum Ca2+-ATPase, Regulation by ADP*. Biol Chem, 2001. **276**(27): p. 25078–87.
26. de Meis, L., *Ca2+-ATPases (SERCA): energy transduction and heat production in transport ATPases*. J Membr Biol, 2002. **188**(1): p. 1–9.
27. de Meis, L., A.P. Arruda, and D.P. Carvalho, *Role of sarco/endoplasmic reticulum Ca(2+)-ATPase in thermogenesis*. Biosci Rep, 2005. **25**(3–4): p. 181–90.
28. Kjelstrup, S., et al., *Is the Ca2+-ATPase from sarcoplasmic reticulum also a heat pump?* Eur Biophys J, 2008. **38**(1): p. 59–67.
29. Ketzer, L.A. and L. de Meis, *Heat production by skeletal muscles of rats and rabbits and utilization of glucose 6-phosphate as ATP regenerative system by rats and rabbits heart Ca2+-ATPase*. Biochem Biophys Res Commun, 2008. **369**(1): p. 265–9.
30. Turner, N., et al., *Scaling of Na+,K+-ATPase molecular activity and membrane fatty acid composition in mammalian and avian hearts*. Physiol Biochem Zool, 2006. **79**(3): p. 522–33.
31. Hirabayashi, M., et al., *Transformation of Skeletal Muscle from Fast- to Slow-Twitch during Acquisition of Cold Tolerance in the Chick*. Endocrinology, 2005. **146**(1): p. 399–405.
32. Else, P.L., D.J. Windmill, and V. Markus, *Molecular activity of sodium pumps in endotherms and ectotherms*. Am J Phys, 1996. **271**(5 Pt 2): p. R1287–94.
33. Porter, R.K., A.J. Hulbert, and M.D. Brand, *Allometry of mitochondrial proton leak: influence of membrane surface area and fatty acid composition*. Am J Phys, 1996. **271**(6 Pt 2): p. R1550–60.
34. Talbot, D.A., et al., *Uncoupling protein and ATP/ADP carrier increase mitochondrial proton conductance after cold adaptation of king penguins*. J Physiol, 2004. **558**(Pt 1): p. 123–35.
35. Trzcionka, M., et al., *The effects of fasting and cold exposure on metabolic rate and mitochondrial proton leak in liver and skeletal muscle of an amphibian, the cane toad Bufo marinus*. J Exp Biol, 2008. **211** (Pt 12): p. 1911–8.
36. Bal, N.C., et al., *Sarcolipin is a newly identified regulator of muscle-based thermogenesis in mammals*. Nat Med, 2012. **18**(10): p. 1575–9.
37. Bal, N.C. and M. Periasamy, *Uncoupling of sarcoendoplasmic reticulum calcium ATPase pump activity by sarcolipin as the basis for muscle non-shivering thermogenesis*. Philos Trans R Soc Lond Ser B Biol Sci, 2020. **375**(1793): p. 20190135.
38. Bal, N.C., et al., *The Role of Sarcolipin in Muscle Non-shivering Thermogenesis*. Front Physiol, 2018. **9**: p. 1217.
39. Bombardier, E., et al., *Sarcolipin trumps β-adrenergic receptor signaling as the favored mechanism for muscle-based diet-induced thermogenesis*. FASEB J, 2013. **27**(9): p. 3871–8.
40. Periasamy, M. and A. Kalyanasundaram, *SERCA pump isoforms: their role in calcium transport and disease*. Muscle Nerve, 2007. **35**(4): p. 430–42.
41. Xu, H. and H. Van Remmen, *The SarcoEndoplasmic Reticulum Calcium ATPase (SERCA) pump: a potential target for intervention in aging and skeletal muscle pathologies*. 2021. **11**(1): p. 25.
42. Lytton, J., et al., *Functional comparisons between isoforms of the sarcoplasmic or endoplasmic reticulum family of calcium pumps*. J Biol Chem, 1992. **267**(20): p. 14483–9.
43. Gamu, D., et al., *The sarcoplasmic reticulum and SERCA: a nexus for muscular adaptive thermogenesis*. Appl Physiol Nutr Metab, 2020. **45**(1): p. 1–10.
44. Periasamy, M., et al., *Role of SERCA Pump in Muscle Thermogenesis and Metabolism*. Compr Physiol, 2017. **7**(3): p. 879–890.
45. Anderson, D.M., et al., *A micropeptide encoded by a putative long noncoding RNA regulates muscle performance*. Cell, 2015. **160**(4): p. 595–606.

46. Nelson, B.R., et al., *A peptide encoded by a transcript annotated as long noncoding RNA enhances SERCA activity in muscle*. Science, 2016. **351**(6270): p. 271–5.
47. Desmond, P.F., et al., *Interactions between small ankyrin 1 and sarcolipin coordinately regulate activity of the sarco(endo)plasmic reticulum Ca(2+)-ATPase (SERCA1)*. J Biol Chem, 2017. **292**(26): p. 10961–10972.
48. Paran, C.W., et al., *Lipogenesis mitigates dysregulated sarcoplasmic reticulum calcium uptake in muscular dystrophy*. Biochim Biophys Acta, 2015. **1851**(12): p. 1530–8.
49. Shaikh, S.A., S.K. Sahoo, and M. Periasamy, *Phospholamban and sarcolipin: Are they functionally redundant or distinct regulators of the Sarco(Endo)Plasmic Reticulum Calcium ATPase?* J Mol Cell Cardiol, 2016. **91**: p. 81–91.
50. Smith, W.S., et al., *Sarcolipin uncouples hydrolysis of ATP from accumulation of Ca2+ by the Ca2+-ATPase of skeletal-muscle sarcoplasmic reticulum*. Biochem J, 2002. **361**(Pt 2): p. 277–86.
51. Rowland, L.A., et al., *Uncoupling Protein 1 and Sarcolipin Are Required to Maintain Optimal Thermogenesis, and Loss of Both Systems Compromises Survival of Mice under Cold Stress*. J Biol Chem, 2015. **290**(19): p. 12282–9.
52. Sahoo, S.K., et al., *Sarcolipin protein interaction with sarco(endo)plasmic reticulum Ca2+ ATPase (SERCA) is distinct from phospholamban protein, and only sarcolipin can promote uncoupling of the SERCA pump*. J Biol Chem, 2013. **288**(10): p. 6881–9.
53. Glaves, J.P., et al., *Interaction of a Sarcolipin Pentamer and Monomer with the Sarcoplasmic Reticulum Calcium Pump, SERCA*. Biophys J, 2020. **118**(2): p. 518–531.
54. Aguayo-Ortiz, R. and L.M. Espinoza-Fonseca, *Linking Biochemical and Structural States of SERCA: Achievements, Challenges, and New Opportunities*. Int J Mol Sci, 2020. **21**(11).
55. Toyoshima, C., et al., *Crystal structures of the calcium pump and sarcolipin in the Mg2+-bound E1 state*. Nature, 2013. **495**(7440): p. 260–4.
56. Winther, A.M., et al., *The sarcolipin-bound calcium pump stabilizes calcium sites exposed to the cytoplasm*. Nature, 2013. **495**(7440): p. 265–9.
57. Mall, S., et al., *The presence of sarcolipin results in increased heat production by Ca(2+)-ATPase*. J Biol Chem, 2006. **281**(48): p. 36597–602.
58. Maurya, S.K. and M. Periasamy, *Sarcolipin is a novel regulator of muscle metabolism and obesity*. Pharmacol Res, 2015. **102**: p. 270–5.
59. Xu, H. and H. Van Remmen, *The SarcoEndoplasmic Reticulum Calcium ATPase (SERCA) pump: a potential target for intervention in aging and skeletal muscle pathologies*. Skelet Muscle, 2021. **11**(1): p. 25.
60. Meizoso-Huesca, A., L. Pearce, C.J. Barklay, B.S. Launikonis, *Ca(2+) leak through ryanodine receptor 1 regulates thermogenesis in resting skeletal muscle*. Proc Natl Acad Sci USA, 2022. **119**(4): e2119203119.
61. Periasamy, M., J.L. Herrera, and F.C.G. Reis, *Skeletal Muscle Thermogenesis and Its Role in Whole Body Energy Metabolism*. Diabetes Metab J, 2017. **41**(5): p. 327–336.
62. Autry, J.M., D.D. Thomas, and L.M. Espinoza-Fonseca, *Sarcolipin promotes uncoupling of the SERCA Ca2+ pump by inducing a structural rearrangement in the energy-transduction domain*. Biochemistry, 2016. **55**(44): p. 6083–6086.
63. Quan, C., et al., *A PKB-SPEG signaling nexus links insulin resistance with diabetic cardiomyopathy by regulating calcium homeostasis*. Nat Commun, 2020. **11**(1): p. 2186.
64. Fredersdorf, S., et al., *Increased myocardial SERCA expression in early type 2 diabetes mellitus is insulin dependent: In vivo and in vitro data*. Cardiovasc Diabetol, 2012. **11**(1): p. 57.
65. Seth, M., et al., *Dynamic regulation of sarcoplasmic reticulum Ca(2+) stores by stromal interaction molecule 1 and sarcolipin during muscle differentiation*. Dev Dyn, 2012. **241**(4): p. 639–47.
66. Babu, G.J., et al., *Differential expression of sarcolipin protein during muscle development and cardiac pathophysiology*. J Mol Cell Cardiol, 2007. **43**(2): p. 215–22.
67. Sopariwala, D.H., et al., *Sarcolipin overexpression improves muscle energetics and reduces fatigue*. J Appl Physiol (1985), 2015. **118**(8): p. 1050–8.
68. Bal, N.C., et al., *Increased Reliance on Muscle-based Thermogenesis upon Acute Minimization of Brown Adipose Tissue Function*. J Biol Chem, 2016. **291**(33): p. 17247–57.
69. Kaspari, R.R.and A. Reyna-Neyra, *The paradoxical lean phenotype of hypothyroid mice is marked by increased adaptive thermogenesis in the skeletal muscle. Proc Natl Acad Sci USA* 2020. **117**(36): p. 22544–22551.

70. Hofmann, W.E., et al., *Effects of genetic background on thermoregulation and fatty acid-induced uncoupling of mitochondria in UCP1-deficient mice*. J Biol Chem, 2001. **276**(15): p. 12460–5.
71. Bal, N.C., et al., *The role of sarcolipin in muscle non-shivering thermogenesis*. Front Physiol, 2018. **9**: p. 1217.
72. Nowack, J. and S.G. Vetter, *Muscle nonshivering thermogenesis in a feral mammal*. 2019. **9**(1): p. 6378.
73. Zhu, H., et al., *Differential Expression of Metabolism-Related Genes in Plateau Pika (Ochotona curzoniae) at Different Altitudes on the Qinghai-Tibet Plateau*. Front Genet, 2021. **12**: p. 784811.
74. Pani, P., et al., *Seasonal cold induces divergent structural/biochemical adaptations in different skeletal muscles of Columba livia: evidence for nonshivering thermogenesis in adult birds Biochem J*, 2023. **480**(17): p. 1397–1409.
75. Stager, M. and Z.A. Cheviron, *Is there a role for sarcolipin in avian facultative thermogenesis in extreme cold?* Biol Lett, 2020. **16**(6): p. 20200078.
76. Maurya, S.K., et al., *Sarcolipin Signaling Promotes Mitochondrial Biogenesis and Oxidative Metabolism in Skeletal Muscle*. Cell Rep, 2018. **24**(11): p. 2919–2931.
77. Braun, J.L., et al., *The effects of neurogranin knockdown on SERCA pump efficiency in soleus muscles of female mice fed a high fat diet*. Front Endocrinol (Lausanne), 2022. **13**: p. 957182.
78. Morales-Alamo, D., M. Martinez-Canton, and M. Gelabert-Rebato, *Sarcolipin expression in human skeletal muscle: Influence of energy balance and exercise*. Scand J Med Sci Sports, 2020. **30**(3): p. 408–420.
79. Son, J.S., et al., *Maternal exercise intergenerationally drives muscle-based thermogenesis via activation of apelin-AMPK signaling*. EBioMedicine, 2022. **76**: p. 103842.
80. Chambers, P.J., E.S. Juracic, and V.A. Fajardo, *Role of SERCA and sarcolipin in adaptive muscle remodeling*. Am J Physiol Cell Physiol, 2022. **322**(3): p. C382–c394.
81. Pani, P. and N.C. Bal, *Avian adjustments to cold and non-shivering thermogenesis: whats, wheres and hows*. Biol Rev, 2022. **97**(6): p. 2106–2126.
82. Swalsingh, G., P. Pani, and N.C. Bal, *Structural functionality of skeletal muscle mitochondria and its correlation with metabolic diseases*. Clin Sci (Lond), 2022. **136**(24): p. 1851–1871.
83. Wu, H., et al., *Regulation of mitochondrial biogenesis in skeletal muscle by CaMK*. Science, 2002. **296**(5566): p. 349–52.
84. Taylor, D.F. and D.J. Bishop, *Transcription factor movement and exercise-induced mitochondrial biogenesis in human skeletal muscle: Current knowledge and future perspectives*. Int J Mol Sci, 2022. **23**(3).
85. Franck, J.P.C., E. Slight-Simcoe, and N.C. Wegner, *Endothermy in the smalleye opah (Lampris incognitus): A potential role for the uncoupling protein sarcolipin*. Comp Biochem Physiol A Mol Integr Physiol, 2019. **233**: p. 48–52.
86. Tattersall, G.J., C.A. Leite, and C.E. Sanders, *Seasonal reproductive endothermy in tegu lizards*. Sci Adv, 2016. **2**(1): p. e1500951.
87. Hutchison, V.H., H.G. Dowling, and A. Vinegar, *Thermoregulation in a brooding female Indian python, Python molurus bivittatus*. Science, 1966. **151**(3711): p. 694–6.
88. Harlow, P. and G. Grigg, *Shivering thermogenesis in a brooding diamond python, Python spilotes spilotes*. Copeia, 1984, 18: p. 959–965.
89. Sup, D. and R. Shine, *Reptilian endothermy: a field study of thermoregulation by brooding diamond pythons*. J Zool, 1988. **216**(2): p. 367–78.
90. Grigg, G.C., L.A. Beard, and M.L. Augee, *The evolution of endothermy and its diversity in mammals and birds*. Physiol Biochem Zool, 2004. **77**(6): p. 982–97.
91. Woltering, J.M., *From lizard to snake; behind the evolution of an extreme body plan*. Curr Genomics, 2012. **13**(4): p. 289–99.

10 Recruitment of Non-Shivering Thermogenesis in Skeletal Muscle

Importance of Environmental Context

Colleen M. Novak and Christina A. Watts
Kent State University, Kent, Ohio, USA

10.1 MUSCLE THERMOGENESIS AND ALLOSTASIS

Skeletal muscle can generate heat through shivering as well as non-shivering thermogenesis. This encompasses a range of mechanisms [1–5], some of which serve temperature homeostasis (see Chapters 4 and 6). For example, sea otters have a highly elevated metabolic rate primarily stemming from muscle mitochondrial proton leak; this heat helps otters sustain body temperature in cold water [6, 7]. In humans, skeletal muscle is the predominant site of thermogenesis with adaptation to a cold environment [8–10]. Not all muscle thermogenesis may serve temperature homeostasis, however. While the homeostatic process realigns the balance of the internal environment by actively coordinating control centers to maintain a set point and defend these parameters [11], allostasis, in contrast, refers to an active adaptation process to defend against potential threat [12]. Allostasis, a derivative of homeostasis, relies on similar terminology to expand the understanding of purposeful physiological changes required to meet the demands of a challenge, specifically focusing on predicting harm to elicit appropriate mechanisms of defense. Its general acceptance to delineate short-term "loans" of energetic resources to adapt to acute stressors has provided a platform for understanding reactivity to challenge. This emergency response encompasses energetic demands and coping mechanisms for stress, including predator threat [12]. For example, exposure to predator threat induces a rapid, high-magnitude muscle thermogenic response [13]. Exposure to the odor of a predator (e.g., ferret) increases muscle temperature in rats and mice (Figure 10.1A and B) [13]. Harnessing this mechanism could be advantageous in amplifying energy expenditure, even in the absence of exercise.

Predator threat in laboratory animal models like rats and mice can be simulated using odors of predators including cat, fox, and ferret [14–17]. Most commonly, the odorant trimethylthiazoline (TMT), found in fox urine, has been utilized to evoke a stress response and anti-predator behaviors [15, 16, 18]. Behavioral and brain activity responses to these odors are not uniform, however, even within species [15, 16, 19]. More naturalistic odors like those from ferrets or cats have been postulated to evoke a more robust innate behavioral reaction to predator threat than TMT [15–17, 19]. Stimuli that might be considered even more representative of natural predator threat include the presentation of an actual or simulated predator like a rat (to a mouse), cat, or snake, and exposure to a context with learned associations with a predator [20–27]. Looming or hovering stimuli also evoke defensive or escape behaviors; these stimuli are processed, not surprisingly, through the visual or photic system [28, 29]. Though brain pathways mediating the defensive response to a looming

DOI: 10.1201/9781003391470-12

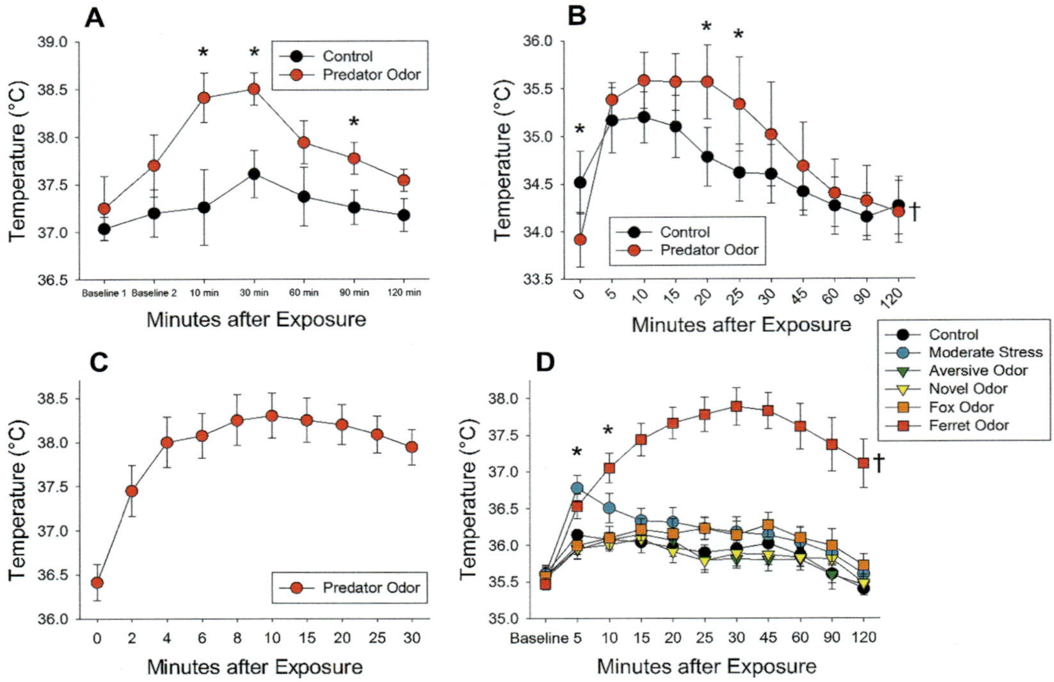

FIGURE 10.1 Exposure to predator (ferret) odor significantly increases skeletal muscle temperature in rats and mice. Measured using transponders implanted against the hindlimb gastrocnemius, muscle temperature increases sharply after exposure to ferret odor in rats compared to rats exposed to control odor (A), as well as in mice (B), compared to control-odor exposure. (C) The escalation in temperature can be detected as quickly as 2 min after initial exposure. (D) In rats, ferret odor induces muscle thermogenesis distinct from other odors or stressors, including fox odor and brief moderate stress (restraint). *Significantly elevated relative to control-odor exposure; †significantly elevated relative to all other conditions; $p<0.05$. Adapted from Gorrell et al. (2020), figures 1 and 4 [13].

object may differ from the neural processing for odor cues [30, 31], the brain regions processing the visual threat stimuli (e.g., midbrain superior colliculus) may also influence autonomic outflow [32–34].

Exposure to predator threat leads to a range of behavioral and physiological responses [27, 35–38]. Animals may attempt to escape or become motionless (i.e., defensive immobility or freezing response) [39, 40]. Through learned associations, animals become fearful or avoidant of locations or contexts where predators or their odors were encountered [41, 42]. Not surprisingly, facing a predator or its odor evokes a robust endocrine stress response [26, 41–44]. Exposure to predator threat also decreases food intake, and long-term exposure can induce weight loss [27, 45–50].

A rapid increase in skeletal muscle temperature is part of the array of physiological reactions to predator threat [13]. When rats encounter the odor of a ferret, one of its prominent natural predators, muscle temperature increases by nearly 2°C by 12–30 min after exposure (Figure 10.1C) [13]. This thermogenic induction is rapid, detectable as quickly as 2 min after initial exposure (Figure 10.1C) [13]. Temperature typically peaks within the first 30 min of exposure and then declines over the course of an hour (in mice) or several hours (in rats) [51]. Interestingly, removal of the odor stimulus after 10 min of exposure does not reverse or hinder the muscle thermogenic response [51].

The thermogenesis evoked by ferret odor in rats appears to be qualitatively and quantitatively distinct from the response to stimuli that are not directly threatening (Figure 10.1D). First, a novel odor, 2-methylbenzoxazole, induced a minor change in muscle temperature indistinguishable from what is seen after control stimulus presentation in rats. Similarly, an aversive odor, butyric acid, did not induce significant thermogenesis. Even the odor of another predator (fox) yielded only a relatively minor muscle thermogenic response. Though this study used Tink's Red Fox-P Cover Scent, the constituent of fox urine, TMT is a commonly used predator-odor stimulus meant to simulate predator threat and provoke ecologically relevant behavioral responses [14–16, 18, 52–56]. Lastly, predator threat also elicits a predictable physiological response, including an endocrine stress response where adrenocortical hormones are elevated [19, 26, 41, 43, 44, 53, 57–65]. These hormones could potentially contribute to the muscle thermogenic response to predator threat as well. Restraint or immobilization stress is commonly used to provoke a robust stress response, elevating adrenal hormones including corticosterone [44, 60]. While restraint for 5 min increased muscle temperature in rats, the thermogenesis was short lived, declining to control levels within less than an hour after the restraint concluded. Altogether, predator odor exerts an influential, distinct effect on muscle thermogenesis separable from its aversive or stressful consequences.

Not surprisingly, this increase in muscle temperature is accompanied by elevated energy expenditure, where rats burn an extra ~40% kcal over the first hour after exposure (Figure 10.2A) [13]. No difference was found in substrate utilization, however [13]. Providing insight on substrate utilization, respiratory exchange ratio (RER), similar to respiratory quotient (RQ), can be calculated during respirometry by comparing oxygen consumption to carbon dioxide production (VCO_2/VO_2). Lower RER signifies a relative increase of fat oxidation (e.g., during fat mobilization from negative energy balance, illness, cold exposure, or high-fat diet). In contrast, a higher RER approaching 1.0 is seen when carbohydrate (glucose) is the primary metabolic fuel. Exposure to predator threat altered RER only during a maximal exercise test, where exposure to predator threat lowered RER during running but not at VO_{2max}. In no other conditions was RER altered, however, implying no predominant bias between glucose and fatty acid utilization in association with muscle thermogenic induction [13].

The caloric expenditure needed to generate heat in muscle is relevant to energy balance (see Chapter 12). Male and female rats exposed to ferret odor chronically for 7 days experienced weight loss and decreased adiposity [13]. This partially stemmed from lower food intake, especially in male rats; however, female rats, which showed minimal change in energy intake, also lost weight when chronically exposed to predator odor [13]. To factor out the confounding effect of appetite suppression, rats were food restricted (25% fewer kcal/day) concurrent with chronic exposure to predator odor or control odor daily for 15 days. Again, rats chronically exposed to the ferret odor lost significantly more weight than rats housed with the control odor [13]. When examining habituation, rats displayed a significant muscle thermogenic response to the presentation of a new ferret-odor stimulus even after a week of chronic exposure [13]. However, after 2 weeks of chronic exposure, the muscle thermogenic response was dampened; interestingly, rats living with the ferret odor retained higher muscle temperatures in their home cages when measured before the next daily odor was introduced. The survival value of vigilance and readiness to respond to a predator would not diminish with repeated exposure to the predator-threat cues, provided the cues are fresh [27, 62]. In contrast, voles chronically exposed to cat feces showed habituation in their behavioral and endocrine response to the odor, yet showed sensitization in their expression of brain DeltaFosB, a transcription factor with cumulative or sustained activity in response to neuronal activation [26, 63]; interestingly, this suggests concurrent habituation and sensitization of different processes.

Exposure to predator threat induces an array of behavioral and physiological responses [26, 63]; many behavioral responses to predator odor, like flight or escape, entail increased physical activity [29, 52, 66]. Because muscle contraction generates heat, physical activity or locomotion could be the source of some of the heat generated upon exposure to predator odor, even in animals housed in laboratory environments. To consider the contribution of contractile thermogenesis to predator odor–induced muscle thermogenesis, rats were exposed to the odor while walking on a treadmill. Even when contractile thermogenesis is held constant, the presence of the ferret odor significantly increased caloric expenditure (Figure 10.2B) and muscle temperature (Figure 10.2C) [13].

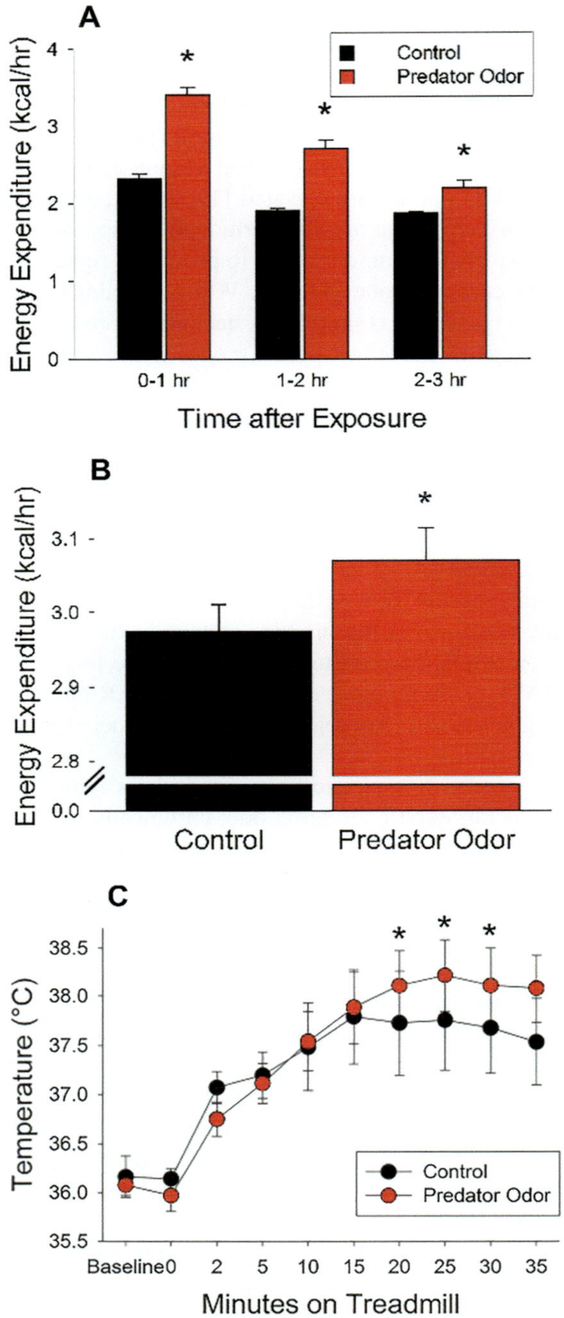

FIGURE 10.2 When locomotor activity is held constant, exposure to predator odor significantly increases energy expenditure (EE), and muscle thermogenesis. (A) Compared to control odor, exposure to ferret odor significantly increased EE in freely moving rats. Significant increases in EE (B) and muscle temperature (C) were evident with predator-odor compared to control-odor exposure when locomotion was held constant using treadmill walking. *Significantly elevated relative to control-odor exposure, $p<0.05$. Adapted from figures 1 and 2 of Gorrell et al. (2020) [13].

When freely moving rats are stimulated with predator odor, their physical activity increases [13]; however, the same magnitude of physical activity induced in other contexts without predator odor present does not increase energy expenditure to the same extent [13, 67]. Altogether, contextual predator threat induces significant heat generation in muscle outside of contractile thermogenesis.

Predator threat activates innate pathways to evoke conserved behavioral responses [19, 65, 68, 69]. Nonetheless, the behavioral reaction is modified by context. For example, escape can be suppressed if a cue is learned to be nonthreatening [28]. Conversely, a context can be aversive because of its association with predator-threat cues [23, 30, 42, 60, 66, 70, 71]. This includes the muscle thermogenic response to predator odor [13]. Rats exposed to ferret odor in a salient context will show elevated muscle temperature the following day when placed in the same context with a control stimulus [13]. Altogether, cues indicating predator threat induce innate behavioral and physiological—including thermogenic—responses, but the context and history of the danger can potentially amplify these responses subsequent to learning or conditioning.

10.2 STRESS-INDUCED HYPERTHERMIA, PSYCHOGENIC FEVER, AND SOCIAL HYPERTHERMIA

Muscle thermogenesis in the context of predator threat may overlap with the phenomenon of stress-induced hyperthermia. In laboratory animals, exposure to a stressful context or stimulus increases core temperature as measured using rectal probes [70–72]. Multiple stimuli have been reported to increase core temperature ranging from mild to more intense, including experimenter handling, an air jet, engagement with a novel environment or new cage, moving the animal to new cage, or more intense stimuli such as footshock or startle, as reviewed by Bouwknecht et al. [72]. There is substantial overlap in the neural pathways mediating the thermogenic and other autonomic effects of stress (e.g., respiratory, cardiovascular), as well as overlap in the mechanisms through which different challenges—stress, immune (e.g., fever), cold—impact thermogenesis. These have been reviewed thoroughly elsewhere [73–76]. Briefly, for the thermogenic response to cold stress, the preoptic area (POA) sends projections to the dorsomedial nucleus of the hypothalamus (DMH), which then projects to the raphe pallidus, part of the rostral medullary raphe nucleus (rMR) [77–80]. Similarly, the DMH is involved in stress-induced hyperthermia through its glutamatergic projections to the rMR [74, 78]. While stress-induced hyperthermia mediated through brown adipose tissue (BAT) thermogenesis has been documented, evidence indicates that the phenomenon does not require BAT. Using infrared thermography, Marks et al. identified hyperthermia with contextual fear in rats that persisted in the absence of shivering. More notably, this hyperthermic response was observed upon the removal of interscapular BAT (iBAT), the primary BAT depot [81]. This hyperthermia relied on beta-adrenergic receptors, as the beta-blocker propranolol suppressed the hyperthermia [81]. Similarly, hyperthermia after emotional stress was detected using telemetry in mice lacking uncoupling protein-1 (UCP1) [82]. Altogether, evidence supports the role of sympathetic nervous system (SNS) outflow in mediating the elevated temperature seen after stress or arousal. The neural pathways through which stressful stimuli increase BAT thermogenesis are established. It is likely, however, that increases in temperature also stem from tissues other than BAT, with skeletal muscle as the most likely culprit.

Stress-induced hyperthermia is also evident in people and, in extreme cases, can present as psychogenic fever [78, 83–90]. Increased temperature can occur in a normal stress response, but prolonged hyperthermia has also been noted in case reports and clinical populations [84–86, 88–92]. A case report also supports the mechanistic separation between psychogenic fever and fevers stemming from an immune response [83], consistent with evidence from preclinical models [73]. Data suggestive of BAT involvement in human stress-induced hyperthermia have been reported [93]. No consideration of skeletal muscle as a potential site of thermogenesis with stress-induced hyperthermia or psychogenic fever is in evidence, however.

A related concept to stress- or fear-induced hyperthermia is social hyperthermia [78, 94–96]. Indeed, negative social interactions or defeat have been considered as one way to induce emotional hyperthermia [96, 97]. Accordingly, substantial attention has been paid to the role of BAT in generating hyperthermia seen with social threat or interactions, as well as the importance of the DMH in mediating this effect [78, 94–96]. Other neural pathways have been recognized recently, including orexin neuronal projections [97]. The ventromedial hypothalamus (VMH), specifically a subpopulation of cells in the ventrolateral VMH (vlVMH), is critical for the behavioral response to social fear [23]. As described above for fear-induced hyperthermia, the possible role of muscle thermogenesis in social hyperthermia is unknown. At least some evidence (Figure 10.1D) implies that thermogenic induction stemming from general stressors or aversive stimuli is at least somewhat distinct from thermogenesis following predator threat. This is reinforced by the divergence in the neural pathways that spur thermogenesis in different contexts. Still, it is likely that the processes of predator odor–induced muscle thermogenesis and stress-induced hyperthermia from BAT thermogenesis are at least parallel, possibly interacting, processes.

10.3 BRAIN AND SYMPATHETIC CONTROL OF CONTEXTUAL INDUCTION OF MUSCLE THERMOGENESIS

Animal brains are innately tuned to respond to predator threat [27]. Exposure to predator odor or other predator-threat stimuli evokes characteristic changes in brain activity [17, 20, 26, 27, 53, 69, 71, 98–100]. Chemosensory stimuli are sensed through the main olfactory system along with the accessory olfactory system, including the vomeronasal organ, and the Grueneberg ganglion; these pathways converge on the medial amygdala (MeA) and VMH [68]. Optogenetic and chemogenetic manipulation of cell activity has teased out the roles of VMH cell subtypes in the behavioral response to predator-threat stimuli. The resulting evidence highlights the importance of a group of cells in the VMH containing the transcription factor steroidogenic factor 1 (SF1). This cell group is critically important in the ability of predator-threat stimuli to evoke an adaptive behavioral response. Predator-threat cues increase SF1 cell neural activity, with activity specifically depending on the modality of the threat stimulus [31]. Optogenetic excitation of SF1 neurons increased defensive behaviors specific to the pathway targeted. This evidence implicates projections to the dorsolateral periaqueductal gray (dlPAG) in defensive immobility (freezing) and projections to the anterior hypothalamic nucleus (AHN) in avoidance or escape [40]. Similarly, optogenetic stimulation of SF1 cells by Kunwar et al. immediately spurred defense behaviors, including flight and immobility, and resulted in learned avoidance of areas or contexts where the stimulation occurred [30]. Silva [23] reported that diverse threatening stimuli—predator, social, or shock stimuli—produced divergent patterns of neuronal activation within the VMH despite a similar behavioral profile. Only the predator threat could significantly increase the number of cells immunolabeled with the immediate-early gene c-Fos, marker of neuronal activity, in the dorsolateral and central VMH [23], the region containing the majority of SF1 cells. Chemogenetic inhibition of SF1 cells significantly decreased defensive and locomotor behaviors elicited by predator threat but not social threat or foot shock. Altogether, VMH SF1 cells are critically important for the expression of the array of defensive behaviors specifically in the context of predator-threat cues.

In addition to the large body of evidence implicating VMH SF1 cells in the behavioral response to predator threat, this cell group also plays a role in energy balance, particularly modulating peripheral metabolism. VMH SF1 cells respond to alterations in metabolic fuels like glucose and endocrine cues like leptin and can alter fuel allocation and energy balance [101–124]. Importantly, chemogenetic activation of SF1 neurons increased energy expenditure and enhanced peripheral (including skeletal muscle) insulin sensitivity [114]. SF1 cells also play a role in BAT thermogenesis [101, 104, 118, 125]. Within the larger VMH, optogenetic activation of a population of cold-sensitive neurons expressing prodynorphin increases BAT and core temperatures [95]. Given the growing body of evidence convincingly linking VMH and SF1 cell function to each predator response and energy balance, it follows that this cell group would mediate the physiological and metabolic effects of predator threat. Relatively few direct connections have been made in this direction, however. Viskaitis et al. demonstrated that VMH SF1 cells are a key node for enabling a threat to suppress appetite [126]. With respect to skeletal muscle thermogenesis, mouse gastrocnemius temperature was significantly increased with the chemogenetic activation of VMH SF1 cells (Figure 10.3A) [127]. Consistent with Coutinho et al., mouse energy expenditure was also significantly increased after SF1 cell activation (Figure 10.3B) [114]. The VMH, specifically the SF1 cell population, may be the key node linking threat cues to muscle thermogenesis. SF1 cells are composed of several subpopulations, not all of which, however, may be involved in modulating muscle thermogenesis [128].

Currently, a rich body of work provides clues regarding the most likely brain pathways mediating muscle thermogenesis. Sensory input relaying predator threat converges on the VMH [20], which is an important player in energy balance. Indeed, the VMH was once termed the "satiety center" [129]. Within the VMH, SF1 cell activity is relevant to the behavioral response to predation [23, 31, 130]. SF1 cells also appear to have a prominent role in modulating autonomic outflow to affect peripheral metabolism, including muscle insulin sensitivity [114, 119]. This includes altering muscle metabolism. Coutinho et al. reported elevated muscle glucose uptake with acute chemogenetic activation of SF1 neurons [114], along with increased liver glucose production, thereby maintaining glucose availability and potentially fueling muscle contraction for actions such as escape. VMH influences on muscles are also affected by the circadian clock [131]. Much of the ability of SF1 cells to impact peripheral metabolism and thermogenesis depends on autonomic—especially sympathetic—outflow [115, 118, 122, 132]. Using conditional tract tracing, Lindberg demonstrated that SF1 cells project to multiple brain regions important in behavior (e.g., PAG) as well as to pre-autonomic regions, including the rostral ventrolateral medulla (RVLM) and the nucleus of the solitary tract (NTS) [133]. Physiologic sympathoexcitatory responses to stress, including some thermogenic outputs, are partly mediated through the RVLM [74]. Tract tracing has identified projections to muscle originating in the RVLM and NTS, including leptin receptor-containing cells in the NTS [134].

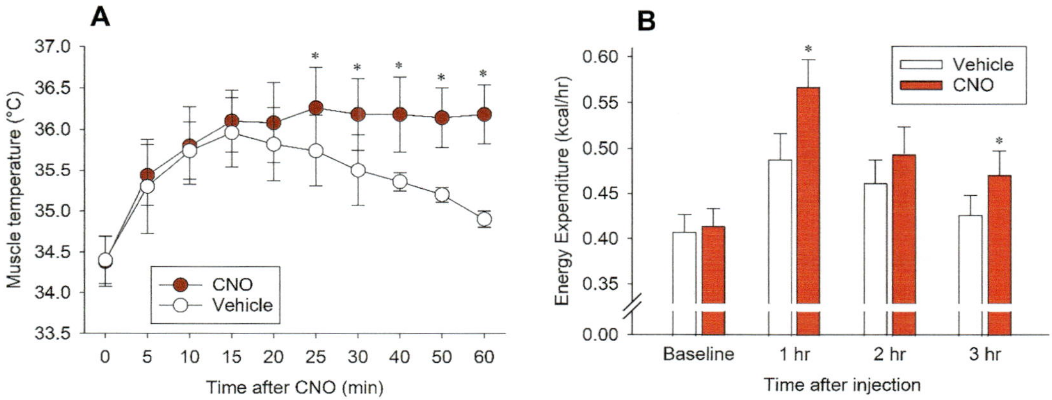

FIGURE 10.3 Chemogenetic activation of steroidogenic factor 1 (SF1) neurons in the ventromedial hypothalamus (VMH) elevates muscle thermogenesis and energy expenditure in mice. Mice expressing Cre recombinase in steroidogenic factor 1 neurons (SF1-Cre) received neural injections of an excitatory viral vector (pAAV-hSyn-DIO-hM3D(Gq)-mCherry (AAV8)) into the ventromedial hypothalamus. (A) Chemogenetic activation of SF1 neurons via intraperitoneal (IP) injections with the agonist clozapine-N-oxide (CNO) resulted in a significant increase in muscle (gastrocnemius) temperature. (B) Increased energy expenditure was also observed 1 and 3 hours after injection of CNO in mice exposed to control odor at 1 hour. *Significantly elevated relative to vehicle injection, $p<0.05$. Adapted from Watts (2023) [127].

Furthermore, noxious stimuli can prompt glucose mobilization, and leptin acts in the PAG to modulate this [135]. Altogether, this evidence outlines a hypothetical neural circuit through which predator threat impacts skeletal muscle.

The SNS innervates skeletal muscle, where postganglionic cells release norepinephrine (NE) [136–139]. Exposure to ferret odor significantly increases SNS outflow to skeletal muscle [140]. With respect to thermogenesis, both sympathetic nerves and NE signaling are needed for ferret odor to exert its full thermogenic effects in rats. Blocking beta-adrenergic receptors using the mixed antagonist nadolol suppresses predator odor-induced muscle thermogenesis (Figure 10.4A) [13]. Moreover, induction of muscle thermogenesis appears to rely primarily on the sympathetic nerves rather than circulating hormones. Unilateral lumbar sympathetic denervation decreased the ability of ferret odor to provoke thermogenesis in the gastrocnemius muscle group ipsilateral to the denervation (SNS neuron innervation does not decussate). Interestingly, the contralateral muscle, served by the intact lumbar sympathetic nerve, showed a robust thermogenic response to predator odor simultaneously in the same rat (Figure 10.4B) [13]. In other words, the muscle relies on the sympathetic nerve to induce muscle thermogenesis. Factors that affect the rat as a whole—locomotion, circulating hormones including catecholamines and glucocorticoids, or other potential heat sources—would affect both legs. Importantly, the muscles (contralateral and ipsilateral to the denervation) showed indistinguishable changes in temperature during treadmill walking, indicating no discernable effect of denervation on contractile thermogenesis [13]. Altogether, this evidence supports a pathway in which the brain activates SNS outflow and releases NE to activate beta-adrenergic receptors on the skeletal myocyte to induce thermogenesis (Figures 10.4 and 10.5).

Evidence suggests mechanisms including ATP-gated K channels [1, 2] and uncoupling of the sarco/endoplasmic reticulum calcium ATPase (SERCA) channel for heat generation in muscle fibers [5, 141–144] (see Chapters 8 and 9). Heemstra et al. (2022) reported that exposure to predator odor uncoupled SERCA in rat soleus muscle, where less calcium is transported into the sarcoplasmic

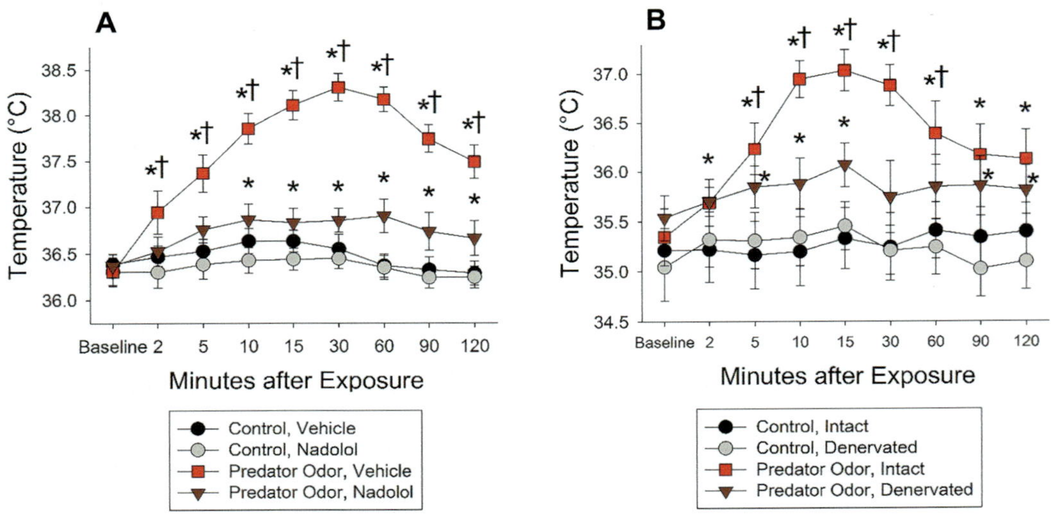

FIGURE 10.4 The ability of predator odor to induce muscle thermogenesis in rats relies on an intact sympathetic nervous system connection. (A) The peripherally acting beta-blocker nadolol significantly suppressed the ability of ferret-odor exposure to increase muscle (gastrocnemius) temperature relative to vehicle control. *Significantly elevated relative to control-odor exposure; †vehicle>nadolol within condition; $p<0.05$. (B) Predator odor–induced muscle thermogenesis was also significantly suppressed in the gastrocnemius ipsilateral to the denervated lumbar sympathetic nerve, with a large induction in the contralateral gastrocnemius concurrently in the same rat. *Significantly elevated relative to control-odor exposure; †greater than all other conditions; $p<0.05$. Adapted from figure 3 of Gorrell et al. (2020) [13].

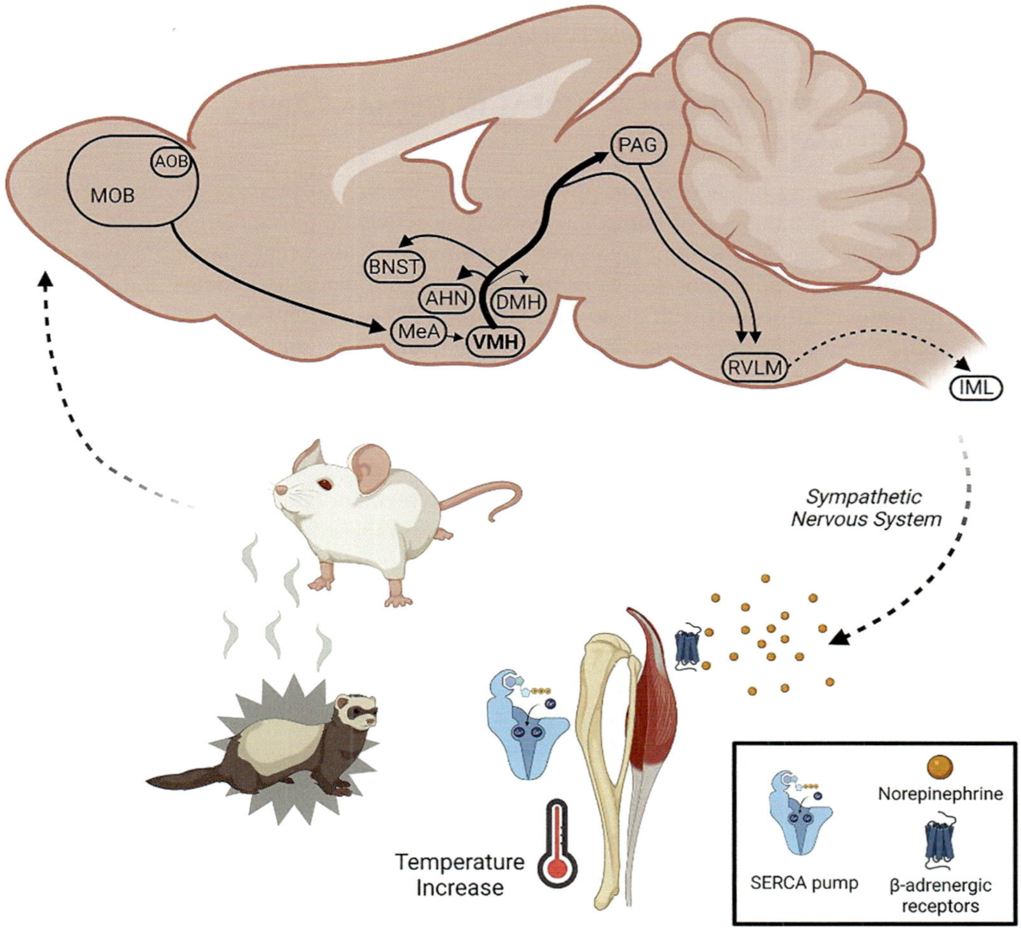

FIGURE 10.5 Hypothesized brain pathway mediating predator threat–induced muscle thermogenesis. Odor is sensed by the main and accessory olfactory systems and the signal is relayed to multiple brain regions, including the medial amygdala (MeA), which projects to the ventromedial hypothalamus (VMH), including the steroidogenic factor 1 (SF1) cell population. The VMH projects to multiple sites to modulate predator defense behaviors; VMH SF1 cells also project to pre-autonomic centers, including the rostral ventrolateral medulla (RVLM), to increase sympathetic outflow to multiple tissues including skeletal muscle. *AHA*, anterior hypothalamic nucleus; *AOB*, accessory olfactory bulb; *BNST*, bed nucleus of stria terminalis; *DMH*, dorsomedial hypothalamus; *IML*, intermediolateral nucleus; *MOB*, main olfactory bulb; *NE*, norepinephrine; *PAG*, periaqueductal gray; *SERCA*, sarco/endoplasmic reticulum calcium ATPase. Created with BioRender.com.

reticulum for a given amount of ATP hydrolysis [145]. Given the relevance of SERCA uncoupling to whole-body metabolism and adaptation to cold environments in mammals [146–150] (Chapter 9), it is not surprising that this may also underlie contextual induction of muscle thermogenesis. This does not rule out potential contributions by mitochondrial uncoupling or ATP-gated potassium channels, however [1, 2, 7]. Aside from the role of the inhibitor sarcolipin in uncoupling SERCA, there is an incomplete description of the intracellular signals linking beta-adrenergic receptor activation and SERCA uncoupling in skeletal muscle [151, 152].

Altogether, evidence supports a pathway whereby sensory cues of predator threat converge on the VMH, with a likely role of SF1 cells, to increase SNS outflow to skeletal muscle. Release of NE and activation of beta-adrenergic receptors on skeletal myocytes lead to uncoupling of SERCA, likely through sarcolipin-mediated inhibition, to generate heat (Figure 10.5). The ability of a cold environment to activate this brain pathway is unknown, however.

10.4 RESPONSES TO PREDATOR THREAT IN HUMANS AND NONHUMAN PRIMATES

The pathways mediating behavioral and physiological responses to predator threat appear to be highly conserved. The induction of immediate-early gene activation throughout the rodent (mouse, rat, vole) brain has been described in detail using multiple predator-threat stimuli, including exposure to predators (e.g., rat, cat, or ferret) or their odors—both naturalistic odors and TMT—and exposure to contexts associated with these cues [16, 17, 19, 23, 25, 26, 45, 53, 60, 62, 68, 69, 71, 98–100, 153]. Brain activation by predator threat consistently encompasses an array of regions included in the medial hypothalamic defense circuit including the dmVMH, AHN, and dorsal premammillary nucleus (PMd), along with upstream areas including the MeA and downstream regions including the PAG [16, 26, 153]. This differs from the pattern of activation seen after general stressors (e.g., elevated plus maze, foot shock) with respect to the degree of activation and the regions affected [26, 53, 98, 99]. Interestingly, chronic exposure to predator odors induces a distinct pattern of DeltaFosB expression [26, 45, 63, 100, 154]. Much of the brain activation pattern appears specific to predator threat and generalizes across species. Marmosets, nonhuman primates, were exposed to an artificial moving snake stimulus [20]. This induced a robust escape response in the marmosets along with c-fos activation in the VMH, MeA, and PAG. In these same marmosets, a dense projection from the MeA to the VMH was identified, with anterograde projections from the VMH to the PAG, similar to what has been identified in rodents [20, 40]. Visual pathways, including the medial pretectal nucleus and superior colliculus, provide a pathway through which visual cues can relay information regarding predator threat to the medial hypothalamic defense circuit [20].

While less is known regarding predator-threat detection by the human brain, it is plausible to predict that a similar multimodal sensory pathway could relay predator-threat cues from the environment to the medial hypothalamic defense circuit. One case report identified a lack of fear experience after a bilateral lesion of the amygdala in one patient [155]. Downstream, the activation of the VMH can induce panic attacks in humans and panic behavior in marmosets [156, 157]. Like other species, humans also appear to be innately responsive to predator-threat stimuli. Infants show a brain response to snakes that is distinct from other, nonthreatening animals [22]. Used as a contextual cue, a threat-relevant odor, 2-methyl-2-thiazoline, enhanced fear learning in human participants [158]. Altogether, this evidence argues for an innate, conserved circuit through which sensory stimuli indicating predator threat affect behavior and potentially other physiological effects, including hormone release [158] and muscle thermogenesis.

10.5 MUSCLE THERMOGENESIS, ENERGETIC EFFICIENCY, AND PERFORMANCE

The question arises regarding the adaptive significance of contextually induced muscle thermogenesis. In other words, what advantage might be incurred by the animal after detection of predator threat increases muscle temperature? One possibility is that the underlying process increases the resistance to fatigue, enhancing aerobic endurance capacity and, in turn, the ability to outrun predators. This is consistent with evidence that beta-adrenergic stimulation of skeletal muscle increases performance—for example, contraction force [152, 159–161]. Adaptation to a cold environment has an overlapping relationship to aerobic capacity or training as well [162–164]. For example, cold adaptation increases maximal running performance, enhances oxidative capacity, and alters muscle metabolism [148, 162, 163]; conversely, exercise training increases cold tolerance [165]. Demonstrating a connection between muscle thermogenesis and fatigue in the context of predator threat, rats were able to run on a treadmill >5 min longer after exposure to ferret odor compared to control odor, suggesting a survival value of fatigue resistance and escape from predation in this context [13]. This resistance to fatigue may stem from enhanced endurance at the muscle level, through actions in the brain that increase motivation to run in the presence of predator threat, or a combination of these factors.

Multiple aspects of the hypothetical neural pathway mediating predator odor–induced muscle thermogenesis (Figure 10.6) have been implicated in modulating running performance, endurance, or fatigue. With respect to central modulation, some skeletal muscle responses to exercise and training (e.g., glycemic control, training performance) are mediated by the brain SF1 cell population within the VMH [166–168]. Adrenergic and sympathetic stimulation sustains endurance capacity and muscle force generation [152, 161, 169]. During locomotion in mice, muscle-specific overexpression of the thermogenic protein sarcolipin reduces fatigue at the level of isolated skeletal muscle as well as at the level of overall exertion in the whole animal [170]. Interestingly, the enhanced performance seen in association with thermogenic activation appears to be connected with an altered economy of activity, alternatively termed muscle work efficiency or energy cost of locomotion or transport (i.e., kcal needed to fuel the muscle work of transporting the animal based on size and distance) [163, 171]. This relationship between muscle performance and muscle energetic efficiency persists in humans as well, where elevated oxidative capacity is consistently associated with a lower economy of exercise or activity [172]. Along with increasing performance, cold adaptation makes locomotion more energetically costly, implicating muscle efficiency [163, 173, 174]. For example, the caloric cost of activity is doubled in mice housed at 22°C compared to 30°C [174]. Not surprisingly, muscle thermogenesis is also associated with muscle energetic inefficiency—increased caloric requirement for the same amount of work (Figure 10.6B) [13]. It is conceivable that the same

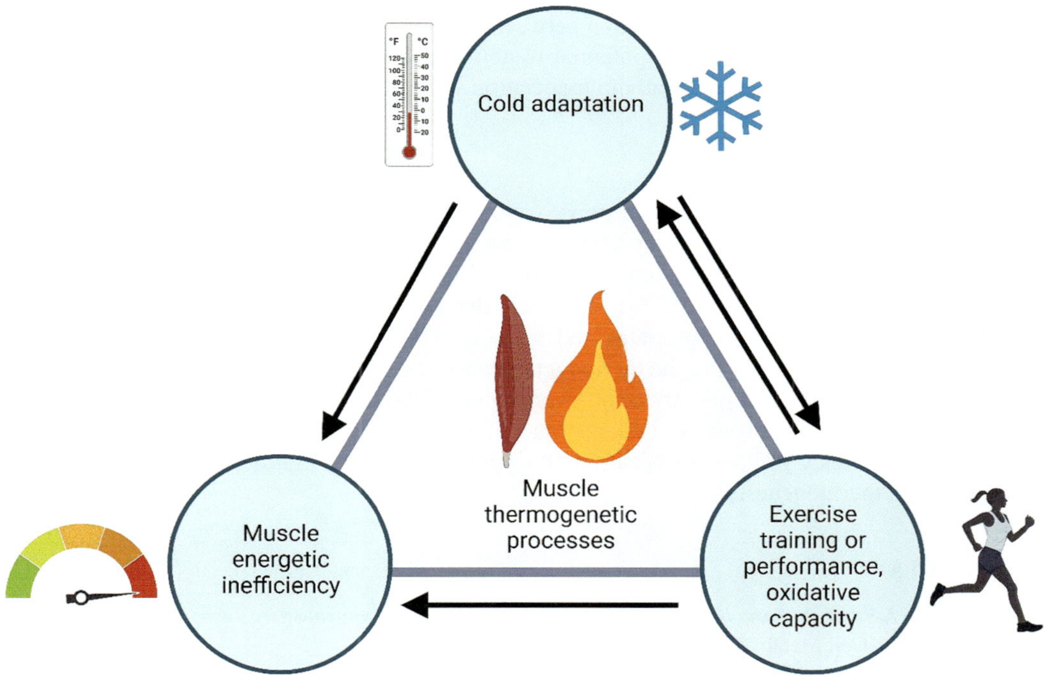

FIGURE 10.6 Interconnection of muscle thermogenic processes with muscle performance and fuel inefficiency. Adaptation to cold both increases running performance and decreases fuel economy of activity. Similarly, exercise training increases cold tolerance and decreases muscle energetic efficiency. All three of these processes may be mediated at least in part by muscle thermogenic mechanisms. Created with BioRender.com.

mechanism underlies each of these processes, leading to elevated aerobic capacity or performance and reduced energetic efficiency, generating heat as a result (Figure 10.6). If these processes stem from the same underlying mechanism, these facets of muscle function would be inexorably linked with heat generation and therefore inseparable. Altogether, this evidence is consistent with the hypothesis that the adaptive significance of this pathway is to increase animal survival by facilitating escape from predators. In this scenario, the heat generated by muscle could be considered a by-product of the underlying mechanism, likely SERCA uncoupling, which confers the survival advantage through enhanced muscle performance (Figure 10.6) (see Chapter 9). Though this hypothesis has not been explicitly examined, it raises questions regarding potential co-optation of energy transfer processes in the evolution of endothermy [175, 176] (see Chapter 6).

One example of the convergence of muscle thermogenesis, elevated performance (i.e., resistance to running fatigue), and muscle inefficiency is the contrasting rat model of high and low intrinsic aerobic capacity developed by Britton and Koch [177–180]. Briefly, innate aerobic fitness was assessed using a treadmill running test, and the rats with the highest and lowest aerobic capacity were independently selected for breeding. Subsequent generations of artificial selection yielded strains of rats with reliably high or low aerobic fitness [177]. Along with this, a multitude of physiological and behavioral traits have segregated, where the high-capacity runners (HCR) are lean and have higher longevity compared to their low-capacity runner counterparts (LCR), which have higher adiposity and are susceptible to metabolic disease [181–183]. The HCR are not only resistant to fatigue but are also less fuel efficient while active, burning significantly more kcal during activity [183, 184]. Moreover, the HCR also have greater muscle thermogenesis, dissipating more heat during treadmill walking [183]. HCR also show significantly elevated muscle temperatures relative to LCR after exposure to predator odor [185]. Lastly, HCR display more muscle (soleus) SERCA uncoupling relative to LCR [145]. Altogether, evidence from this model of high and low aerobic capacity or fitness reinforces the connection between energetic muscle thermogenesis, inefficiency, and running capacity and points to a potential route through which elevated muscle thermogenic attributes may have evolved, specifically in association with enhanced aerobic capacity.

10.6 CONCLUSION

A robust skeletal muscle thermogenic response can be induced reliably by environmental predator threat, specifically predator odor. After detecting the predator-threat cue, the brain increases SNS outflow to muscle, which then alters muscle SERCA efficiency [13, 145]. Though we have demonstrated this in rats and mice [13, 51, 185], it is likely that similar responses may be evident in other mammals, including nonhuman primates and humans. The brain-SNS-muscle pathway through which predator threat activates muscle thermogenesis can be utilized to enhance muscle thermogenesis and energy expenditure in the service of reversing positive energy balance and promoting weight loss. Lastly, the overlap between cold adaptation, oxidative capacity and muscle performance, and muscle fuel inefficiency yields clues regarding the potential adaptive significance of predator odor–induced muscle thermogenesis.

REFERENCES

1. Alekseev, A.E., et al., *Sarcolemmal ATP-sensitive K(+) channels control energy expenditure determining body weight*. Cell Metab, 2010. **11**(1): p. 58–69.
2. Reyes, S., et al., *K(ATP) channels process nucleotide signals in muscle thermogenic response*. Crit Rev Biochem Mol Biol, 2010. **45**(6): p. 506–19.
3. Barclay, C.J. and Launikonis, B.S., *A mathematical model to quantify RYR Ca2+ leak and associated heat production in resting human skeletal muscle fibers*. J Gen Physiol, 2022. **154**. (9).
4. Meizoso-Huesca, A., et al., *Ca(2+) leak through ryanodine receptor 1 regulates thermogenesis in resting skeletal muscle*. Proc Natl Acad Sci U S A, 2022. **119**. (4).

5. Bal, N.C. and Periasamy, M., *Uncoupling of sarcoendoplasmic reticulum calcium ATPase pump activity by sarcolipin as the basis for muscle non-shivering thermogenesis*. Philos Trans R Soc Lond B Biol Sci, 2020. **375**(1793): p. 0135.2019
6. Wright, T. and Sheffield-Moore, M., *Skeletal muscle plasticity and thermogenesis: Insights from sea otters*. Temperature (Austin), 2022. **9**(2): p. 119–121.
7. Wright, T., et al., *Skeletal muscle thermogenesis enables aquatic life in the smallest marine mammal*. Science, 2021. **373**(6551): p. 223–225.
8. Ivanova, Y.M. and Blondin, D.P., *Examining the benefits of cold exposure as a therapeutic strategy for obesity and type 2 diabetes*. J Appl Physiol (1985), 2021. **130**(5): p. 1448–1459.
9. Gordon, K., et al., *Seven days of cold acclimation substantially reduces shivering intensity and increases nonshivering thermogenesis in adult humans*. J Appl Physiol (1985), 2019. **126**(6): p. 1598–1606.
10. Blondin, D.P. and F. Haman, *Shivering and nonshivering thermogenesis in skeletal muscles*. Handb Clin Neurol, 2018. **156**: p. 153–173.
11. Ramsay, D.S. and S.C. Woods, *Clarifying the roles of homeostasis and allostasis in physiological regulation*. Psychol Rev, 2014. **121**(2): p. 225–47.
12. McEwen, B.S. and P.J. Gianaros, *Stress- and allostasis-induced brain plasticity*. Annu Rev Med, 2011. **62**: p. 431–45.
13. Gorrell, E., et al., *Skeletal muscle thermogenesis induction by exposure to predator odor*. J Exp Biol, 2020. **223**(Pt 8).
14. Pagani, J.H. and Rosen, J.B., *The medial hypothalamic defensive circuit and 2,5-dihydro-2,4,5-trimethylthiazoline (TMT) induced fear: comparison of electrolytic and neurotoxic lesions*. Brain Res, 2009. **1286**: p. 133–46.
15. McGregor, I.S., et al., *Not all 'predator odours' are equal: cat odour but not 2,4,5 trimethylthiazoline (TMT; fox odour) elicits specific defensive behaviours in rats*. Behav Brain Res, 2002. **129**(1–2): p. 1–16.
16. Dielenberg, R.A. and McGregor, I.S., *Defensive behavior in rats towards predatory odors: a review*. Neurosci Biobehav Rev, 2001. **25**(7–8): p. 597–609.
17. Dielenberg, R.A., Hunt, G.E., and McGregor, I.S., *"When a rat smells a cat": the distribution of Fos immunoreactivity in rat brain following exposure to a predatory odor*. Neuroscience, 2001. **104**(4): p. 1085–97.
18. Takahashi, L.K., et al., *The smell of danger: a behavioral and neural analysis of predator odor-induced fear*. Neurosci Biobehav Rev, 2005. **29**(8): p. 1157–67.
19. Masini, C.V., Sauer, S, and Campeau, S, *Ferret odor as a processive stress model in rats: neurochemical, behavioral, and endocrine evidence*. Behav Neurosci, 2005. **119**(1): p. 280–92.
20. Montardy, Q., et al., *Mapping the neural circuitry of predator fear in the nonhuman primate*. Brain Struct Funct, 2021. **226**(1): p. 195–205.
21. Coimbra, N.C., et al., *Critical neuropsychobiological analysis of panic attack- and anticipatory anxiety-like behaviors in rodents confronted with snakes in polygonal arenas and complex labyrinths: a comparison to the elevated plus- and T-maze behavioral tests*. Rev Bras Psiquiatr, 2017. **39**(1): p. 72–83.
22. Bertels, J., et al., *Snakes elicit specific neural responses in the human infant brain*. Sci Rep, 2020. **10**(1): p. 7443.
23. Silva, B.A., et al., *Independent hypothalamic circuits for social and predator fear*. Nat Neurosci, 2013. **16**(12): p. 1731–3.
24. Barros, M., et al., *Measuring fear and anxiety in the marmoset (Callithrix penicillata) with a novel predator confrontation model: effects of diazepam*. Behav Brain Res, 2000. **108**(2): p. 205–11.
25. Ribeiro-Barbosa, E.R., et al., *An alternative experimental procedure for studying predator-related defensive responses*. Neurosci Biobehav Rev, 2005. **29**(8): p. 1255–63.
26. Hegab, I.M. and Wei, W, *Neuroendocrine changes upon exposure to predator odors*. Physiol Behav, 2014. **131**: p. 149–55.
27. de Araujo Salgado, I., et al., *Toggling between food-seeking and self-preservation behaviors via hypothalamic response networks*. Neuron, 2023. **111**(18): p. 2899–2917 e6.
28. Lenzi, S.C., et al., *Threat history controls flexible escape behavior in mice*. Curr Biol, 2022. **32**(13): p. 2972–2979 e3.
29. Evans, D.A., et al., *Cognitive Control of Escape Behaviour*. Trends Cogn Sci, 2019. **23**(4): p. 334–348.
30. Kunwar, P.S., et al., *Ventromedial hypothalamic neurons control a defensive emotion state*. Elife, 2015. **4**.
31. Kennedy, A., et al., *Stimulus-specific hypothalamic encoding of a persistent defensive state*. Nature, 2020. **586**(7831): p. 730–734.

32. Lynch, E., et al., *Descending pathways from the superior colliculus mediating autonomic and respiratory effects associated with orienting behaviour*. J Physiol, 2022. **600**(24): p. 5311–5332.
33. Carr, J.A., *I'll take the low road: the evolutionary underpinnings of visually triggered fear*. Front Neurosci, 2015. **9**: p. 414.
34. Dampney, R.A., et al., *Role of dorsolateral periaqueductal grey in the coordinated regulation of cardiovascular and respiratory function*. Auton Neurosci, 2013. **175**(1–2): p. 17–25.
35. Zoladz, P.R. and Diamond, D.M., *Predator-based psychosocial stress animal model of PTSD: Preclinical assessment of traumatic stress at cognitive, hormonal, pharmacological, cardiovascular and epigenetic levels of analysis*. Exp Neurol, 2016. **284**(Pt B): p. 211–219.
36. Bulos, E.M., Pobbe, R.L., and Zangrossi, Jr. H., *Behavioral consequences of predator stress in the rat elevated T-maze*. Physiol Behav, 2015. **146**: p. 28–35.
37. Storchi, R., et al., *A High-Dimensional Quantification of Mouse Defensive Behaviors Reveals Enhanced Diversity and Stimulus Specificity*. Curr Biol, 2020. **30**(23): p. 4619–4630 e5.
38. Cattarelli, M. and Chanel, J, *Influence of some biologically meaningful odorants on the vigilance states of the rat*. Physiol Behav, 1979. **23**(5): p. 831–8.
39. Faber, C.L., et al., *Distinct neuronal projections from the hypothalamic yentromedial nucleus mediate glycemic and behavioral effects*. Diabetes, 2018. **67**(12): p. 2518–2529.
40. Wang, L., Chen, I.Z., and Lin, D, *Collateral pathways from the ventromedial hypothalamus mediate defensive behaviors*. Neuron, 2015. **85**(6): p. 1344–58.
41. Wang, Q., et al., *Predator stress-induced persistent emotional arousal is associated with alterations of plasma corticosterone and hippocampal steroid receptors in rat*. Behav Brain Res, 2012. **230**(1): p. 167–74.
42. Burghardt, P.R., et al., *Risk-assessment and coping strategies segregate with divergent intrinsic aerobic capacity in rats*. Neuropsychopharmacology, 2011. **36**(2): p. 390–401.
43. Perrot-Sinal, T.S., Ossenkopp, K.P., and Kavaliers, M, *Brief predator odour exposure activates the HPA axis independent of locomotor changes*. Neuroreport, 1999. **10**(4): p. 775–80.
44. Munoz-Abellan, C., et al., *Behavioral and endocrine consequences of simultaneous exposure to two different stressors in rats: interaction or independence?* PLoS One, 2011. **6**(6): p. e21426.
45. Genne-Bacon, E.A., Trinko, J.R., and DiLeone, R.J., *Innate fear-induced weight regulation in the C57BL/6J mouse*. Front Behav Neurosci, 2016. **10**: p. 132.
46. Monarca, R.I., et al., *Predation risk modulates diet-induced obesity in male C57BL/6 mice*. Obesity (Silver Spring), 2015. **23**(10): p. 2059–65.
47. Carlsen, M., et al., *The effect of predation risk on body weight in the field vole, Microtus agrestis*. Oikos, 1999. **87**(2): p. 277–285.
48. Sundell, J., et al., *Variation in predation risk and vole feeding behaviour: a field test of the risk allocation hypothesis*. Oecologia, 2004. **139**(1): p. 157–62.
49. Speakman, J.R., *A nonadaptive scenario explaining the genetic predisposition to obesity: the "predation release" hypothesis*. Cell Metab, 2007. **6**(1): p. 5–12.
50. Lima, S.L. and Bednekoff, P.A., *Temporal Variation in Danger Drives Antipredator Behavior: The Predation Risk Allocation Hypothesis*. Am Nat, 1999. **153**(6): p. 649–659.
51. Watts, C.A., et al., *Measuring Skeletal Muscle Thermogenesis in Mice and Rats*. J Vis Exp, 2022(**185**).
52. Blanchard, D.C., Griebel, G, and Blanchard, R.J., *Conditioning and residual emotionality effects of predator stimuli: some reflections on stress and emotion*. Prog Neuropsychopharmacol Biol Psychiatry, 2003. **27**(8): p. 1177–85.
53. Day, H.E., Masini, C.V., and Campeau, S, *The pattern of brain c-fos mRNA induced by a component of fox odor, 2,5-dihydro-2,4,5-trimethylthiazoline (TMT), in rats, suggests both systemic and processive stress characteristics*. Brain Res, 2004. **1025**(1–2): p. 139–51.
54. Galliot, E., et al., *Fear-like behavioral responses in mice in different odorant environments: Trigeminal versus olfactory mediation under low doses*. Behav Processes, 2012. **90**(2): p. 161–6.
55. Horii, Y., et al., *Exposure to TMT odor affects adrenal sympathetic nerve activity and behavioral consequences in rats*. Behav Brain Res, 2010. **214**(2): p. 317–22.
56. Morrow, B.A., Elsworth, J.D., and Roth, R.H., *Fear-like biochemical and behavioral responses in rats to the predator odor, TMT, are dependent on the exposure environment*. Synapse, 2002. **46**(1): p. 11–8.
57. Adamec, R., et al., *Neural plasticity, neuropeptides and anxiety in animals--implications for understanding and treating affective disorder following traumatic stress in humans*. Neurosci Biobehav Rev, 1998. **23**(2): p. 301–18.

58. Adamec, R.E., Blundell, J, and Burton, P, *Relationship of the predatory attack experience to neural plasticity, pCREB expression and neuroendocrine response*. Neurosci Biobehav Rev, 2006. **30**(3): p. 356–75.
59. Cohen, H. and Zohar, J, *An animal model of posttraumatic stress disorder: the use of cut-off behavioral criteria*. Ann N Y Acad Sci, 2004. **1032**: p. 167–78.
60. Figueiredo, H.F., et al., *Stress integration after acute and chronic predator stress: differential activation of central stress circuitry and sensitization of the hypothalamo-pituitary-adrenocortical axis*. Endocrinology, 2003. **144**(12): p. 5249–58.
61. File, S.E., et al., *Dissociation between behavioral and corticosterone responses on repeated exposures to cat odor*. Physiol Behav, 1993. **54**(6): p. 1109–11.
62. Hegab, I.M., et al., *Defensive responses of Brandt's voles (Lasiopodomys brandtii) to stored cat feces*. Physiol Behav, 2014. **123**: p. 193–9.
63. Hegab, I.M., et al., *Defensive responses of Brandt's voles (Lasiopodomys brandtii) to chronic predatory stress*. Physiol Behav, 2014. **126**: p. 1–7.
64. Masini, C.V., et al., *Disruption of neuroendocrine stress responses to acute ferret odor by medial, but not central amygdala lesions in rats*. Brain Res, 2009. **1288**: p. 79–87.
65. Masini, C.V., et al., *Non-associative defensive responses of rats to ferret odor*. Physiol Behav, 2006. **87**(1): p. 72–81.
66. Blanchard, D.C., R.J. Blanchard, and G. Griebel, *Defensive responses to predator threat in the rat and mouse*. Curr Protoc Neurosci, 2005. **Chapter 8**: p. Unit **8** 19.
67. Gavini, C.K., Jones, W.C. 2nd, and Novak, C.M., *Ventromedial hypothalamic melanocortin receptor activation: regulation of activity energy expenditure and skeletal muscle thermogenesis*. J Physiol, 2016. **594**(18): p. 5285–301.
68. Perez-Gomez, A., et al., *Innate predator odor aversion driven by parallel olfactory subsystems that converge in the ventromedial hypothalamus*. Curr Biol, 2015. **25**(10), 1340–1346.
69. Martinez, R.C., et al., *Investigation of the hypothalamic defensive system in the mouse*. Behav Brain Res, 2008. **192**(2): p. 185–90.
70. Staples, L.G., *Predator odor avoidance as a rodent model of anxiety: learning-mediated consequences beyond the initial exposure*. Neurobiol Learn Mem, 2010. **94**(4): p. 435–45.
71. Staples, L.G., et al., *Neural activation during cat odor-induced conditioned fear and 'trial 2' fear in rats*. Neurosci Biobehav Rev, 2005. **29**(8): p. 1265–77.
72. Adriaan Bouwknecht, J., Olivier, B, and Paylor, R.E., *The stress-induced hyperthermia paradigm as a physiological animal model for anxiety: a review of pharmacological and genetic studies in the mouse*. Neurosci Biobehav Rev, 2007. **31**(1): p. 41–59.
73. Nakamura, K., Nakamura, Y, and Kataoka, N, *A hypothalamomedullary network for physiological responses to environmental stresses*. Nat Rev Neurosci, 2022. **23**(1): p. 35–52.
74. Nakamura, K. and Morrison, S.F., *Central sympathetic network for thermoregulatory responses to psychological stress*. Auton Neurosci, 2022. **237**: p. 102918.
75. Morrison, S.F. and Nakamura, K, *Central mechanisms for thermoregulation*. Annu Rev Physiol, 2019. **81**: p. 285–308.
76. Dampney, R.A., *Central mechanisms regulating coordinated cardiovascular and respiratory function during stress and arousal*. Am J Physiol Regul Integr Comp Physiol, 2015. **309**(5): p. R429–43.
77. Munzberg, H., et al., *Neural control of energy expenditure*. Handb Exp Pharmacol, 2016. **233**: p. 173–94.
78. Kataoka, N., et al., *Psychological stress activates a dorsomedial hypothalamus-medullary raphe circuit driving brown adipose tissue thermogenesis and hyperthermia*. Cell Metab, 2014. **20**(2): p. 346–58.
79. Nakamura, K., *Neural circuit for psychological stress-induced hyperthermia*. Temperature (Austin), 2015. **2**(3): p. 352–61.
80. Zhang, W., et al., *Orexin neurons are indispensable for stress-induced thermogenesis in mice*. J Physiol, 2010. **588**(Pt 21): p. 4117–29.
81. Marks, A., Vianna, D.M., and Carrive, P, *Nonshivering thermogenesis without interscapular brown adipose tissue involvement during conditioned fear in the rat*. Am J Physiol Regul Integr Comp Physiol, 2009. **296**(4): p. R1239–47.
82. Eskilsson, A., et al., *The generation of immune-induced fever and emotional stress-induced hyperthermia in mice does not involve brown adipose tissue thermogenesis*. FASEB J, 2020. **34**(4): p. 5863–5876.
83. Hiramoto, T., et al., *Pyrogenic cytokines did not mediate a stress interview-induced hyperthermic response in a patient with psychogenic fever: a case report*. Psychosom Med, 2009. **71**(9): p. 932–6.

84. Oka, T., *Psychogenic fever: how psychological stress affects body temperature in the clinical population*. Temperature (Austin), 2015. **2**(3): p. 368–78.
85. Timmerman, R.J., et al., *Psychogenic periodic fever*. Neth J Med, 1992. **41**(3–4): p. 158–60.
86. Oka, T., *Stress-induced hyperthermia and hypothermia*. Handb Clin Neurol, 2018. **157**: p. 599–621.
87. Olivier, B., *Psychogenic fever, functional fever, or psychogenic hyperthermia?* Temperature (Austin), 2015. **2**(3): p. 324–5.
88. Oka, T. and Oka, K, *Age and gender differences of psychogenic fever: a review of the Japanese literature*. Biopsychosoc Med, 2007. **1**: p. 11.
89. Oka, T., Oka, K, and Hori, T, *Mechanisms and mediators of psychological stress-induced rise in core temperature*. Psychosom Med, 2001. **63**(3): p. 476–86.
90. Richardson, J.S., *Pyrexia of uncertain origin and psychogenic fever*. Practitioner, 1953. **170**(1015): p. 61–6.
91. Nozu, T. and Uehara, A, *The diagnoses and outcomes of patients complaining of fever without any abnormal findings on diagnostic tests*. Intern Med, 2005. **44**(8): p. 901–2.
92. Marazziti, D., Di Muro, A, and Castrogiovanni, P, *Psychological stress and body temperature changes in humans*. Physiol Behav, 1992. **52**(2): p. 393–5.
93. Robinson, L.J., et al., *Brown adipose tissue activation as measured by infrared thermography by mild anticipatory psychological stress in lean healthy females*. Exp Physiol, 2016. **101**(4): p. 549–57.
94. Lkhagvasuren, B., et al., *Social defeat stress induces hyperthermia through activation of thermoregulatory sympathetic premotor neurons in the medullary raphe region*. Eur J Neurosci, 2011. **34**(9): p. 1442–52.
95. Feng, C., et al., *Cold-sensitive ventromedial hypothalamic neurons control homeostatic thermogenesis and social interaction-associated hyperthermia*. Cell Metab, 2022. **34**(6): p. 888–901 e5.
96. Mohammed, M., Ootsuka, Y., and Blessing, W., *Brown adipose tissue thermogenesis contributes to emotional hyperthermia in a resident rat suddenly confronted with an intruder rat*. Am J Physiol Regul Integr Comp Physiol, 2014. **306**(6): p. R394–400.
97. Mohammed, M., et al., *Reduced brown adipose tissue thermogenesis during environmental interactions in transgenic rats with ataxin-3-mediated ablation of hypothalamic orexin neurons*. Am J Physiol Regul Integr Comp Physiol, 2014. **307**(8): p. R978–89.
98. Butler, R.K., et al., *Comparison of the activation of somatostatin- and neuropeptide Y-containing neuronal populations of the rat amygdala following two different anxiogenic stressors*. Exp Neurol, 2012. **238**(1): p. 52–63.
99. Baisley, S.K., Cloninger, C.L., and Bakshi, V.P., *Fos expression following regimens of predator stress versus footshock that differentially affect prepulse inhibition in rats*. Physiol Behav, 2011. **104**(5): p. 796–803.
100. Staples, L.G., McGregor, I.S., and Hunt, G.E., *Long-lasting FosB/DeltaFosB immunoreactivity in the rat brain after repeated cat odor exposure*. Neurosci Lett, 2009. **462**(2): p. 157–61.
101. Seoane-Collazo, P., et al., *SF1-Specific AMPKalpha1 Deletion Protects Against Diet-Induced Obesity*. Diabetes, 2018. **67**(11): p. 2213–2226.
102. Felsted, J.A., et al., *Alpha2delta-1 in SF1(+) Neurons of the Ventromedial Hypothalamus Is an Essential Regulator of Glucose and Lipid Homeostasis*. Cell Rep, 2017. **21**(10): p. 2737–2747.
103. Ramos-Lobo, A.M., et al., *SOCS3 ablation in SF1 cells causes modest metabolic effects during pregnancy and lactation*. Neuroscience, 2017. **365**: p. 114–124.
104. Martinez-Sanchez, N., et al., *Hypothalamic AMPK-ER Stress-JNK1 Axis Mediates the Central Actions of Thyroid Hormones on Energy Balance*. Cell Metab, 2017. **26**(1): p. 212–229 e12.
105. Sohn, J.W., et al., *Leptin and insulin engage specific PI3K subunits in hypothalamic SF1 neurons*. Mol Metab, 2016. **5**(8): p. 669–79.
106. Kamitakahara, A., Xu, B., and Simerly, R., *Ventromedial hypothalamic expression of Bdnf is required to establish normal patterns of afferent GABAergic connectivity and responses to hypoglycemia*. Mol Metab, 2016. **5**(2): p. 91–101.
107. Orozco-Solis, R., et al., *SIRT1 relays nutritional inputs to the circadian clock through the Sf1 neurons of the ventromedial hypothalamus*. Endocrinology, 2015: **156**(6), 2174–2184.
108. Zhang, R., et al., *Selective inactivation of Socs3 in SF1 neurons improves glucose homeostasis without affecting body weight*. Endocrinology, 2008. **149**(11): p. 5654–61.
109. Dhillon, H., et al., *Leptin directly activates SF1 neurons in the VMH, and this action by leptin is required for normal body-weight homeostasis*. Neuron, 2006. **49**(2): p. 191–203.

110. Berger, A., et al., *Gsalpha Deficiency in the Ventromedial Hypothalamus Enhances Leptin Sensitivity and Improves Glucose Homeostasis in Mice on a High-Fat Diet*. Endocrinology, 2016. **157**(2): p. 600–10.
111. Brix, L.M., et al., *Contribution of the co-chaperone FKBP51 in the ventromedial hypothalamus to metabolic homeostasis in male and female mice*. Mol Metab, 2022. **65**: p. 101579.
112. Cardinal, P., et al., *CB1 cannabinoid receptor in SF1-expressing neurons of the ventromedial hypothalamus determines metabolic responses to diet and leptin*. Mol Metab, 2014. **3**(7): p. 705–16.
113. Cheung, C.C., et al., *Sex-dependent changes in metabolism and behavior, as well as reduced anxiety after eliminating ventromedial hypothalamus excitatory output*. Mol Metab, 2015. **4**(11): p. 857–66.
114. Coutinho, E.A., et al., *Activation of SF1 neurons in the ventromedial hypothalamus by DREADD technology increases insulin sensitivity in peripheral tissues*. Diabetes, 2017. **66**(9), 2372–2386.
115. Fagan, M.P., et al., *Essential and sex-specific effects of mGluR5 in ventromedial hypothalamus regulating estrogen signaling and glucose balance*. Proc Natl Acad Sci U S A, 2020. **117**(32): p. 19566–19577.
116. Fosch, A., et al., *New Insights of SF1 Neurons in Hypothalamic Regulation of Obesity and Diabetes*. Int J Mol Sci, 2021. **22**(12).
117. Kaneko, K., et al., *Rap1 in the VMH regulates glucose homeostasis*. JCI Insight, 2021. **6**(11).
118. Milbank, E., et al., *Small extracellular vesicle-mediated targeting of hypothalamic AMPKalpha1 corrects obesity through BAT activation*. Nat Metab, 2021. **3**(10): p. 1415–1431.
119. Ramadori, G., et al., *SIRT1 deacetylase in SF1 neurons protects against metabolic imbalance*. Cell Metab, 2011. **14**(3): p. 301–12.
120. Rouabhi, M., et al., *BBSome ablation in SF1 neurons causes obesity without comorbidities*. Mol Metab, 2021. **48**: p. 101211.
121. Senn, S.S., et al., *Unsilencing of native LepRs in hypothalamic SF1 neurons does not rescue obese phenotype in LepR-deficient mice*. Am J Physiol Regul Integr Comp Physiol, 2019. **317**(3): p. R451–R460.
122. Shimazu, T. and Minokoshi, Y., *Systemic Glucoregulation by Glucose-Sensing Neurons in the Ventromedial Hypothalamic Nucleus (VMH)*. J Endocr Soc, 2017. **1**(5): p. 449–459.
123. Xu, Y., Faulkner, L.D., and Hill, J.W., *Cross-Talk between Metabolism and Reproduction: The Role of POMC and SF1 Neurons*. Front Endocrinol (Lausanne), 2011. **2**: p. 98.
124. Yang, H., et al., *Regulation of Energy Balance via BDNF Expressed in Nonparaventricular Hypothalamic Neurons*. Mol Endocrinol, 2016. **30**(5): p. 494–503.
125. Milbank, E., et al., *Small extracellular vesicle targeting of hypothalamic AMPKalpha1 promotes weight loss in leptin receptor deficient mice*. Metabolism, 2023. **139**: p. 155350.
126. Viskaitis, P., et al., *Modulation of SF1 neuron activity coordinately regulates both feeding behavior and associated emotional states*. Cell Rep, 2017. **21**(12): p. 3559–3572.
127. Watts, C.A., *Investigating the Role of Steroidogenic Factor 1 (SF1) Neurons in Energy Balance: An Analysis of Predator Odor-Induced Skeletal Muscle Thermogenesis in Mice Utilizing DREADD Technology*. 2023, Kent State University.
128. Khodai, T. and Luckman, S.M., *Ventromedial Nucleus of the Hypothalamus Neurons Under the Magnifying Glass*. Endocrinology, 2021. **162**(10).
129. King, B.M., *The rise, fall, and resurrection of the ventromedial hypothalamus in the regulation of feeding behavior and body weight*. Physiol Behav, 2006. **87**(2): p. 221–44.
130. Esteban Masferrer, M., et al., *Differential Encoding of Predator Fear in the Ventromedial Hypothalamus and Periaqueductal Grey*. J Neurosci, 2020. **40**(48): p. 9283–9292.
131. Aras, E., et al., *Light Entrains Diurnal Changes in Insulin Sensitivity of Skeletal Muscle via Ventromedial Hypothalamic Neurons*. Cell Rep, 2019. **27**(8): p. 2385–2398 e3.
132. Labbe, S.M., et al., *Hypothalamic control of brown adipose tissue thermogenesis*. Front Syst Neurosci, 2015. **9**: p. 150.
133. Lindberg, D., Chen, P., and Li, C., *Conditional viral tracing reveals that steroidogenic factor 1-positive neurons of the dorsomedial subdivision of the ventromedial hypothalamus project to autonomic centers of the hypothalamus and hindbrain*. J Comp Neurol, 2013. **521**(14): p. 3167–90.
134. Babic, T., et al., *Innervation of skeletal muscle by leptin receptor-containing neurons*. Brain Res, 2010. **1345**: p. 146–55.
135. Flak, J.N., et al., *A leptin-regulated circuit controls glucose mobilization during noxious stimuli*. J Clin Invest, 2017. **127**(8): p. 3103–3113.
136. Kerman, I.A., Akil, H., and Watson, S.J., *Rostral elements of sympatho-motor circuitry: a virally mediated transsynaptic tracing study*. J Neurosci, 2006. **26**(13): p. 3423–33.

137. Barker, D. and Saito, M., *Autonomic innervation of receptors and muscle fibres in cat skeletal muscle*. Proc R Soc Lond B Biol Sci, 1981. **212**(1188): p. 317–32.
138. Baron, R., Janig, W., and Kollmann, W., *Sympathetic and afferent somata projecting in hindlimb nerves and the anatomical organization of the lumbar sympathetic nervous system of the rat*. J Comp Neurol, 1988. **275**(3): p. 460–8.
139. Almundarij, T.I., Gavini, C.K., and Novak, C.M., *Suppressed sympathetic outflow to skeletal muscle, muscle thermogenesis, and activity energy expenditure with calorie restriction*. Physiol Rep, 2017. **5**(4).
140. Gorrell, E.E., *Induction of skeletal muscle thermogenesis by exposure to predator odor*. 2020, Kent State University.
141. Bal, N.C., et al., *Sarcolipin is a newly identified regulator of muscle-based thermogenesis in mammals*. Nat Med, 2012. **18**(10): p. 1575–9.
142. Bal, N.C., et al., *The Role of Sarcolipin in Muscle Non-shivering Thermogenesis*. Front Physiol, 2018. **9**: p. 1217.
143. Gamu, D., et al., *The sarcoplasmic reticulum and SERCA: a nexus for muscular adaptive thermogenesis*. Appl Physiol Nutr Metab, 2020. **45**(1): p. 1–10.
144. Maurya, S.K. and Periasamy, M., *Sarcolipin is a novel regulator of muscle metabolism and obesity*. Pharmacol Res, 2015. **102**: p. 270–5.
145. Heemstra, L.A., et al., *Altered skeletal muscle sarco-endoplasmic reticulum Ca(2+)-ATPase calcium transport efficiency after a thermogenic stimulus*. Am J Physiol Regul Integr Comp Physiol, 2022. **323**(5): p. R628–R637.
146. Nowack, J., et al., *Muscle nonshivering thermogenesis in a feral mammal*. Sci Rep, 2019. **9**(1): p. 6378.
147. Periasamy, M., Herrera, J.L., and Reis, F.C.G., *Skeletal muscle thermogenesis and its role in whole body energy metabolism*. Diabetes Metab J, 2017. **41**(5): p. 327–336.
148. Sepa-Kishi, D.M., et al., *Cold acclimation causes fiber type-specific responses in glucose and fat metabolism in rat skeletal muscles*. Sci Rep, 2017. **7**(1): p. 15430.
149. Liu, M., et al., *Recruitment of Muscle Genes as an Effect of Brown Adipose Tissue Ablation in Cold-Acclimated Brandt's Voles (Lasiopodomys brandtii)*. Int J Mol Sci, 2022. **24**(1): 342.
150. Bal, N.C., et al., *Both brown adipose tissue and skeletal muscle thermogenesis processes are activated during mild to severe cold adaptation in mice*. J Biol Chem, 2017. **292**(40): p. 16616–16625.
151. Schertzer, J.D., et al., *Beta2-agonist administration increases sarcoplasmic reticulum Ca2+-ATPase activity in aged rat skeletal muscle*. Am J Physiol Endocrinol Metab, 2005. **288**(3): p. E526–33.
152. Cairns, S.P. and Borrani, F., *beta-Adrenergic modulation of skeletal muscle contraction: key role of excitation-contraction coupling*. J Physiol, 2015. **593**(21): p. 4713–27.
153. Canteras, N.S., *The medial hypothalamic defensive system: hodological organization and functional implications*. Pharmacol Biochem Behav, 2002. **71**(3): p. 481–91.
154. Velasco, E.R., et al., *PACAP-PAC1R modulates fear extinction via the ventromedial hypothalamus*. Nat Commun, 2022. **13**(1): p. 4374.
155. Feinstein, J.S., et al., *The human amygdala and the induction and experience of fear*. Curr Biol, 2011. **21**(1): p. 34–8.
156. Wilent, W.B., et al., *Induction of panic attack by stimulation of the ventromedial hypothalamus*. J Neurosurg, 2010. **112**(6): p. 1295–8.
157. Lipp, H.P. and Hunsperger, R.W., *Threat, attack and flight elicited by electrical stimulation of the ventromedial hypothalamus of the marmoset monkey Callithrix jacchus*. Brain Behav Evol, 1978. **15**(4): p. 260–93.
158. Taylor, J.E., et al., *An Evolutionarily Threat-Relevant Odor Strengthens Human Fear Memory*. Front Neurosci, 2020. **14**: p. 255.
159. Ryall, J.G., et al., *Beta 2-agonist fenoterol has greater effects on contractile function of rat skeletal muscles than clenbuterol*. Am J Physiol Regul Integr Comp Physiol, 2002. **283**(6): p. R1386–94.
160. Blackwood, S.J. and Katz, A., *Isoproterenol enhances force production in mouse glycolytic and oxidative muscle via separate mechanisms*. Pflugers Arch, 2019. **471**(10): p. 1305–1316.
161. Schaeffer, P.J., Villarin, J.J., and Lindstedt, S.L., *Chronic cold exposure increases skeletal muscle oxidative structure and function in Monodelphis domestica, a marsupial lacking brown adipose tissue*. Physiol Biochem Zool, 2003. **76**(6): p. 877–87.
162. Schaeffer, P.J., et al., *Cold exposure increases running VO(2max) and cost of transport in goats*. Am J Physiol Regul Integr Comp Physiol, 2001. **280**(1): p. R42–7.

163. Harri, M.N. and Valtola, J., *Comparison of the effects of physical exercise, cold acclimation and repeated injections of isoprenaline on rat muscle enzymes*. Acta Physiol Scand, 1975. **95**(4): p. 391–9.
164. Ostman, I. and Sjostrand, N., *Reduced urinary noradrenaline excretion during rest, exercise and cold stress in trained rats: a comparison between physically-trained rats, cold-acclimated rats and warm-acclimated rats*. Acta Physiol Scand, 1975. **95**(3): p. 209–18.
165. Knuth, C.M., et al., *Prior exercise training improves cold tolerance independent of indices associated with non-shivering thermogenesis*. J Physiol, 2018. **596**(18): p. 4375–4391.
166. Fujikawa, T., et al., *SF-1 expression in the hypothalamus is required for beneficial metabolic effects of exercise*. Elife, 2016. **5**.
167. Pedroso, J.A.B., et al., *SOCS3 expression in SF1 cells regulates adrenal differentiation and exercise performance*. J Endocrinol, 2017. **235**(3): p. 207–222.
168. Pedroso, J.A.B., et al., *Deletion of growth hormone receptor in hypothalamic neurons affects the adaptation capacity to aerobic exercise*. Peptides, 2021. **135**: p. 170426.
169. Rizvi, M.H., Abdul Azeem, M., and Savanur, A., *Effects of adrenaline on contractility and endurance of isolated mammalian soleus with different calcium concentrations*. J Muscle Res Cell Motil, 2019. **40**, 373–378.
170. Sopariwala, D.H., et al., *Sarcolipin overexpression improves muscle energetics and reduces fatigue*. J Appl Physiol (1985), 2015. **118**(8): p. 1050–8.
171. Pontzer, H., *A unified theory for the energy cost of legged locomotion*. Biol Lett, 2016. **12**(2): p. 20150935.
172. Hunter, G.R., et al., *Inverse relationship between exercise economy and oxidative capacity in muscle*. Eur J Appl Physiol, 2005. **94**(5–6): p. 558–68.
173. Schaeffer, P.J., et al., *Cost of transport is increased after cold exposure in Monodelphis domestica: training for inefficiency*. J Exp Biol, 2005. **208**(Pt 16): p. 3159–67.
174. Even, P.C. and Blais, A., *Increased Cost of Motor Activity and Heat Transfer between Non-Shivering Thermogenesis, Motor Activity, and Thermic Effect of Feeding in Mice Housed at Room Temperature - Implications in Pre-Clinical Studies*. Front Nutr, 2016. **3**: p. 43.
175. Nowack, J., et al., *Muscle Non-shivering Thermogenesis and Its Role in the Evolution of Endothermy*. Front Physiol, 2017. **8**: p. 889.
176. Rowland, L.A., Bal, N.C., and Periasamy, M., *The role of skeletal-muscle-based thermogenic mechanisms in vertebrate endothermy*. Biol Rev Camb Philos Soc, 2015. **90**(4): p. 1279–97.
177. Koch, L.G. and Britton, S.L., *Rat Models of Exercise for the Study of Complex Disease*. Methods Mol Biol, 2019. **2018**: p. 309–317.
178. Koch, L.G. and Britton, S.L., *Theoretical and Biological Evaluation of the Link between Low Exercise Capacity and Disease Risk*. Cold Spring Harb Perspect Med, 2018. **8**(1).
179. Koch, L.G. and Britton, S.L., *Biology: Motion is Function*. Function (Oxf), 2022. **3**(4): p. zqac030.
180. Koch, L.G. and Britton, S.L., *Evolution, atmospheric oxygen, and complex disease*. Physiol Genomics, 2007. **30**(3): p. 205–8.
181. Wisloff, U., et al., *Cardiovascular risk factors emerge after artificial selection for low aerobic capacity*. Science, 2005. **307**(5708): p. 418–20.
182. Noland, R.C., et al., *Artificial selection for high-capacity endurance running is protective against high-fat diet-induced insulin resistance*. Am J Physiol Endocrinol Metab, 2007. **293**(1): p. E31–41.
183. Gavini, C.K., et al., *Leanness and heightened nonresting energy expenditure: role of skeletal muscle activity thermogenesis*. Am J Physiol Endocrinol Metab, 2014. **306**(6): p. E635–47.
184. Novak, C.M., et al., *Spontaneous activity, economy of activity, and resistance to diet-induced obesity in rats bred for high intrinsic aerobic capacity*. Horm Behav, 2010. **58**(3): p. 355–67.
185. Shemery, A.M., et al., *Reduced contextually induced muscle thermogenesis in rats with calorie restriction and lower aerobic fitness but not monogenic obesity*. Temperature (Austin), 2023. **10**(3): p. 379–393.

Unit 3

Biomedical Relevance of Non-Shivering Thermogenesis

OVERVIEW

Obesity has increasingly become a burden on the human health care system worldwide. Effective treatment of obesity greatly reduces the morbidity of other metabolic diseases. Obesity results from storage of the excess metabolic energy. Energy homeostasis is closely related to the active function of adipose tissue and skeletal muscle. Adipose tissue in mammals consists mainly of brown adipose tissue (BAT) and white adipose tissue (WAT). BAT includes classical BAT and recently discovered beige adipocyte aggregates, which resists cold and obesity through adaptive heat production, known as non-shivering thermogenesis (NST). NST is regulated by multiple signaling pathways. Thermogenic adipocytes, including brown adipocytes and beige adipocytes, have been shown to be abundant in human infants, and recent research has detected it in adults as well. It is becoming increasingly clear that BAT and BAT-like tissues have significant effects on total body metabolism in humans. These factors, along with the role of mitochondrial dysfunction in these tissues, are discussed in Chapter 11.

Humans have evolved complex neural circuits that drive behavioral, somatic, and autonomic thermoregulatory responses to maintain their body temperature. While they are well adapted to dissipate heat in warm climates, they have a reduced capacity to preserve it in cold environments. Consequently, heat production is critical to maintaining their core temperature. As in other large mammals, skeletal muscles are the primary source of heat production recruited in cold-exposed humans. This is achieved voluntarily in the form of contractions from exercising muscles or involuntarily in the form of contractions from shivering muscles and the recruitment of NST mechanisms. Chapters 12 and 13 of this unit describe our current understanding of shivering and non-shivering thermogenesis in skeletal muscles, from the neural circuitry driving their recruitment to the metabolic substrates that fuel them. More recent investigations have focused on understanding the metabolic processes that produce the heat from both contracting and noncontracting mechanisms. With the growing interest in the potential therapeutic benefits of shivering and NST in

DOI: 10.1201/9781003391470-13

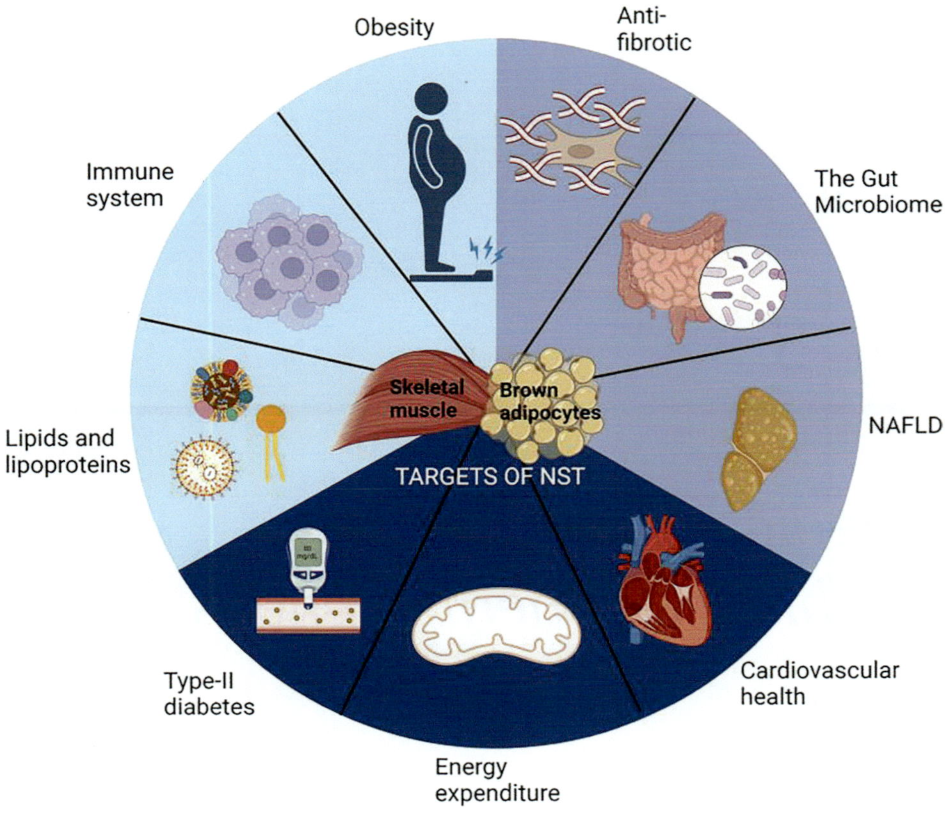

skeletal muscle to counter the effects of neuromuscular, cardiovascular, and metabolic diseases, we expect this field to continue its growth in the coming years.

To realize and fully utilize the potential of these cellular metabolic processes, the evolutionary history and different regulatory mechanisms of thermogenesis should be explored. The chapters of this unit also illustrate the current findings on all the thermogenesis-related signaling pathways that have been studied in recent years. There has been some excitement in the field to target these processes directly and/or indirectly by pharmacological agents to increase futile energy expenditure. Chapter 14 illustrates the potential therapeutic applications, mainly focusing on natural products and their active ingredients that may decrease obesity by affecting related signaling pathways.

11 Interrogating the Nature of Mitochondrial Bioenergetic Dysfunction with Aging

Sam D. Houman, Julius J. Larion and Abraham Karimi
Touro University California College of Osteopathic Medicine, Vallejo, California, USA

Shona A. Mookerjee
Touro University California College of Osteopathic Medicine, Vallejo, California, United States
Buck Institute for Research on Aging, Novato, California, USA

11.1 INTRODUCTION

Mitochondrial dysfunction appears to be a common denominator in aging, and is defined in a broad variety of ways—so many that understanding what "mitochondrial dysfunction" is and how we measure it is almost as complicated as defining and measuring the aging process itself. Decades of literature support aging-associated changes to mitochondrial components: genome integrity, copy number, and variability, RNA transcripts, membrane protein composition and modification, as well as changes to processes related to mitochondrial morphology and turnover and mitochondrial interaction with other organelles, and finally changes to the system fluxes of substrate import and oxidation, oxidative phosphorylation to generate ATP, and generation of reactive oxygen species (ROS). Reaching outside the cell, mitochondrial signaling to pathways including inflammation and insulin signaling connect mitochondrial status and function to systemic signaling and metabolism. Because these activities are so varied, an expectation that "mitochondrial dysfunction" can be collected into one or a few mechanisms is probably incorrect. Yet, the use of this term often implies such an expectation. To clarify how we think about mitochondrial dysfunction during aging, we can start at the beginning, with the free radical theory of aging.

The free radical theory of aging was published in 1956 by Denham Harman, and subsequently updated, by him and many others, through the present [1–5]. Arguably the first molecular theory of aging (foundational evidence for another early contender came shortly after with the publication of the Hayflick limit in 1961 [6]), the free radical theory of aging posited that oxidation of cellular components led to accumulating damage and cellular dysfunction [7, 8]. With the publication of Mitchell's chemiosmotic theory [9] and the subsequent identification of the mitochondrial electron transport chain as a major source of intracellular reactive oxygen species (ROS) [10, 11], mainly superoxide and hydrogen peroxide [12], the free radical theory of aging was adapted accordingly.

The mitochondrial free radical theory of aging posits a "downward spiral" of oxidative damage within the mitochondrial compartment leading to decreased ATP-generating capacity due to mutated and damaged machinery, which in turn drives increased superoxide production that further damages mitochondria. Acceptance of this theory has shifted over time, and it is now

DOI: 10.1201/9781003391470-14

considered less a universal explanation and more one of multiple likely mechanisms that drive cellular and organismal aging [13–16]. In addition, despite an abundance of correlative and indirect evidence, more progress in sorting out the specific relationships between mitochondrial damage, mitochondrial loss of function, and measurements of aging is needed before direct mechanisms emerge.

The purpose of this review is to focus on an assumed common outcome of most, if not all, mechanisms of mitochondrial dysfunction: impairment of mitochondrial respiration and oxidative phosphorylation [17]. Here, we revisit a crucial question: Is defective mitochondrial energy production a hallmark of aging [18]? To do so, we explore the relationships between mitochondrial dysfunction as described mostly by changes observed in mitochondrial diseases and aging-related diseases and the demonstrated or potential role that energy insufficiency may play in connecting these observations to aging.

11.2 MITOCHONDRIAL DNA DAMAGE—RED HERRING OR MOLECULAR AGING ORIGIN STORY?

Accumulation of damage to the mitochondrial DNA (mtDNA) is a central premise of the mitochondrial free radical theory of aging. Historical lines of evidence supporting this theory are the many inherited mtDNA mutations that underlie various mitochondrial disease syndromes, and the many observations of mtDNA mutation accumulation with age in humans and various model organisms. The mitochondrial genome in humans and most animals encodes 13 proteins, all subunits of mitochondrial electron transport chain complexes. Mutations to these genes, and the resulting expression of nonfunctioning or poorly functioning respiratory complexes, certainly or plausibly decreases ATP-generating capacity by mitochondria [19].

MtDNA has also historically been considered to be more vulnerable to damage and mutation than nuclear DNA, through greater exposure to toxins and mutagens, higher replication error rate, and lack of DNA repair mechanisms [20]. Over the last two decades, though, the identification of robust systems of mtDNA structural organization and repair have weakened many of these assertions [21–25].

Despite the evidence that mtDNA mutations cause pathological bioenergetic deficits, and that mtDNA mutations accumulate with age, a consistent and direct role for the types (or distribution) of mtDNA mutations that accumulate during aging actually causing aging-related energy insufficiency remains elusive. For example, the 5-kb "common deletion" of the mitochondrial genome increases in frequency with age and was proposed to drive aging pathology [26, 27]. This region contains 7 of the 13 mitochondrially encoded respiratory complex subunits. In a 1992 analysis of human skeletal muscle, the aging-related increase in this common deletion was determined to be too sparse to drive defects in respiration [28]. More recent analyses have associated the common deletion convincingly with photoaging and decreased mtDNA copy number [29–31], while in platelets, accumulation of the common deletion was associated with increased copy number and no measurable effect on ATP content [32]. Importantly, overall mitochondrial genome copy number, rather than frequency of the common deletion, appears predominantly correlated with cellular respiratory activity in muscle and brain [33, 34]

But what about smaller mutations, driven by oxidative or other damage? Other researchers have noted considerably higher concentrations of 8-hydroxy-2'-deoxyguanosine (8-OHdG) in mtDNA than in nuclear DNA in aged human tissue samples. 8-OHdG, a common marker of oxidative DNA damage, was found to be 10-fold higher in mtDNA than in nuclear DNA in brain tissue samples from healthy adults in their 40s. This increased to 15-fold in adults over 70, possibly reflecting increased oxidative damage, either by increased production of reactive oxygen species or by increased exposure of mtDNA to damage, or by decreased repair, or a combination of these processes [35]. Separately, analysis of heart and liver samples of Wistar rats showed that from 6 to 23 months of age, the level of 8-OHdG increased 2.5-fold in mtDNA, while in nuclear DNA it was unchanged. Comparable to human brain, rat heart and liver mtDNA showed 5-fold more 8-OHdG in mtDNA than nuclear DNA in adulthood (6 months) that increased to 12-fold by 23 months of age [36].

Much of the foundational research in DNA damage and mutation accumulation with age used 8-OHdG as a direct reporter. Sequencing approaches that assess mitochondrial mutations more broadly have since become available, and have shown that replication errors like SNP conversions, insertions, and deletions may drive the predominant mutations that differentiate young and old mitochondrial DNA, rather than oxidative damage [37]. In fact, an interesting mouse model with a mitochondrial-specific knockout of DNA repair glycosylases OGG1 and MUTYH, and a heart-specific knockout of superoxide dismutase (SOD2), was generated to test the combined effect of decreased base excision repair and increased superoxide level on mtDNA integrity. Sequencing analysis of this model showed no accumulation of mtDNA mutations relative to control mice, calling into question the idea that proximity to superoxide production drives high rates of mtDNA mutation [38].

The effects of accumulating mtDNA replication errors are perhaps best known from the progeroid POLG "mutator" mouse, in which the exonuclease proofreading function of the mitochondrial DNA polymerase gamma (POLG) is impaired. These mice exhibit the same phenotype as their wild-type counterparts at birth and when young, but by 9 months of age they display significantly accelerated age-related phenotypes such as hair loss, anemia, hearing loss, decreased bone and muscle mass, and decreased overall life span. Despite the increased mitochondrial mutation load, there was no evidence of increased oxidative stress. This further suggests that ROS and oxidative stress may play a less significant role in some mechanisms of aging than was previously thought [39, 40].

Another group bred heterozygous POLG mutants for several generations and found that even offspring who were homozygous wild-type for POLG but had inherited their mitochondria from a heterozygous mother exhibited accelerated aging and shortened life span. Enzyme activity analysis showed that mice that were deficient in POLG and inherited mutant mitochondria had lower respiratory complex activity (specifically, complexes 1 and 4) than mice who were also deficient in POLG but inherited wild-type mitochondria [41]. While this is one for the few studies to directly assess respiratory chain component activity, and supports that respiratory capacity may be impaired, it falls short of demonstrating under what conditions cellular ATP supply and demand are meaningfully disrupted. Overall, however, these findings support that the developmental timing of mitochondrial mutation burden makes a difference, with mutations that occur earlier in life having a more deleterious impact than those that occur later in life. Contrasting these findings, a separate study found no life span deficit in POLG heterozygous mutants [42].

POLG inhibition and impairment of respiratory chain complex activity is also demonstrated in patients prescribed nucleoside reverse transcriptase inhibitors, or NRTIs, to treat human immunodeficiency virus (HIV) [43, 44]. HIV+ patients taking NRTIs can display premature aging phenotypes such as frailty [45], degenerative cardiovascular disease [46], and neurodegeneration [47]. Analysis of skeletal muscle biopsies from HIV+ patients taking NRTIs, alongside HIV+ patients not taking NRTIs, and healthy HIV– control individuals (all under the age of 50) revealed deficits in cytochrome oxidase (COX) respiratory chain activity only in the group taking NRTIs, demonstrating that it was independent of HIV status. Furthermore, the severity of the COX defect strongly correlated with total lifetime treatment duration with NRTI, and not with recency of treatment, suggesting that the mtDNA mutations acquired with treatment persisted over time, possibly through clonal expansion [48].

Clonal expansion of mtDNA mutations that disrupt cytochrome oxidase activity are well studied in intestinal crypt cells, where the cells derived from individual mutation events are easily visualized by immunohistochemistry [49, 50]. This phenomenon is apparent in the POLG-D257A mouse model, and is suggested to be similar in pattern to the accumulation observed in human aging [51]. This model was further analyzed to study how mtDNA mutations contribute to changes in cell function within the small intestine [52]. Bioenergetic analysis showed that the cells of the villus but not the crypts had decreased ATP and a higher ADP:ATP ratio compared to controls. Even at a "young" 3 months of age, morphological changes to the small intestine were observed, and attributed to cell cycle dysregulation, evident within the crypts and villi of the small intestine, and the significant increase in apoptosis of stem cell progenitors [52]. This mitochondrial dysfunction and bioenergetic deficit correlated with impaired ability to absorb lipids, consistent with widely reported observations of nutrient malabsorption in aging [53, 54]. The study by Fox et al. reported one of relatively few attempts in the literature to associate mtDNA mutation accumulation with a directly measured bioenergetic deficit.

Accumulation of mutations in individual mitochondrial genomes contributes to mitochondrial DNA heteroplasmy, in which the population of mitochondrial genomes within a single cell is heterogenous in sequence. Different cells even within the same tissue may have different ratios of mutant to wild-type mtDNA. Heteroplasmy complicates the pathology of mitochondrial diseases and contributes to wide variation in mitochondrial function between cells, tissues, and patients [55, 56]. It can also modify the appearance of other diseases, including neurodegeneration [57]. The threshold required for a deleterious mutation to cause energetic deficits depends on the mutation, cell type, energy demand, mtDNA copy number, and other factors, but overall it might be around 60–80% [58, 59]. Mitochondrial DNA heteroplasmy, and the clonal expansion of mutated mitochondrion, is thought to be a potential contributor to the aging phenotype.

So, what do we make of the relative lack of evidence that, aside from mutations that appear to abolish respiration, the gradual accumulation of mtDNA mutation and heteroplasmy with age does not appear to yield a gradual decline of bioenergetic function that might be expected to result? This is not to ask whether mtDNA integrity is important for bioenergetic function; the faithful replication and timely repair of the mitochondrial genome are clearly required to maintain a functioning respiratory chain. But what is the overall relationship between mutational burden and energetic insufficiency? Why does most of our evidence support a mostly catastrophic relationship at a certain threshold?

One possibility is biological; it could be that accumulating mutations under a catastrophic threshold (perhaps buffered by multiple other factors, including mtDNA copy number, e.g., [60]) makes little difference to the economy of ATP supply and demand under most biological conditions, while exceeding that mutation threshold causes respiratory collapse. Or, a subthreshold mutation burden accommodates basal function but capacity is decreased, and while basal function is preserved, energetic stress occurs at lower energy demand, as we speculated previously [61].

Another possibility is artifactual; the histological approaches employed in tissues are generally good at identifying energetic collapse but may be poorly suited for identifying transient or small energy deficits whose effects accumulate over time. Such a measurement bias would yield a large amount of evidence that energy deficits that associated with aging appear to be the stochastic bioenergetic collapse in individual cells that contribute to gradual tissue and physiological malfunction. The identification of more subtle bioenergetic deficits on a cell-by-cell or tissue basis would require more sophisticated bioenergetic analysis, e.g., of the forces and fluxes of ATP production rates and capacities, which is technically challenging but has become more addressable with newer technologies, including fluorescent probes for determining mitochondrial membrane potential, ATP levels, and ATP:ADP ratios in whole cells, as well as the Seahorse XF Analyzer [62–65].

11.3 MTDNA EPIGENETIC MODIFICATION AND TRANSCRIPTION SOMETIMES SHOW CHANGES WITH AGE, BUT WITH NO CURRENT EVIDENCE FOR DIRECT ATTENUATION OF MITOCHONDRIAL ENERGY PRODUCTION

While enzymic methylation of mitochondrial DNA has long been recognized [66–68], our understanding of how mtDNA methylation affects its packaging, replication, and transcription is incomplete [69]. Some findings indicate that it may modulate the affinity of the mitochondrial histone-like protein TFAM for mtDNA [68, 70] or the processing of polycistronic transcripts [68, 71]. Hypomethylation of CpG sites M1215 and M1313 of the 12S ribosomal RNA gene in mtDNA has shown to be correlated with increasing age [72, 73]. However, others found that increased methylation of the same gene was associated with increased mortality risk [73, 74], suggesting that perhaps there are locus-specific differences that need to be further investigated. Many researchers have noted epigenetic modifications to the D-loop of the mitochondrial genome. Although the effects of these modifications are not yet known, there does seem to be a correlation with age-related diseases like Alzheimer's and Parkinson's [73, 75]. The influence of "mitoepigenetics" on cellular energy homeostasis in the context of aging is an area of research that appears unexplored yet would be of high value to the field.

Early studies of the impact of aging on global transcription of the mitochondrial genome show variable reports of whether mitochondrial transcript levels decline with age or remain constant [76]. Multiple studies have shown, however, that mitochondrial DNA content increases with age, and thus the mtRNA:mtDNA ratio consistently decreases with age across various models of aging [77–79]. Increased mtDNA is speculated to correct for decrease transcriptional activity and maintain mRNA transcript levels. As in nuclear transcriptional regulation, noncoding micro-RNAs (miRNAs) also regulate mtDNA transcriptional activity. These short RNA sequences are primarily nuclear-encoded, but a few mtDNA-encoded miRNAs have also recently been identified [80]. MiRNAs associate with core proteins like AGO/AGO2 to form the RNA-induced silencing complex (RISC). MiRNAs and RISC components have been shown also to be active in the mitochondria and target mitochondrial transcripts [81]. Some of these miRNAs may modulate senescence and autophagy, as well as ATP and ROS production [82]. MiR-210 and miR-494 are two similar senescence-promoting miRNAs that are induced by DNA damage and oxidative stress [83], and that have been shown to decrease mitochondrial respiration. MiRNA-210 is asserted to regulate cellular responses to hypoxia [84]. In a human pulmonary arterial endothelial cell model, upregulating miRNA-210 expression was shown to decrease free ATP by 18% [85]. Another study comparing the brain tissue of different-aged mice showed that miRNAs were differentially upregulated with age, and were computationally predicted to target complexes 3 and 4 as well as the mitochondrial ATP synthase [86]. Overall, however, while RNA-level changes of both protein-coding and regulatory RNA species correlate with aging in some cases, the role that these changes play in inducing bioenergetic activity changes, and particularly whether those changes are adaptive or deleterious, remains unclear.

11.4 MITOCHONDRIAL PROTEOSTASIS IS DISRUPTED BY AGING-RELATED DISEASE

Protein homeostasis, or proteostasis, refers to the processes by which protein synthesis, folding, modification, positioning, and degradation are regulated to maintain functional proteins at the levels needed by the cell under given conditions. Like other systems of homeostasis, these so-called "quality control" mechanisms require extensive coordination. Diseases of protein accumulation are frequently framed as a loss of proteostasis, including Alzheimer's, Parkinson's, and other neurodegenerative diseases marked by protein misfolding and accumulation. Similarly, the role of proteostasis in non-pathologic aging has been broadly investigated through longevity studies in a range of eukaryotic organisms [87]. Disruption of proteostasis at multiple different points may be expected to impair mitochondrial bioenergetic activity. Identifying how disruption of different steps affects (or does not affect) bioenergetic activity is important to clarify how a decline in proteostasis may contribute to a loss of coordinated bioenergetic activity between mitochondria and the rest of the cell.

Mitochondria depend on import of proteins encoded in the nucleus for biogenesis and proper function. During aging, the basal expression and/or activity of proteins involved in proper protein folding and import into the mitochondria decline in basal expression and activity, notably the outer membrane translocase TOM70. In addition to its role in mitochondrial protein import, TOM70 also appears to regulate mitochondrial transcription [88] and be involved in Ca^{2+} transfer between mitochondrial and the endoplasmic reticulum [89]. In this latter role, TOM70 is asserted to maintain bioenergetic function, as its depletion decreased both basal and maximal respiration in cells [88]. Conversely, when the translocase of the outer membrane complex (TOM20) was overexpressed in neuronal models of Parkinson's disease, mitochondrial membrane potential increased to near control levels, promoting membrane potential–dependent mitochondrial protein import [90, 91]. Experimental manipulation of other chaperone proteins including HSP70 and PDIA3 also affects mitochondrial membrane potential and basal respiration, supporting a broader but as yet unclear relationship between chaperone function and bioenergetic maintenance [92, 93].

A well-studied mitochondrial stress response is the unfolded protein response (UPR) [76, 94, 95]. Expression of the transcription factor CHOP is upregulated in response to protein aggregation and mtDNA damage, and in turn upregulates the expression of target chaperones HSP60, HSP90, and others [96]. A consequence of UPR in the mitochondria (mtUPR) is the preservation of import machinery expression, which was shown to correspond with increased protein import and longer life span in a model of *C. elegans* aging [97]. In mammalian models, neurodegeneration is strongly associated with increased mtUPR activity, where it is thought to be activated in response to protein aggregation [98–100]. In a cellular model of Alzheimer's disease, increased UPR activity due to Aβ and tau overexpression was associated with decreased ATP content (probably reflecting decreased adenine nucleotide pool size) and lower mitochondrial membrane potential. It remained unclear, however, if these decreases were a cause or effect of UPR activation, or the degree to which they were related to Aβ and tau overexpression rather than UPR activity [101]. These results are consistent with potential bioenergetic deficits, but causality remains to be determined by careful determination of bioenergetic relationships under controlled manipulation of individual UPR components.

Altogether, these and other studies identify both mitochondrial membrane potential and chaperone activity as important components of mitochondrial protein import, which promotes overall function and adaptation in response to cellular and metabolic signals. What is lacking, however, is a clear understanding of cause and effect among protein import, mitochondrial membrane potential, and aging-related phenotypes.

11.5 PROTEIN TURNOVER AND MITOPHAGY SLOW DURING AGING

Degradation of individual mitochondrial proteins is mediated by a suite of proteases in the matrix (m-) and intermembrane space (i-), called the m- and i-AAA proteases, for <u>A</u>TPases <u>A</u>ssociated with diverse cellular <u>A</u>ctivities [102–104]. Mitochondrial protein turnover is observed to be slower than cytosolic protein turnover in age-matched samples, irrespective of age. However, additional slowing of mitochondrial protein turnover, particularly of chaperone and respiratory chain components, is seen in aging models [105]. As in protein import, mitochondrial membrane potential is also suggested to play a role in protein turnover by the AAA proteases. In one example, membrane hyperpolarization was shown to directly inhibit m-AAA protease activity through interaction with the m-AAA protease inhibitor TMBIM5 [91]. Longer-term hyperpolarization degraded TMBIM5 and promoted m-AAA activity. Notably, AAA protease activity appears to be particularly important for mitochondrial remodeling during adaptation to altered substrate conditions [106].

Larger-scale mitochondrial turnover in the form of mitochondrial autophagy, or mitophagy, also takes place. Like individual protein turnover, attenuation of mitophagy is associated both with aging and with specific aging-related diseases, particularly neurodegeneration. Mitophagy, rather than targeting individual proteins, involved the engulfment and autophagic degradation of entire mitochondrial fragments. The mechanisms of mitophagy elucidated over the last 15–20 years identify the "quality control signal" of low mitochondrial membrane potential as trigger for mitophagy. Regions of the mitochondrial reticulum displaying lower membrane potential undergo fission and recruit the PTEN-induced kinase (PINK1) and Parkin to "tag" these particles for autophagic degradation [74–76].

In some cell types this is sufficient to activate mitochondrial engulfment and autophagy. In contrast, a human osteosarcoma cell model required the additional inhibition of mammalian target of rapamycin (mTOR) in order to recruit Parkin to depolarized mitochondria and initiate mitophagy [107].

Recent studies continue efforts to elucidate whether age-related bioenergetic decline triggers mitophagy. When respiratory chain proteins (Complex I-IV) were inhibited in cerebellar granule cells, mitochondrial membrane potential decreased, and PINK1 was observed to localize and aggregate to the mitochondrial outer membrane. However, there were differential results regarding MMP changes and activation of mitophagy depending on which complex was inhibited. It was concluded that loss of MMP itself was neither necessary nor sufficient for mitophagy [108], in agreement with prior work mentioned earlier [107]. Together, these and more recent findings suggest that multiple signals can initiate mitophagy other than decreased mitochondrial membrane potential and PINK1 recruitment [109–113].

Other studies have attempted to clarify the relationship between mitophagy and mitochondrial respiration using gene knockout and substrate manipulation. In one, inhibition of PINK1 and Parkin in renal podocytes decreased maximal respiration and increased glycolytic flux, suggesting compensatory ATP production from glycolysis as mitochondrial ATP production decreased. Conversely, hyperglycemia exposure inhibited PINK1 and Parkin expression while upregulating fission markers [114, 115]. Similar decreases in mitophagy and increases in fission in the presence of high glucose have been observed in other studies [115].

Mitophagy is the terminal stage in mitochondrial dynamics, an essential homeostatic process. Reduction in its activity is implicated as a factor of aging, with plausible explanations of waste buildup and proliferation of damaged mitochondria. However, uncertainty remains regarding exactly how much and under what conditions mitochondrial membrane potential contributes to initiating mitophagy. While longevity studies correlate positively with mitophagy upregulation, it remains unclear whether mitophagy downregulates results in aging.

11.6 FISSION AND FUSION DYNAMICS

The correct balance between mitochondrial fission and fusion, and the resulting mitochondrial morphology, depends on factors including cell type, substrate, energy demand, and signaling. Broadly, fusion predominates under ATP-depleted (and high-energy-demand) states, when O_2 and non-sugar substrates are present. Fusion promotes mitochondrial elongation, which is associated with aerobic growth in yeast [116] and higher respiratory capacity in other models [117]. In contrast, fission and subsequent fragmentation of the mitochondrial network occurs when respiratory demand is lower, when cells are dividing, or cellular damage has occurred [119]. Fission-driven fragmentation of networks has been observed in *C. elegans* and other models of aging and is asserted to be "pro-aging" [118, 119]. Fusion-driven tubular and highly interconnected networks are more prominent in young than in old *C. elegans* [120].

While fission is correlated to aging phenotypes, fusion has been associated with increased longevity [121]. Interestingly, though loss of fission might be expected to disrupt mitophagy by preventing generation of fragments to be "tagged" by PINK1/Parkin, it did not appear to do so in a *C. elegans* model [122]. Because of this finding, the authors proposed a role for fragmentation in protecting mitochondrial regions of higher membrane potential, rather than only sequestering regions of lower membrane potential. Induction of mitochondrial fusion (with the prediction that mitochondrial function will "improve") also yields mixed results. Aged rat leydig cells (LC) treated with M1, a small molecule fusion promoter [123], demonstrated increased maximal respiration and ATP production compared to young LC levels. In contrast, overexpression of OPA1 resulted in a decrease in respiration [124]. Notably, not all mitochondrial fusion is associated with increased expression of OPA1. In a cardiomyocyte model of doxorubicin-induced senescence, elongated mitochondrial networks were observed, along with decreased matrix density and decreased respiration rate. However, these changes were not accompanied by a change in OPA1 transcription [125].

Genetic manipulation of fission and fusion machinery has conclusively demonstrated the importance of these processes in mitochondrial and cellular function, and the association of deranged mitochondrial dynamics with various diseases and with aging [126–133]. The detectable consequences of this derangement on respiration appear to be more gradual than derangement of mtDNA, although differences in experimental approaches may contribute to this impression. In addition, contradictory effects in different models and conditions likely reflect both an optimum of fusion activity (balanced by fission as required for the conditions present) and that titration of this activity may be difficult to achieve experimentally; carefully controlling both experimental conditions and the degree of effect on mitochondrial dynamics would allow better interpretation the effects of aging-related changes to mitochondrial protein turnover bioenergetic activity on one another.

11.7 "UNCOUPLING TO SURVIVE" DOES NOT CONFLICT WITH MEETING EVERYDAY ENERGY DEMAND, BUT CAN DECREASE RESPIRATORY CAPACITY

Mitochondrial uncoupling refers to the separation of "oxidative" and "phosphorylation"—that is, mitochondrial uncoupling is the "leaking" of protons down the proton gradient without passing through the mitochondrial ATP synthase, so that mitochondrial protonmotive force (established by substrate oxidation and electron transport) is uncoupled from phosphorylation of ADP to make ATP. Mitochondrial uncoupling occurs constitutively [134, 135], can be stimulated by the activation of endogenous uncoupling proteins [136], or can be mediated by pharmacological uncouplers like dinitrophenol (DNP) and carbonyl cyanide-p-trifluoromethoxyphenylhydrazone (FCCP). The degree of uncoupling, resulting protonmotive force, and biological conditions all determine whether mitochondrial uncoupling promotes or disrupts its proper function.

No mitochondrial system studied to date is completely coupled, indicating inherent endogenous proton leak. In isolated mitochondrial and whole animals, proton leak is estimated to drive 25–50% of respiration rate; the rest is coupled to ATP production [134, 135]. The "uncoupling to survive" hypothesis formulated by Brand and colleagues [137, 138] states that a biological benefit of this endogenous proton leak is the maintenance of mitochondrial membrane potential within a range of voltages that minimized ROS production while maintaining energy sufficiency. Accordingly, individual mice with long life spans had higher degrees of uncoupling [138].

What mediates this endogenous proton leak? In Speakman et al. [138], the adenine nucleotide translocase and uncoupling protein 3 were identified as the main conduits for proton reentry. Together with uncoupling protein 2, these "non-canonical" uncoupling proteins are both strongly implicated in modulating life span [139–143], although not all reports agree [144]. The "canonical" uncoupling protein, UCP1 [145], is more narrowly expressed in brown fat and a few other tissues, and is indirectly linked to life span through its influence on the development of metabolic conditions [146]. Knockout studies of UCP3 support the assertion that mild uncoupling decreases ROS production; mice lacking skeletal muscle UCP3 expression showed greater mitochondrial coupling and higher rates of ROS production, though measured physiological markers (appearance, body weight, exercise tolerance, fatty acid oxidation, and cold-induced thermogenesis) remained unchanged [147]. Additional examples of life span extension in mammalian models of aging by manipulating mitochondrial proton leak have been recently reviewed [148].

Pharmacological uncoupling has a long history of use in food additives, unregulated supplements, and as a treatment for obesity [149], but historical agents like dinitrophenol (DNP) confer significant toxicity [150–152]. Despite these drawbacks, therapeutic uncoupling remains a compelling target in drug discovery, with focus on specific conditions including cancers, fatty liver disease, and obesity [150, 153–155]. Recent progress in this area includes a reformulation of DNP into a controlled-release pro-drug form (controlled-release mitochondrial protonophore, CRMP) that is metabolized by liver CYP450 enzymes into the active compound and has its subsequent effects primarily in the liver. When fed orally to mice subjected to a high-fat diet, CRMP did not alter life span but did improve numerous hepatic markers, including decreasing lipid accumulation, increasing insulin sensitivity, and decreasing inflammatory signaling [156]. A separate uncoupling compound, identified through high-throughput screening for uncoupled respiration and currently named OPC-163493, is similarly liver-targeting. Treatment of mice with this compound lowered HbA1C levels and improved glucose tolerance in treated mice [154]. These and other examples suggest that a key to using uncouplers effectively as therapeutic agents may be to target their action to specific tissues.

Both correlative and experimental studies find that small degrees of uncoupling are associated with life span extension and decreases in measurements of aging-related decline, with potentially different underlying mechanisms in different tissues [157]. But does "mild uncoupling" necessarily cause energy insufficiency? The short answer is no—as demonstrated in Speakman et al., it was not only that longer-lived animals displayed higher uncoupling; they also had higher metabolic rates overall, consistent with meeting energy demand despite a "leakier" system. Though it is straightforward to demonstrate *in vitro* that respiratory capacity can be attenuated with low concentrations of uncoupler, more measurements of respiratory capacity *in vivo* are needed to understand its significance to physiology and life span in whole organisms.

One notable exception to mild uncoupling improving functional decline comes from work in neuronal models, where mild uncoupling with FCCP led to predicted decreases in spare respiratory capacity (defined as maximum respiratory capacity minus basal respiration rate; this can be a useful concept but can also make the source of a change ambiguous, e.g., whether a decrease in spare capacity results from decreased maximal rate, increased basal rate, or both). However, mild uncoupling did not decrease intracellular superoxide levels or improve tolerance to oxidative stress in these models [158, 159], perhaps because mild uncoupling in neurons also triggers glutamate excitotoxicity [160]. A later study in *C. elegans* found that treating whole worms with DNP or CCCP (cyanide m-chlorophenyl hydrazone, closely related to FCCP) was neuroprotective [161]. The proposed mechanism of this neuroprotection was not decreased ROS production but increased mitophagy, as deletion of PINK1 abrogated the effect.

These findings highlight the point that the degree of uncoupling is critically important in interpreting the result of experimental uncoupling in a biological system. Uncoupling that maintains energy sufficiency might depolarize mitochondrial membrane potential much less than the degree that activates PINK1/Parkin-mediated mitophagy [162, 163], which may be similar to or less than the depolarization that leads to energetic collapse. While it is not always possible to report changes to mitochondrial membrane potential in mV units, papers that do so make it easier to understand what degree of uncoupling is present in a given experimental system, and therefore make it easier to interpret and compare the results of different experiments.

11.8 ROS AND AGING: IT'S COMPLICATED

Recent reviews on the mitochondrial generation of ROS [164, 165] and its association with aging [166–168] already exist, and our goal is not to replicate them here. Instead, we focus on how ROS production might attenuate ATP production and alter mitochondrial activity, and how this might be an underlying factor in aging. Across the literature, a broadly negative correlation exists between measured rates of mitochondrial ROS production and life span in many mammalian species [167]. Even a notable exception, the naked mole rat, supports the correlational conclusion that low rates of oxidative damage, either by low rates of ROS production, or high rates of ROS neutralization and/or clearance and repair of oxidative damage, promote longer life span [169].

The increasing recognition that ROS production may be a source of beneficial signaling as well as oxidative damage further complicates the prediction that ROS production should be negatively correlated with life span [167]. These potentially competing biological roles may help explain the current literature, from which it is difficult to assume simple relationships for ROS and antioxidants such as increased ROS leading to a shorter life span. Treatment of *C. elegans* with 2-deoxy-D-glucose to inhibit glycolysis was shown to increase mitochondrial respiration and extend life span [170], but as faster electron transport leads to overall reduction of the respiratory complexes and therefore decreases ROS production [164], faster respiration cannot explain the concurrent increase in ROS as measured by DCF-DA assay [170]. Importantly, however, these investigators also demonstrated that antioxidant supplementation with N-acetyl cysteine (NAC) prevented the increase in longevity. This supports a role for increased ROS levels (either through increased production or decreased neutralization) in promoting increased life span, as the life span extension effect abolished with supplemented antioxidants.

Overall, supplemental antioxidants have mixed effects on life span, and the effects of their use on mitochondrial respiration and ATP sufficiency are unclear [167, 171–173]. An earlier meta-analysis found an overall negative correlation between supplemental antioxidant level and life span [172]. Although it may initially seem counterintuitive for antioxidant levels to be negatively correlated with life span, this is speculated to result from "over-detoxification" of ROS, implying an optimum level that either too much or too little intracellular ROS fails to achieve [167]. In contrast, supplementation with glycine and N-acetylcysteine (GlyNAC) can reverse aging-related decreases in intracellular glutathione levels, and is associated with increased life span, decreased accumulation of protein carbonyls in cardiac tissue, and preservation of mitochondrial biogenesis [173]. While GlyNAC does have some direct antioxidant properties, it is a relatively poor scavenger, and its main biological activity is speculated to be stimulation of glutathione synthesis. The possibility that glutathione may be better regulated as a part of cellular redox homeostasis may contribute to the beneficial effects of GlyNAC in an experimental system, in contrast to other antioxidant molecules. However, no human studies to date have shown that antioxidant supplementation slows aging phenotypes or decreases mortality [171, 174, 175].

ROS-mediated oxidation of TCA cycle enzymes can inhibit their activity, primarily aconitase [176–178] and oxoglutarate dehydrogenase (α-ketoglutarate dehydrogenase) [179]. But does this meaningfully contribute to energy insufficiency *in vivo* during aging? Predictions based on kinetic modeling of neuronal cell bioenergetics suggested that about 60% inhibition of OGDHC/αKGDHC was needed to suppress ATP production, which occurred precipitously when this threshold was reached, while aconitase suppression had little predicted effect on ATP production activity [180]. *In silico* modeling of SDH inhibition (which is not ROS mediated, but can result from exposure to, e.g., pesticides) also suggests little impact on ATP production *in vivo* [181], although biological confirmation is needed. From these and other data it is not clear that energy insufficiency is a certain outcome of ROS production, or that the mechanisms of respiratory attenuation described above are deleterious to the cell. Some evidence supports a role for SDH inhibition, ROS-induced or experimental, in improving, for example, ischemia-reperfusion injury [182], in which rapid succinate oxidation during reperfusion may drive rapid ROS production through mitochondrial membrane hyperpolarization [183].

11.9 MAKING ATP TO MEET ORDINARY AND EXTRAORDINARY DEMAND

Decreasing ATP-generating capacity in aging skeletal muscle is perhaps one of the best examples of bioenergetic decline with age [105, 119, 184]. Even here, though, it is not clear what kind of dysfunction this decreased capacity might represent, or if the broad interpretation that muscles with lower respiratory capacity are dysfunctional can be applied wholesale to other cell types and organisms. Not all muscle models of aging show respiratory decline with age [185]. Much of the muscle tissue decline in respiratory capacity is due to decreased mitochondrial content, not necessarily decreased mitochondrial oxidative capacity, though indirect estimation does support decreased oxidative capacity in some cases [186]. These factors make it important to consider the scale of the question (mitochondrial, cellular, or tissue-level respiration) and the control of ATP production under given conditions when considering whether a measured decrease in ATP production represents dysfunction (i.e., energy insufficiency), adaptation (e.g., fewer mitochondria), or simply lower ATP demand (e.g., decreased organism activity).

Studies have also compared the relationship between mitochondrial respiration and physical fitness. In one example, mitochondrial respiration rates of permeabilized skeletal muscle fibers from volunteer subject biopsies were compared to their performance on a treadmill and other tests of cardiorespiratory fitness or muscle performance. This study found a decline in respiratory capacity with age in both men and women but little to no difference in basal respiration, no relationship to gait speed, and a statistical decline in muscle strength of the upper and lower extremities that weakened after adjusting for age and BMI [187].

ATP production from oxidative phosphorylation has been shown to decrease with age across multiple studies in skeletal muscle and cardiac muscle [184, 188, 189]. However, not all the studies agree whether there is a decrease in the amount of mitochondria and a functional impairment or just a functional impairment alone that occurs with age [187].

On the level of system supply and demand, mechanisms that could decrease energy sufficiency are decreased substrate uptake and oxidation, decreased mitochondrial membrane potential, and decreased coupling. We must, therefore, look for examples of these in models of aging to support a system-level insufficiency in ATP production with aging. Altered substrate uptake and oxidation, as well as increased mitochondrial uncoupling, are associated with aging, especially in skeletal muscle models [190–192]. Lower resting mitochondrial membrane potentials have been associated with aging [193–195]. In one example, lower mitochondrial membrane potential was more predictive of aging phenotypes than chronology, in hematopoietic stem cells [196], although using flow cytometry to measure membrane potential, as was done in this report, can yield artifacts that must be carefully controlled for [63].

These changes might be expected to decrease ATP-generating capacity and limit the ability of the mitochondrial machinery to meet extraordinary demand (i.e., maximum demand), but not the ability to meet ordinary (i.e., basal) energy demand. Accumulation of damage to the respiratory chain or ATP synthase might also plausibly do this. Some evidence for accumulation of such damage is the aging-associated carbonylation of multiple subunits of the mitochondrial ATP synthase, associated with impaired synthase dimerization, decreased coupled respiration rate, and increased likelihood of energetic collapse, in mitochondria isolated from cardiac tissue in an aging mouse model [188]. This finding is consistent with some [197, 198] but not all [199] earlier work examining carbonylation of mitochondrial proteins.

But is lower mitochondrial membrane potential good (i.e., adaptive, or promoting proper function), bad (i.e., diminishing function), or inconsequential? Is it a cause or effect of broader aging-related changes, from tissue alteration to behavior? Perhaps most importantly, would reversing it promote an antiaging effect or ameliorate risk or symptoms of aging-related disease? In yeast, depolarization of the mitochondrial membrane potential has been shown to extend chronological life span [200]. In *C. elegans*, increasing mitochondrial membrane potential extends life span [195]. The apparent contradictions presented by these and other examples may not be surprising, as the relationship of mitochondrial membrane potential to both ROS production (occurs at high membrane potential with reduced redox centers in respiratory complexes) and ATP production (redox centers are more oxidized due to faster electron current) is highly condition dependent.

11.10 CONCLUSIONS AND FUTURE DIRECTIONS

Different models present conflicting evidence for the role of substantial bioenergetic alteration in aging. From multiple lines of evidence, most data associated with mitochondrial dysfunction from *in vivo* systems is of energetic collapse in individual cells rather than the more subtle decline. How would energy insufficiency be properly measured to achieve this? How would an observed decrease in respiration be supported as energy insufficiency? One criterion is that it should be accompanied by an acute decrease in mitochondrial membrane potential, indicating a "pull" of ATP demand. Another is that it should yield phenotypes that are abrogated by alternative supply of ATP. A recent example of this approach used a xanthine oxidase inhibitor in a *C. elegans* model to show that the increased intracellular ATP levels resulting from this treatment corresponded with improved muscle cell integrity during chronological aging. When combined with ascorbic acid treatment, ATP levels increased further and corresponded with a life span increase [201]. Crucially, the authors measured ATP levels following rapid sample freezing to avoid equilibration of the nucleotide pool. Though it sounds simple, energy insufficiency in biological systems may not always be easy to demonstrate, which is part of why we have relatively little evidence of it to support and understand the role of mitochondrial alteration in aging.

A final suggestion for building a better evidence base for the role of bioenergetic alteration in aging is to do more bioenergetic analysis in different models of aging. As one example, comparing rates of respiration before and after (or with and without) an experimental manipulation yields one pair of values. This is insufficient to build a meaningful relationship that allows causality and mechanism to be determined, because respiration rate is not a fixed property of a cell or tissue. Like an individual enzyme whose activity follows Michaelis-Menten kinetics, a rate measured under given conditions is a function of one or more other factors. Varying the conditions (e.g., substrate concentration) may change the rate, revealing a curve that describes the activity. Since mitochondrial respiration is a function of multiple enzymes operating together in a system, its rate is a function not just of the activities of individual enzymes but also of the kinetic control of the system [202]. For more robust bioenergetic conclusions, determining, for example, a force-flux curve can distinguish whether an intervention changes mitochondrial status without changing the underlying relationships of the energy supply and demand system (i.e., changes the position on a given force-flux curve) or changes the underlying relationship. The latter result would allow identification of mechanisms of aging that meaningfully alter the bioenergetic function. More data like these would allow much greater clarity on how bioenergetic activities affect, and are affected by, aging.

REFERENCES

1. Harman D. The free radical theory of aging. *Antioxidants and Redox Signaling*. 2003. doi:10.1089/152308603770310202
2. Harman D. Free radical theory of aging. *Mutation Research DNAging*. 1992;275. doi:10.1016/0921-8734(92)90030-S
3. Ziada AS, Smith MSR, Côté HCF. Updating the Free Radical Theory of Aging. *Front Cell Dev Biol*. 2020;8. doi:10.3389/fcell.2020.575645
4. Pomatto LCD, Davies KJA. Adaptive homeostasis and the free radical theory of ageing. *Free Radical Biology and Medicine*. 2018. doi:10.1016/j.freeradbiomed.2018.06.016
5. Gladyshev VN. The free radical theory of aging is dead. Long live the damage theory! *Antioxid Redox Signal*. 2014;20. doi:10.1089/ars.2013.5228
6. Hayflick L. The limited in vitro lifetime of human diploid cell strains. *Exp Cell Res*. 1965;37. doi:10.1016/0014-4827(65)90211-9
7. Harman D. The Biologic Clock: The Mitochondria? *J Am Geriatr Soc*. 1972;20. doi:10.1111/j.1532-5415.1972.tb00787.x
8. Harman D. Free radical theory of aging: Consequences of mitochondrial aging. *Age (Omaha)*. 1983;6. doi:10.1007/BF02432509
9. Mitchell P. Coupling of phosphorylation to electron and hydrogen transfer by a chemi-osmotic type of mechanism. *Nature*. 1961;191. doi:10.1038/191144a0
10. Jensen PK. Antimycin-insensitive oxidation of succinate and reduced nicotinamide-adenine dinucleotide in electron-transport particles II. Steroid effects. *BBA - Enzymology and Biological Oxidation*. 1966;122. doi:10.1016/0926-6593(66)90058-0
11. Jensen PK. Antimycin-insensitive oxidation of succinate and reduced nicotinamide-adenine dinucleotide in electron-transport particles I. pH dependency and hydrogen peroxide formation. *BBA - Enzymology and Biological Oxidation*. 1966;122. doi:10.1016/0926-6593(66)90057-9

12. Boveris A, Chance B. The mitochondrial generation of hydrogen peroxide. General properties and effect of hyperbaric oxygen. *Biochemical Journal*. 1973;134. doi:10.1042/bj1340707
13. Son JM, Lee C. Aging: All roads lead to mitochondria. *Semin Cell Dev Biol*. 2021;116. doi:10.1016/j.semcdb.2021.02.006
14. Stuart JA, Maddalena LA, Merilovich M, Robb EL. A midlife crisis for the mitochondrial free radical theory of aging. *Longev Healthspan*. 2014;3. doi:10.1186/2046-2395-3-4
15. Barja G. The mitochondrial free radical theory of aging. *Progress in Molecular Biology and Translational Science*. 2014. doi:10.1016/B978-0-12-394625-6.00001-5
16. Ziegler D. V., Wiley CD, Velarde MC. Mitochondrial effectors of cellular senescence: Beyond the free radical theory of aging. *Aging Cell*. 2015. doi:10.1111/acel.12287
17. Miwa S, Kashyap S, Chini E, von Zglinicki T. Mitochondrial dysfunction in cell senescence and aging. *Journal of Clinical Investigation*. 2022. doi:10.1172/JCI158447
18. Maklashina E, Ackrell BAC. Is defective electron transport at the hub of aging? *Aging Cell*. 2004. doi:10.1111/j.1474-9728.2003.00078.x
19. Wallace DC. Bioenergetic origins of complexity and disease. *Cold Spring Harb Symp Quant Biol*. 2011;76. doi:10.1101/sqb.2011.76.010462
20. Fleming JE, Miquel J, Cottrell SF, Yengoyan LS, Economos AC. Is cell aging caused by respiration-dependent injury to the mitochondrial genome? *Gerontology*. 1982. doi:10.1159/000212510
21. Stein A, Sia EA. Mitochondrial DNA repair and damage tolerance. *Frontiers in Bioscience - Landmark*. 2017. doi:10.2741/4525
22. Roy A, Kandettu A, Ray S, Chakrabarty S. Mitochondrial DNA replication and repair defects: Clinical phenotypes and therapeutic interventions. *Biochimica et Biophysica Acta - Bioenergetics*. 2022. doi:10.1016/j.bbabio.2022.148554
23. Allkanjari K, Baldock RA. Beyond base excision repair: An evolving picture of mitochondrial DNA repair. *Bioscience Reports*. 2021. doi:10.1042/BSR20211320
24. Ling F, Bradshaw E, Yoshida M. Prevention of mitochondrial genomic instability in yeast by the mitochondrial recombinase Mhr1. *Sci Rep*. 2019;9. doi:10.1038/s41598-019-41699-9
25. Gredilla R, Sánchez-Román I, Gómez A, López-Torres M, Barja G. Mitochondrial base excision repair positively correlates with longevity in the liver and heart of mammals. *Geroscience*. 2020;42. doi:10.1007/s11357-020-00158-4
26. Cortopassi GA, Shibata D, Soong NW, Arnheim N. A pattern of accumulation of a somatic deletion of mitochondrial DNA in aging human tissues. *Proc Natl Acad Sci U S A*. 1992;89. doi:10.1073/pnas.89.16.7370
27. Chinnery PF, Turnbull DM. Mitochondrial DNA mutations in the pathogenesis of human disease. *Molecular Medicine Today*. 2000. doi:10.1016/S1357-4310(00)01805-0
28. Cooper JM, Mann VM, Schapira AHV. Analyses of mitochondrial respiratory chain function and mitochondrial DNA deletion in human skeletal muscle: Effect of ageing. *J Neurol Sci*. 1992;113. doi:10.1016/0022-510X(92)90270-U
29. Berneburg M, Grether-Beck S, Kürten V, Ruzicka T, Briviba K, Sies H, et al. Singlet oxygen mediates the UVA-induced generation of the photoaging-associated mitochondrial common deletion. *Journal of Biological Chemistry*. 1999;274. doi:10.1074/jbc.274.22.15345
30. Diba LZ, Ardebili SMM, Gharesouran J, Houshmand M. Age-related decrease in mtDNA content as a consequence of mtDNA 4977 bp deletion. *Mitochondrial DNA*. 2016;27. doi:10.3109/19401736.2015.1063046
31. Quan C, Cho MK, Perry D, Quan T. Age-associated reduction of cell spreading induces mitochondrial DNA common deletion by oxidative stress in human skin dermal fibroblasts: Implication for human skin connective tissue aging. *J Biomed Sci*. 2015;22. doi:10.1186/s12929-015-0167-6
32. Özaydin A, Kerimova A, Onaran I. The Effects of Mitochondrial Mt4977 Deletion on Platelet Function in Ischemic Heart Disease Patients. *Phoenix Medical Journal*. 2021;3. doi:10.38175/phnx.869878
33. Filograna R, Mennuni M, Alsina D, Larsson NG. Mitochondrial DNA copy number in human disease: the more the better? *FEBS Letters*. 2021. doi:10.1002/1873-3468.14021
34. Sciacco M, Bonilla E, Schon EA, Dimauro S, Moraes CT Distribution of wild-type and common deletion forms of mtDNA in normal and respiration-deficient muscle fibers from patients with mitochondrial myopathy. *Hum Mol Genet*. 1994;3. doi:10.1093/hmg/3.1.13
35. Mecocci P, MacGarvey U, Kaufman AE, Koontz D, Shoffner JM, Wallace DC, et al. Oxidative damage to mitochondrial DNA shows marked age-dependent increases in human brain. *Ann Neurol*. 1993;34. doi:10.1002/ana.410340416

36. Zhao L, Sumberaz P. Mitochondrial DNA Damage: Prevalence, Biological Consequence, and Emerging Pathways. *Chemical Research in Toxicology*. 2020. doi:10.1021/acs.chemrestox.0c00083
37. Kennedy SR, Salk JJ, Schmitt MW, Loeb LA. Ultra-Sensitive Sequencing Reveals an Age-Related Increase in Somatic Mitochondrial Mutations That Are Inconsistent with Oxidative Damage. *PLoS Genet*. 2013;9. doi:10.1371/journal.pgen.1003794
38. Kauppila JHK, Bonekamp NA, Mourier A, Isokallio MA, Just A, Kauppila TES, et al. Base-excision repair deficiency alone or combined with increased oxidative stress does not increase mtDNA point mutations in mice. *Nucleic Acids Res*. 2018;46. doi:10.1093/nar/gky456
39. Kujoth CC, Hiona A, Pugh TD, Someya S, Panzer K, Wohlgemuth SE, et al. Medicine: Mitochondrial DNA mutations, oxidative stress, and apoptosis in mammalian aging. *Science* (1979). 2005;309. doi:10.1126/science.1112125
40. Szczepanowska K, Trifunovic A. Different faces of mitochondrial DNA mutators. *Biochimica et Biophysica Acta - Bioenergetics*. 2015. doi:10.1016/j.bbabio.2015.05.016
41. Ross JM, Stewart JB, Hagström E, Brené S, Mourier A, Coppotelli G, et al. Germline mitochondrial DNA mutations aggravate ageing and can impair brain development. *Nature*. 2013;501. doi:10.1038/nature12474
42. Vermulst M, Bielas JH, Kujoth GC, Ladiges WC, Rabinovitch PS, Prolla TA, et al. Mitochondrial point mutations do not limit the natural lifespan of mice. *Nat Genet*. 2007;39. doi:10.1038/ng1988
43. Sohl CD, Szymanski MR, Mislak AC, Shumate CK, Amiralaei S, Schinazi RF, et al. Probing the structural and molecular basis of nucleotide selectivity by human mitochondrial DNA polymerase γ. *Proc Natl Acad Sci U S A*. 2015;112. doi:10.1073/pnas.1421733112
44. Lim SE, Copeland WC. Differential Incorporation and Removal of Antiviral Deoxynucleotides by Human DNA Polymerase γ. *Journal of Biological Chemistry*. 2001;276. doi:10.1074/jbc.M101114200
45. Desquilbet L, Jacobson LP, Fried LP, Phair JP, Jamieson BD, Holloway M, et al. HIV-1 infection is associated with an earlier occurrence of a phenotype related to frailty. *Journals of Gerontology - Series A Biological Sciences and Medical Sciences*. 2007;62. doi:10.1093/gerona/62.11.1279
46. Guaraldi G, Zona S, Alexopoulos N, Orlando G, Carli F, Ligabue G, et al. Coronary aging in HIV-infected patients. *Clinical Infectious Diseases*. 2009;49. doi:10.1086/648080
47. Valcour VG, Shikuma CM, Watters MR, Sacktor NC. Cognitive impairment in older HIV-1-seropositive individuals: Prevalence and potential mechanisms. *AIDS*. 2004.
48. Payne BAI, Wilson IJ, Hateley CA, Horvath R, Santibanez-Koref M, Samuels DC, et al. Mitochondrial aging is accelerated by anti-retroviral therapy through the clonal expansion of mtDNA mutations. *Nat Genet*. 2011;43. doi:10.1038/ng.863
49. Taylor RW, Barron MJ, Borthwick GM, Gospel A, Chinnery PF, Samuels DC, et al. Mitochondrial DNA mutations in human colonic crypt stem cells. *Journal of Clinical Investigation*. 2003;112. doi:10.1172/JCI19435
50. Greaves LC, Preston SL, Tadrous PJ, Taylor RW, Barron MJ, Oukrif D, et al. Mitochondrial DNA mutations are established in human colonic stem cells, and mutated clones expand by crypt fission. *Proc Natl Acad Sci U S A*. 2006;103. doi:10.1073/pnas.0505903103
51. Baines HL, Stewart JB, Stamp C, Zupanic A, Kirkwood TBL, Larsson NG, et al. Similar patterns of clonally expanded somatic mtDNA mutations in the colon of heterozygous mtDNA mutator mice and ageing humans. *Mech Ageing Dev*. 2014;139. doi:10.1016/j.mad.2014.06.003
52. Fox RG, Magness S, Kujoth GC, Prolla TA, Maeda N. Mitochondrial DNA polymerase editing mutation, PolgD257A, disturbs stem-progenitor cell cycling in the small intestine and restricts excess fat absorption. *Am J Physiol Gastrointest Liver Physiol*. 2012;302. doi:10.1152/ajpgi.00402.2011
53. Drozdowski L, Thomson ABR. Aging and the intestine. *World Journal of Gastroenterology*. 2006. doi:10.3748/wjg.v12.i47.7578
54. Thomson ABR. Small intestinal disorders in the elderly. *Best Pract Res Clin Gastroenterol*. 2009;23. doi:10.1016/j.bpg.2009.10.009
55. Stefano GB, Kream RM. Mitochondrial DNA heteroplasmy in human health and disease (Review). *Biomedical Reports*. 2016. doi:10.3892/br.2016.590
56. Nissanka N, Moraes CT. Mitochondrial DNA heteroplasmy in disease and targeted nuclease-based therapeutic approaches. *EMBO Rep*. 2020;21. doi:10.15252/embr.201949612
57. Trinh J, Hicks AA, König IR, Delcambre S, Lüth T, Schaake S, et al. Mitochondrial DNA heteroplasmy distinguishes disease manifestation in PINK1 / PRKN- linked Parkinson's disease. *Brain*. 2022. doi:10.1093/brain/awac464

58. Stewart JB, Chinnery PF. The dynamics of mitochondrial DNA heteroplasmy: Implications for human health and disease. *Nature Reviews Genetics*. 2015. doi:10.1038/nrg3966
59. Kraytsberg Y, Kudryavtseva E, McKee AC, Geula C, Kowall NW, Khrapko K. Mitochondrial DNA deletions are abundant and cause functional impairment in aged human substantia nigra neurons. *Nat Genet*. 2006;38. doi:10.1038/ng1778
60. Foote K, Reinhold J, Yu EPK, Figg NL, Finigan A, Murphy MP, et al. Restoring mitochondrial DNA copy number preserves mitochondrial function and delays vascular aging in mice. *Aging Cell*. 2018;17. doi:10.1111/acel.12773
61. Mookerjee SA, Divakaruni AS, Jastroch M, Brand MD. Mitochondrial uncoupling and lifespan. *Mech Ageing Dev*. 2010;131. doi:10.1016/j.mad.2010.03.010
62. Mookerjee SA, Gerencser AA, Nicholls DG, Brand MD. Quantifying intracellular rates of glycolytic and oxidative ATP production and consumption using extracellular flux measurements. *Journal of Biological Chemistry*. 2017;292: 7189–7207. doi:10.1074/jbc.M116.774471
63. Lerner CA, Gerencser AA. Unbiased Millivolts Assay of Mitochondrial Membrane Potential in Intact Cells. *Methods in Molecular Biology*. 2022. doi:10.1007/978-1-0716-2309-1_2
64. Berg J, Hung YP, Yellen G. A genetically encoded fluorescent reporter of ATP/ADP ratio. doi:10.1038/nmeth.1288
65. Gerencser AA, Mookerjee SA, Jastroch M, Brand MD. Measurement of the absolute magnitude and time courses of mitochondrial membrane potential in primary and clonal pancreatic beta-cells. *PLoS One*. 2016;11. doi:10.1371/journal.pone.0159199
66. Shock LS, Thakkar P V., Peterson EJ, Moran RG, Taylor SM. DNA methyltransferase 1, cytosine methylation, and cytosine hydroxymethylation in mammalian mitochondria. *Proc Natl Acad Sci U S A*. 2011;108. doi:10.1073/pnas.1012311108
67. Nass MMK. Differential methylation of mitochondrial and nuclear DNA in cultured mouse, hamster and virus-transformed hamster cells In vivo and in vitro methylation. *J Mol Biol*. 1973;80. doi:10.1016/0022-2836(73)90239-8
68. Bacalini MG, D'Aquila P, Marasco E, Nardini C, Montesanto A, Franceschi C, et al. The methylation of nuclear and mitochondrial DNA in ageing phenotypes and longevity. *Mechanisms of Ageing and Development*. 2017. doi:10.1016/j.mad.2017.01.006
69. Chocron ES, Munkácsy E, Pickering AM. Cause or casualty: The role of mitochondrial DNA in aging and age-associated disease. *Biochimica et Biophysica Acta - Molecular Basis of Disease*. 2019. doi:10.1016/j.bbadis.2018.09.035
70. van der Wijst MGP, Rots MG. Mitochondrial epigenetics: An overlooked layer of regulation? *Trends in Genetics*. 2015. doi:10.1016/j.tig.2015.03.009
71. Iacobazzi V, Castegna A, Infantino V, Andria G. Mitochondrial DNA methylation as a next-generation biomarker and diagnostic tool. *Molecular Genetics and Metabolism*. 2013. doi:10.1016/j.ymgme.2013.07.012
72. Mawlood SK, Dennany L, Watson N, Dempster J, Pickard BS. Quantification of global mitochondrial DNA methylation levels and inverse correlation with age at two CpG sites. *Aging*. 2016;8. doi:10.18632/aging.100892
73. Chatterjee D, Das P, Chakrabarti O. Mitochondrial Epigenetics Regulating Inflammation in Cancer and Aging. *Frontiers in Cell and Developmental Biology*. 2022. doi:10.3389/fcell.2022.929708
74. D'Aquila P, Bellizzi D, Passarino G. Mitochondria in health, aging and diseases: the epigenetic perspective. *Biogerontology*. 2015. doi:10.1007/s10522-015-9562-3
75. Blanch M, Mosquera JL, Ansoleaga B, Ferrer I, Barrachina M. Altered mitochondrial DNA methylation pattern in Alzheimer disease-related pathology and in Parkinson disease. *American Journal of Pathology*. 2016;186. doi:10.1016/j.ajpath.2015.10.004
76. Lima T, Li TY, Mottis A, Auwerx J. Pleiotropic effects of mitochondria in aging. *Nat Aging*. 2022;2. doi:10.1038/s43587-022-00191-2
77. Barrientos A, Casademont J, Cardellach F, Estivill X, Urbano-Marquez A, Nunes V. Reduced steady-state levels of mitochondrial RNA and increased mitochondrial DNA amount in human brain with aging. *Molecular Brain Research*. 1997;52. doi:10.1016/S0169-328X(97)00278-7
78. Schwarze SR, Weindruch R, Aiken JM. Decreased mitochondrial RNA levels without accumulation of mitochondrial DNA deletions in aging Drosophila melanogaster. *Mutation Research - Mutation Research Genomics*. 1998;382. doi:10.1016/S1383-5726(97)00013-7

79. Calleja M, Pena P, Ugalde C, Ferreiro C, Marco R, Garesse R. Mitochondrial DNA remains intact during Drosophila aging, but the levels of mitochondrial transcripts are significantly reduced. *Journal of Biological Chemistry*. 1993;268. doi:10.1016/s0021-9258(17)46710-x
80. Kuthethur R, Shukla V, Mallya S, Adiga D, Kabekkodu SP, Ramachandra L, et al. Expression analysis and function of mitochondrial genome-encoded microRNAs. *J Cell Sci*. 2022;135. doi:10.1242/jcs.258937
81. Latronico MVG, Condorelli G. The might of microRNA in mitochondria. *Circulation Research*. 2012. doi:10.1161/CIRCRESAHA.112.271312
82. Lauri A, Pompilio G, Capogrossi MC. The mitochondrial genome in aging and senescence. *Ageing Research Reviews*. 2014. doi:10.1016/j.arr.2014.07.001
83. Faraonio R, Salerno P, Passaro F, Sedia C, Iaccio A, Bellelli R, et al. A set of miRNAs participates in the cellular senescence program in human diploid fibroblasts. *Cell Death Differ*. 2012;19. doi:10.1038/cdd.2011.143
84. Magenta A, Greco S, Gaetano C, Martelli F. Oxidative stress and microRNAs in vascular diseases. *International Journal of Molecular Sciences*. 2013. doi:10.3390/ijms140917319
85. Chan SY, Zhang YY, Hemann C, Mahoney CE, Zweier JL, Loscalzo J. MicroRNA-210 Controls Mitochondrial Metabolism during Hypoxia by Repressing the Iron-Sulfur Cluster Assembly Proteins ISCU1/2. *Cell Metab*. 2009;10. doi:10.1016/j.cmet.2009.08.015
86. Li N, Bates DJ, An J, Terry DA, Wang E. Up-regulation of key microRNAs, and inverse down-regulation of their predicted oxidative phosphorylation target genes, during aging in mouse brain. *Neurobiol Aging*. 2011;32. doi:10.1016/j.neurobiolaging.2009.04.020
87. Powers ET, Morimoto RI, Dillin A, Kelly JW, Balch WE. Biological and chemical approaches to diseases of proteostasis deficiency. *Annual Review of Biochemistry*. 2009. doi:10.1146/annurev.biochem.052308.114844
88. Liu Q, Chang CE, Wooldredge AC, Fong B, Kennedy BK, Zhou C. Tom70-based transcriptional regulation of mitochondrial biogenesis and aging. *Elife*. 2022;11. doi:10.7554/eLife.75658
89. Filadi R, Leal NS, Schreiner B, Rossi A, Dentoni G, Pinho CM, et al. TOM70 Sustains Cell Bioenergetics by Promoting IP3R3-Mediated ER to Mitochondria Ca2+ Transfer. *Current Biology*. 2018;28. doi:10.1016/j.cub.2017.12.047
90. Franco-Iborra S, Cuadros T, Parent A, Romero-Gimenez J, Vila M, Perier C. Defective mitochondrial protein import contributes to complex I-induced mitochondrial dysfunction and neurodegeneration in Parkinson's disease. *Cell Death Dis*. 2018;9. doi:10.1038/s41419-018-1154-0
91. Patron M, Tarasenko D, Nolte H, Kroczek L, Ghosh M, Ohba Y, et al. Regulation of mitochondrial proteostasis by the proton gradient. *EMBO J*. 2022;41. doi:10.15252/embj.2021110476
92. Sato TK, Kawano S, Endo T. Role of the membrane potential in mitochondrial protein unfolding and import. *Sci Rep*. 2019;9. doi:10.1038/s41598-019-44152-z
93. Keasey MP, Razskazovskiy V, Jia C, Peterknecht ED, Bradshaw PC, Hagg T. PDIA3 inhibits mitochondrial respiratory function in brain endothelial cells and C. elegans through STAT3 signaling and decreases survival after OGD. *Cell Communication and Signaling*. 2021;19. doi:10.1186/s12964-021-00794-z
94. Guo Y, Guan T, Shafiq K, Yu Q, Jiao X, Na D, et al. Mitochondrial dysfunction in aging. *Ageing Research Reviews*. 2023. doi:10.1016/j.arr.2023.101955
95. Muñoz-Carvajal F, Sanhueza M. The Mitochondrial Unfolded Protein Response: A Hinge Between Healthy and Pathological Aging. *Frontiers in Aging Neuroscience*. 2020. doi:10.3389/fnagi.2020.581849
96. Suárez-Rivero JM, Pastor-Maldonado CJ, Povea-Cabello S, Álvarez-Córdoba M, Villalón-García I, Talaverón-Rey M, et al. Activation of the Mitochondrial Unfolded Protein Response: A New Therapeutic Target? *Biomedicines*. 2022. doi:10.3390/biomedicines10071611
97. Xin N, Durieux J, Yang C, Wolff S, Kim HE, Dillin A. The UPRmt preserves mitochondrial import to extend lifespan. *Journal of Cell Biology*. 2022;221. doi:10.1083/jcb.202201071
98. Wang CL, Ohkubo R, Mu WC, Chen W, Fan JL, Song Z, et al. The mitochondrial unfolded protein response regulates hippocampal neural stem cell aging. *Cell Metab*. 2023;35. doi:10.1016/j.cmet.2023.04.012
99. Wodrich APK, Scott AW, Shukla AK, Harris BT, Giniger E. The Unfolded Protein Responses in Health, Aging, and Neurodegeneration: Recent Advances and Future Considerations. *Frontiers in Molecular Neuroscience*. 2022. doi:10.3389/fnmol.2022.831116
100. Johns A, Higuchi-Sanabria R, Thorwald MA, Vilchez D. A tale of two pathways: Regulation of proteostasis by UPRmt and MDPs. *Current Opinion in Neurobiology*. 2023. doi:10.1016/j.conb.2022.102673

101. Poirier Y, Grimm A, Schmitt K, Eckert A. Link between the unfolded protein response and dysregulation of mitochondrial bioenergetics in Alzheimer's disease. *Cellular and Molecular Life Sciences*. 2019;76. doi:10.1007/s00018-019-03009-4
102. Langer T. AAA proteases: Cellular machines for degrading membrane proteins. *Trends in Biochemical Sciences*. 2000. doi:10.1016/S0968-0004(99)01541-8
103. Hamon MP, Bulteau AL, Friguet B. Mitochondrial proteases and protein quality control in ageing and longevity. *Ageing Research Reviews*. 2015. doi:10.1016/j.arr.2014.12.010
104. Deshwal S, Fiedler KU, Langer T, Langer T, Langer T. Mitochondrial Proteases: Multifaceted Regulators of Mitochondrial Plasticity. *Annual Review of Biochemistry*. 2020. doi:10.1146/annurev-biochem-062917-012739
105. Kruse SE, Karunadharma PP, Basisty N, Johnson R, Beyer RP, Maccoss MJ, et al. Age modifies respiratory complex I and protein homeostasis in a muscle type-specific manner. *Aging Cell*. 2016;15. doi:10.1111/acel.12412
106. Ahola S, Langer T, Macvicar T. Mitochondrial proteolysis and metabolic control. *Cold Spring Harb Perspect Biol*. 2019;11. doi:10.1101/cshperspect.a033936
107. Gilkerson RW, De Vries, Rla, Lebot P, Wikstrom JD, Torgyekes E, Shirihai OS, et al. Mitochondrial autophagy in cells with mtDNA mutations results from synergistic loss of transmembrane potential and mTORC1 inhibition. *Hum Mol Genet*. 2012;21. doi:10.1093/hmg/ddr529
108. Shin YS, Ryall JG, Britto JM, Lau CL, Devenish RJ, Nagley P, et al. Inhibition of bioenergetics provides novel insights into recruitment of PINK1-dependent neuronal mitophagy. *J Neurochem*. 2019;149. doi:10.1111/jnc.14667
109. Kubli DA, Cortez MQ, Moyzis AG, Najor RH, Lee Y, Gustafsson ÅB. PINK1 is dispensable for mitochondrial recruitment of parkin and activation of mitophagy in cardiac myocytes. *PLoS One*. 2015;10. doi:10.1371/journal.pone.0130707
110. Allen GFG, Toth R, James J, Ganley IG. Loss of iron triggers PINK1/Parkin-independent mitophagy. *EMBO Rep*. 2013;14. doi:10.1038/embor.2013.168
111. Oshima Y, Cartier E, Boyman L, Verhoeven N, Polster BM, Huang W, et al. Parkin-independent mitophagy via drp1-mediated outer membrane severing and inner membrane ubiquitination. *Journal of Cell Biology*. 2021;220. doi:10.1083/jcb.202006043
112. Miyazaki N, Shiratori R, Oshima T, Zhang Z, Valencia R, Kranrod J, et al. PINK1-dependent and Parkin-independent mitophagy is involved in reprogramming of glycometabolism in pancreatic cancer cells. *Biochem Biophys Res Commun*. 2022;625. doi:10.1016/j.bbrc.2022.08.004
113. McWilliams TG, Prescott AR, Montava-Garriga L, Ball G, Singh F, Barini E, et al. Basal Mitophagy Occurs Independently of PINK1 in Mouse Tissues of High Metabolic Demand. *Cell Metab*. 2018;27. doi:10.1016/j.cmet.2017.12.008
114. Audzeyenka I, Rogacka D, Piwkowska A. P0982PTEN-INDUCED KINASE 1 (PINK1) DEPLETION ALTERS MITOCHONDRIAL EFFICIENCY AND GLYCOLYSIS IN HUMAN PODOCYTES. *Nephrology Dialysis Transplantation*. 2020;35. doi:10.1093/ndt/gfaa142.p0982
115. Kobayashi S, Zhao F, Zhang Z, Kobayashi T, Huang Y, Shi B, et al. Mitochondrial Fission and Mitophagy Coordinately Restrict High Glucose Toxicity in Cardiomyocytes. *Front Physiol*. 2020;11. doi:10.3389/fphys.2020.604069
116. Visser W, van Spronsen EA, Nanninga N, Pronk JT, Kuenen JG, van Dijken JP. Effects of growth conditions on mitochondrial morphology in Saccharomyces cerevisiae. *Antonie Van Leeuwenhoek*. 1995;67. doi:10.1007/BF00873688
117. Yao CH, Wang R, Wang Y, Kung CP, Weber JD, Patti GJ. Mitochondrial fusion supports increased oxidative phosphorylation during cell proliferation. *Elife*. 2019;8. doi:10.7554/eLife.41351
118. Jiang HC, Hsu JM, Yen CP, Chao CC, Chen RH, Pan CL. Neural activity and CaMKII protect mitochondria from fragmentation in aging Caenorhabditis elegans neurons. *Proc Natl Acad Sci U S A*. 2015;112. doi:10.1073/pnas.1501831112
119. Leduc-Gaudet JP, Picard M, Pelletier FSJ, Sgarioto N, Auger MJ, Vallée J, et al. Mitochondrial morphology is altered in atrophied skeletal muscle of aged mice. *Oncotarget*. 2015;6. doi:10.18632/oncotarget.4235
120. Weir HJ, Yao P, Huynh FK, Escoubas CC, Goncalves RL, Burkewitz K, et al. Dietary Restriction and AMPK Increase Lifespan via Mitochondrial Network and Peroxisome Remodeling. *Cell Metab*. 2017;26. doi:10.1016/j.cmet.2017.09.024

121. Chaudhari SN, Kipreos ET. Increased mitochondrial fusion allows the survival of older animals in diverse C. Elegans longevity pathways. Nat Commun. 2017;8. doi:10.1038/s41467-017-00274-4
122. Burman JL, Pickles S, Wang C, Sekine S, Vargas JNS, Zhang Z, et al. Mitochondrial fission facilitates the selective mitophagy of protein aggregates. *Journal of Cell Biology*. 2017;216. doi:10.1083/jcb.201612106
123. Wang D, Wang J, Bonamy GMC, Meeusen S, Brusch RG, Turk C, et al. A small molecule promotes mitochondrial fusion in mammalian cells. *Angewandte Chemie - International Edition*. 2012;51. doi:10.1002/anie.201204589
124. Garza S, Chen L, Galano M, Cheung G, Sottas C, Li L, et al. Mitochondrial dynamics, Leydig cell function, and age-related testosterone deficiency. *FASEB Journal*. 2022;36. doi:10.1096/fj.202201026R
125. Morris S, Molina-Riquelme I, Barrientos G, Bravo F, Aedo G, Gómez W, et al. Inner mitochondrial membrane structure and fusion dynamics are altered in senescent human iPSC-derived and primary rat cardiomyocytes. *Biochim Biophys Acta Bioenerg*. 2023;1864. doi:10.1016/j.bbabio.2022.148949
126. Willems PHGM, Rossignol R, Dieteren CEJ, Murphy MP, Koopman WJH. Cell Metabolism Perspective Redox Homeostasis and Mitochondrial Dynamics. *Cell Metab*. 2015;22.
127. Ul Fatima N, Ananthanarayanan V. Mitochondrial movers and shapers: Recent insights into regulators of fission, fusion and transport. *Current Opinion in Cell Biology*. 2023. doi:10.1016/j.ceb.2022.102150
128. Youle RJ, Van Der Bliek AM. Mitochondrial fission, fusion, and stress. *Science*. 2012. doi:10.1126/science.1219855
129. Leduc-Gaudet JP, Hussain SNA, Barreiro E, Gouspillou G. Mitochondrial dynamics and mitophagy in skeletal muscle health and aging. *International Journal of Molecular Sciences*. 2021. doi:10.3390/ijms22158179
130. Sharma A, Smith HJ, Yao P, Mair WB. Causal roles of mitochondrial dynamics in longevity and healthy aging. *EMBO Rep*. 2019;20. doi:10.15252/embr.201948395
131. Seo AY, Joseph AM, Dutta D, Hwang JCY, Aris JP, Leeuwenburgh C. New insights into the role of mitochondria in aging: Mitochondrial dynamics and more. *Journal of Cell Science*. 2010. doi:10.1242/jcs.070490
132. Sebastián D, Palacín M, Zorzano A. Mitochondrial Dynamics: Coupling Mitochondrial Fitness with Healthy Aging. *Trends in Molecular Medicine*. 2017. doi:10.1016/j.molmed.2017.01.003
133. Chan DC. Mitochondrial Dynamics and Its Involvement in Disease. *Annual Review of Pathology: Mechanisms of Disease*. 2020;15. doi:10.1146/annurev-pathmechdis-012419-032711
134. Rolfe DFS, Brand MD. The physiological significance of mitochondrial proton leak in animal cells and tissues. *Bioscience Reports*. 1997. doi:10.1023/A:1027327015957
135. Rolfe DFS, Brand MD. Contribution of mitochondrial proton leak to skeletal muscle respiration and to standard metabolic rate. *Am J Physiol Cell Physiol*. 1996;271. doi:10.1152/ajpcell.1996.271.4.c1380
136. Nicholls DG. The physiological regulation of uncoupling proteins. *Biochimica et Biophysica Acta - Bioenergetics*. 2006. doi:10.1016/j.bbabio.2006.02.005
137. Brand MD. Uncoupling to survive? The role of mitochondrial inefficiency in ageing. *Experimental Gerontology*. 2000. doi:10.1016/S0531-5565(00)00135-2
138. Speakman JR, Talbot DA, Selman C, Snart S, McLaren JS, Redman P, et al. Uncoupled and surviving: Individual mice with high metabolism have greater mitochondrial uncoupling and live longer. *Aging Cell*. 2004;3. doi:10.1111/j.1474-9728.2004.00097.x
139. Andrews ZB, Horvath TL. Uncoupling protein-2 regulates lifespan in mice. *Am J Physiol Endocrinol Metab*. 2009;296. doi:10.1152/ajpendo.90903.2008
140. Halaschek-Wiener J, Amirabbasi-Beik M, Monfared N, Pieczyk M, Sailer C, Kollar A, et al. Genetic variation in healthy oldest-old. *PLoS One*. 2009;4. doi:10.1371/journal.pone.0006641
141. Rose G, Crocco P, de Rango F, Montesanto A, Passarino G. Further support to the uncoupling-to-survive theory: The genetic variation of human UCP genes is associated with longevity. *PLoS One*. 2011;6. doi:10.1371/journal.pone.0029650
142. B. Andrews Z. Uncoupling Protein-2 and the Potential Link Between Metabolism and Longevity. *Current Aging Sciencee*. 2010;3. doi:10.2174/1874609811003020102
143. Hirose M, Schilf P, Lange F, Mayer J, Reichart G, Maity P, et al. Uncoupling protein 2 protects mice from aging. *Mitochondrion*. 2016;30. doi:10.1016/j.mito.2016.06.004
144. Kukat A, Dogan SA, Edgar D, Mourier A, Jacoby C, Maiti P, et al. Loss of UCP2 Attenuates Mitochondrial Dysfunction without Altering ROS Production and Uncoupling Activity. *PLoS Genet*. 2014;10. doi:10.1371/journal.pgen.1004385

145. Nicholls DG. Fifty years on: How we uncovered the unique bioenergetics of brown adipose tissue. *Acta Physiologica*. 2023. doi:10.1111/apha.13938
146. Kontani Y, Wang Y, Kimura K, Inokuma KI, Saito M, Suzuki-Miura T, et al. UCP1 deficiency increases susceptibility to diet-induced obesity with age. *Aging Cell*. 2005;4. doi:10.1111/j.1474-9726.2005.00157.x
147. Vidal-Puig AJ, Grujic D, Zhang CY, Hagen T, Boss O, Ido Y, et al. Energy metabolism in uncoupling protein 3 gene knockout mice. *Journal of Biological Chemistry*. 2000;275. doi:10.1074/jbc.M910179199
148. Klaus S, Ost M. Mitochondrial uncoupling and longevity – A role for mitokines? *Experimental Gerontology*. 2020. doi:10.1016/j.exger.2019.110796
149. Tainter ML, Stockton AB, Cutting WC. Use of dinitrophenol in obesity and related conditions: A progress report. *J Am Med Assoc*. 1933;101. doi:10.1001/jama.1933.02740440032009
150. Shrestha R, Johnson E, Byrne FL. Exploring the therapeutic potential of mitochondrial uncouplers in cancer. *Molecular Metabolism*. 2021. doi:10.1016/j.molmet.2021.101222
151. Parascandola J. Dinitrophenol and bioenergetics: An historical perspective. *Mol Cell Biochem*. 1974;5. doi:10.1007/BF01874175
152. Colman E. Dinitrophenol and obesity: An early twentieth-century regulatory dilemma. *Regulatory Toxicology and Pharmacology*. 2007. doi:10.1016/j.yrtph.2007.03.006
153. Harper JA, Dickinson K, Brand MD. Mitochondrial uncoupling as a target for drug development for the treatment of obesity. *Obesity Reviews*. 2001;2. doi:10.1046/j.1467-789X.2001.00043.x
154. Okamoto T, Shimada T, Matsumura C, Minoshima H, Ban T, Itotani M, et al. New Approach to Drug Discovery of a Safe Mitochondrial Uncoupler: OPC-163493. *ACS Omega*. 2021;6. doi:10.1021/acsomega.1c01993
155. Christoffersen BØ, Sanchez-Delgado G, John LM, Ryan DH, Raun K, Ravussin E. Beyond appetite regulation: Targeting energy expenditure, fat oxidation, and lean mass preservation for sustainable weight loss. *Obesity*. 2022. doi:10.1002/oby.23374
156. Goedeke L, Murt KN, Di Francesco A, Camporez JP, Nasiri AR, Wang Y, et al. Sex- and strain-specific effects of mitochondrial uncoupling on age-related metabolic diseases in high-fat diet-fed mice. *Aging Cell*. 2022;21. doi:10.1111/acel.13539
157. Bellanti F, Romano AD, Giudetti AM, Rollo T, Blonda M, Tamborra R, et al. Many faces of mitochondrial uncoupling during age: Damage or defense? *Journals of Gerontology - Series A Biological Sciences and Medical Sciences*. 2013;68. doi:10.1093/gerona/gls332
158. Johnson-Cadwell LI, Jekabsons MB, Wang A, Polster BM, Nicholls DG. "Mild Uncoupling" does not decrease mitochondrial superoxide levels in cultured cerebellar granule neurons but decreases spare respiratory capacity and increases toxicity to glutamate and oxidative stress. *J Neurochem*. 2007;101. doi:10.1111/j.1471-4159.2007.04516.x
159. Tretter L, Adam-Vizi V. Uncoupling is without an effect on the production of reactive oxygen species by in situ synaptic mitochondria. *J Neurochem*. 2007;103. doi:10.1111/j.1471-4159.2007.04891.x
160. Nicholls DG. Oxidative stress and energy crises in neuronal dysfunction. *Annals of the New York Academy of Sciences*. 2008. doi:10.1196/annals.1427.002
161. Cho I, Song HO, Cho JH. Mitochondrial uncoupling attenuates age-dependent neurodegeneration in C. Elegans. *Mol Cells*. 2017;40. doi:10.14348/molcells.2017.0172
162. Matsuda N, Sato S, Shiba K, Okatsu K, Saisho K, Gautier CA, et al. PINK1 stabilized by mitochondrial depolarization recruits Parkin to damaged mitochondria and activates latent Parkin for mitophagy. *Journal of Cell Biology*. 2010;189. doi:10.1083/jcb.200910140
163. Narendra DP, Jin SM, Tanaka A, Suen DF, Gautier CA, Shen J, et al. PINK1 is selectively stabilized on impaired mitochondria to activate Parkin. *PLoS Biol*. 2010;8. doi:10.1371/journal.pbio.1000298
164. Murphy MP. How mitochondria produce reactive oxygen species. *Biochemical Journal*. 2009. doi:10.1042/BJ20081386
165. Brand MD. Riding the tiger–physiological and pathological effects of superoxide and hydrogen peroxide generated in the mitochondrial matrix. *Critical Reviews in Biochemistry and Molecular Biology*. 2020. pp. 592–661. doi:10.1080/10409238.2020.1828258
166. Balaban RS, Nemoto S, Finkel T. Mitochondria, oxidants, and aging. *Cell*. 2005. doi:10.1016/j.cell.2005.02.001
167. Shields HJ, Traa A, Van Raamsdonk JM. Beneficial and Detrimental Effects of Reactive Oxygen Species on Lifespan: A Comprehensive Review of Comparative and Experimental Studies. *Frontiers in Cell and Developmental Biology*. 2021. doi:10.3389/fcell.2021.628157

168. Gómez J, Mota-Martorell N, Jové M, Pamplona R, Barja G. Mitochondrial ROS production, oxidative stress and aging within and between species: Evidences and recent advances on this aging effector. *Experimental Gerontology*. 2023. doi:10.1016/j.exger.2023.112134
169. Munro D, Baldy C, Pamenter ME, Treberg JR. The exceptional longevity of the naked mole-rat may be explained by mitochondrial antioxidant defenses. *Aging Cell*. 2019;18. doi:10.1111/acel.12916
170. Schulz TJ, Zarse K, Voigt A, Urban N, Birringer M, Ristow M. Glucose Restriction Extends Caenorhabditis elegans Life Span by Inducing Mitochondrial Respiration and Increasing Oxidative Stress. *Cell Metab*. 2007;6. doi:10.1016/j.cmet.2007.08.011
171. Dolara P, Bigagli E, Collins A. Antioxidant vitamins and mineral supplementation, life span expansion and cancer incidence: A critical commentary. *European Journal of Nutrition*. 2012. doi:10.1007/s00394-012-0389-2
172. Selman C, McLaren JS, Collins AR, Duthie GG, Speakman JR. Deleterious consequences of antioxidant supplementation on lifespan in a wild-derived mammal. *Biol Lett*. 2013;9. doi:10.1098/rsbl.2013.0432
173. Kumar P, Osahon OW, Sekhar R V. GlyNAC (Glycine and N-Acetylcysteine) Supplementation in Mice Increases Length of Life by Correcting Glutathione Deficiency, Oxidative Stress, Mitochondrial Dysfunction, Abnormalities in Mitophagy and Nutrient Sensing, and Genomic Damage. *Nutrients*. 2022;14. doi:10.3390/nu14051114
174. Conti V, Izzo V, Corbi G, Russomanno G, Manzo V, De Lise F, et al. Antioxidant supplementation in the treatment of aging-associated diseases. *Frontiers in Pharmacology*. 2016. doi:10.3389/fphar.2016.00024
175. Sadowska-Bartosz I, Bartosz G. Effect of antioxidants supplementation on aging and longevity. *Biomed Res Int*. 2014;2014. doi:10.1155/2014/404680
176. Murakami K, Yoshino M. Inactivation of aconitase in yeast exposed to oxidative stress. *Biochem Mol Biol Int*. 1997;41. doi:10.1080/15216549700201501
177. Cantu D, Schaack J, Patel M. Oxidative inactivation of mitochondrial aconitase results in iron and H2O2-mediated neurotoxicity in rat primary mesencephalic cultures. *PLoS One*. 2009;4. doi:10.1371/journal.pone.0007095
178. Yan LJ, Levine RL, Sohal RS. Oxidative damage during aging targets mitochondrial aconitase. *Proc Natl Acad Sci U S A*. 1997;94. doi:10.1073/pnas.94.21.11168
179. Tretter L, Adam-Vizi V. Alpha-ketoglutarate dehydrogenase: A target and generator of oxidative stress. *Philosophical Transactions of the Royal Society B: Biological Sciences*. 2005. doi:10.1098/rstb.2005.1764
180. Berndt N, Bulik S, Holzhütter HG. Kinetic modeling of the mitochondrial energy metabolism of neuronal cells: The impact of reduced α-Ketoglutarate dehydrogenase activities on ATP production and generation of reactive oxygen species. *Int J Cell Biol*. 2012. doi:10.1155/2012/757594
181. Kamp H, Wahrheit J, Stinchcombe S, Walk T, Stauber F, Ravenzwaay B V. Succinate dehydrogenase inhibitors: in silico flux analysis and in vivo metabolomics investigations show no severe metabolic consequences for rats and humans. *Food and Chemical Toxicology*. 2021;150. doi:10.1016/j.fct.2021.112085
182. Beach TE, Prag HA, Pala L, Logan A, Huang MM, Gruszczyk A V., et al. Targeting succinate dehydrogenase with malonate ester prodrugs decreases renal ischemia reperfusion injury. *Redox Biol*. 2020;36. doi:10.1016/j.redox.2020.101640
183. Mookerjee SA, Gerencser AA, Watson MA, Brand MD. Controlled power: how biology manages succinate-driven energy release. *Biochemical Society Transactions*. 2021. doi:10.1042/BST20211032
184. Porter C, Hurren NM, Cotter M V., Bhattarai N, Reidy PT, Dillon EL, et al. Mitochondrial respiratory capacity and coupling control decline with age in human skeletal muscle. *Am J Physiol Endocrinol Metab*. 2015;309. doi:10.1152/ajpendo.00125.2015
185. DiStefano G, Standley RA, Dubé JJ, Carnero EA, Ritov VB, Stefanovic-Racic M, et al. Chronological age does not influence Ex-vivo mitochondrial respiration and quality control in skeletal muscle. *Journals of Gerontology - Series A Biological Sciences and Medical Sciences*. 2017;72. doi:10.1093/gerona/glw102
186. Conley KE, Jubrias SA, Esselman PC. Oxidative capacity and ageing in human muscle. *Journal of Physiology*. 2000;526. doi:10.1111/j.1469-7793.2000.t01-1-00203.x
187. Gonzalez-Freire M, Scalzo P, D'Agostino J, Moore ZA, Diaz-Ruiz A, Fabbri E, et al. Skeletal muscle ex vivo mitochondrial respiration parallels decline in vivo oxidative capacity, cardiorespiratory fitness, and muscle strength: The Baltimore Longitudinal Study of Aging. *Aging Cell*. 2018;17. doi:10.1111/acel.12725
188. Bou-Teen D, Fernandez-Sanz C, Miro-Casas E, Nichtova Z, Bonzon-Kulichenko E, Casós K, et al. Defective dimerization of FoF1-ATP synthase secondary to glycation favors mitochondrial energy deficiency in cardiomyocytes during aging. *Aging Cell*. 2022;21. doi:10.1111/acel.13564

189. Short KR, Bigelow ML, Kahl J, Singh R, Coenen-Schimke J, Raghavakaimal S, et al. Decline in skeletal muscle mitochondrial function with aging in humans. *Proc Natl Acad Sci U S A*. 2005;102. doi:10.1073/pnas.0501559102
190. Befroy DE, Petersen KF, Dufour S, Mason GF, Rothman DL, Shulman GI. Increased substrate oxidation and mitochondrial uncoupling in skeletal muscle of endurance-trained individuals. *Proc Natl Acad Sci U S A*. 2008;105. doi:10.1073/pnas.0808889105
191. Petersen KF, Morino K, Alves TC, Kibbey RG, Dufour S, Sono S, et al. Effect of aging on muscle mitochondrial substrate utilization in humans. *Proc Natl Acad Sci U S A*. 2015;112. doi:10.1073/pnas.1514844112
192. Aas V, Thoresen GH, Rustan AC, Lund J. Substrate oxidation in primary human skeletal muscle cells is influenced by donor age. *Cell Tissue Res*. 2020;382. doi:10.1007/s00441-020-03275-w
193. Sugrue MM, Tatton WG. Mitochondrial membrane potential in aging cells. *Neurosignals*. 2001;10. doi:10.1159/000046886
194. Nicholls DG. Mitochondrial membrane potential and aging. *Aging Cell*. 2004. pp. 35–40. doi:10.1111/j.1474-9728.2003.00079.x
195. Berry BJ, Vodičková A, Müller-Eigner A, Meng C, Ludwig C, Kaeberlein M, et al. Optogenetic rejuvenation of mitochondrial membrane potential extends C. elegans lifespan. Nat *Aging*. 2023;3. doi:10.1038/s43587-022-00340-7
196. Mansell E, Sigurdsson V, Deltcheva E, Brown J, James C, Miharada K, et al. Mitochondrial Potentiation Ameliorates Age-Related Heterogeneity in Hematopoietic Stem Cell Function. *Cell Stem Cell*. 2021;28. doi:10.1016/j.stem.2020.09.018
197. Prokai L, Yan LJ, Vera-Serrano JL, Stevens SM, Forster MJ. Mass spectrometry-based survey of age-associated protein carbonylation in rat brain mitochondria. *Journal of Mass Spectrometry*. 2007. doi:10.1002/jms.1345
198. Das N, Jana CK. Age-associated oxidative modifications of mitochondrial α-subunit of F1 ATP synthase from mouse skeletal muscles. *Free Radic Res*. 2015;49. doi:10.3109/10715762.2015.1017477
199. Davies SMK, Poljak A, Duncan MW, Smythe GA, Murphy MP. Measurements of protein carbonyls, ortho- and meta-tyrosine and oxidative phosphorylation complex activity in mitochondria from young and old rats. *Free Radic Biol Med*. 2001;31. doi:10.1016/S0891-5849(01)00576-7
200. De Souza-Guerreiro TC, Asally M. Seeking Insights into Aging through Yeast Mitochondrial Electrophysiology. *Bioelectricity*. 2021. doi:10.1089/bioe.2021.0011
201. Yoshina S, Izuhara L, Kamatani N, Mitani S. Regulation of aging by balancing mitochondrial function and antioxidant levels. *Journal of Physiological Sciences*. 2022;72. doi:10.1186/s12576-022-00853-1
202. Fell D. *Understanding the Control of Metabolism*. London: Portland Press; 1997.

12 Muscle-Based Thermogenesis and Its Physiological Implications for Obesity and Muscle Health

Jessica L. Braun, Bianca M. Marcella*, Mia S. Geromella*, and Val A. Fajardo*
Brock University, Canada

12.1 SKELETAL MUSCLE AND METABOLISM

Skeletal muscle is the largest organ in the body that is best known for its contractile activity that supports posture, movement, and breathing. However, the importance of this tissue for mammalian physiology extends well beyond locomotion. Skeletal muscle is a metabolically active tissue and is largely involved in energy expenditure both at rest and during physical activity.[1] Furthermore, and by virtue of its size (40–50% of body mass), skeletal muscle accounts for 80% of whole-body insulin-stimulated glucose uptake[2] and encompasses metabolic processes highly involved in fuel combustion to meet the energy demands of activity, such as fatty acid metabolism and glycogen synthesis/breakdown. To support both oxidative (fat) and glycolytic (carbohydrates) metabolism, skeletal muscles can be made up of two general fiber types, type I and type II, though there are various subclasses of type II fibers.[3] Type I fibers are oxidative fibers that take advantage of a dense mitochondrial network to produce adenosine triphosphate (ATP) via oxidative phosphorylation. These fibers are also rich in insulin receptors and glucose uptake machinery, which explains why these fibers are relatively more insulin sensitive. Conversely, type II glycolytic fibers rely more on cytosolic/anaerobic metabolism to synthesize ATP and are relatively less insulin sensitive[3,4]. Indeed, skeletal muscle is highly plastic such that it is able to adapt in response to caloric intake, physical demands, or metabolic stress—one adaptation being a change in fiber type composition.

Skeletal muscle can also act as an endocrine organ, meaning that it can send signals that alter the properties of other tissues and organs such as the brain, adipose tissue, bone, liver, gut, pancreas, and vasculature.[5] This interorgan cross-talk further exemplifies the importance of skeletal muscle to whole-body homeostasis. In terms of energy expenditure and metabolism, various myokines (e.g., irisin and brain-derived neurotrophic factor) have been identified that can direct whole-body metabolism, particularly by altering adipose-based thermogenesis, which is primarily regulated by uncoupling protein 1 in brown/beige adipocytes. However, skeletal muscle itself is a thermogenic tissue, having implications for metabolic disorders such as obesity and type II diabetes. Herein lies the purpose of this chapter, which is to elaborate on the role of skeletal muscle as a thermogenic organ and its implications for health and disease.

* These authors contributed equally.

DOI: 10.1201/9781003391470-15

12.2 SKELETAL MUSCLE AND ITS CONTRIBUTION TO ENERGY EXPENDITURE AND ADAPTIVE THERMOGENESIS

All biological work is powered by ATP. This high-energy molecule consists of a nitrogenous base, adenosine, a ribose sugar, and three phosphate groups bound to ribose through phosphoanhydride bonds.[6] It is the presence of these three phosphate groups that naturally want to repel each other (i.e., electrostatic repulsion) that provides the ATP molecule with instability and thus a high amount of captured free energy.[6] Taken together, ATP "wants" to release a phosphate group to become more stable, and in doing so releases 30.5 $\frac{kJ}{mol}$ of free energy[6]. Through a process called *thermodynamic coupling*, this free energy fuels most if not all of the endergonic reactions required for biological work, of which there are three forms: mechanical, transport, and chemical.[6,7] Mechanical work refers to processes that generate motion, such as muscle contraction or the movement of other motor proteins.[7] Transport work refers to the movement of substances across membranes, such as the active transport of ions.[7] Finally, chemical work refers to any process involving chemical reactions, as seen in processes such as glucose breakdown, protein synthesis, and other aspects of metabolism.[7] On a daily basis our bodies consume a large amount of ATP, and on any given day, the human body can turn over its equivalent in body weight (e.g., a 75-kg individual could hydrolyze up to 75 kg of ATP per day).[8] This is referred to as daily energy expenditure.

Daily energy expenditure can be broken down into several components, including basal metabolic rate (BMR), the thermic effect of feeding, physical activity, and finally adaptive thermogenesis.[9] BMR represents all cellular reactions occurring at rest to maintain cellular homeostasis and accounts for 60% of total daily energy expenditure. The thermic effect of feeding refers to the thermogenic response to caloric intake and is required for proper nutrient digestion, accounting for 10–15% of daily energy expenditure. Likely the most variable component (15–30%), physical activity, refers to energy expenditure for normal daily activities and in response to exercise.[10] Finally, adaptive thermogenesis is the cellular phenomenon of increasing heat production in response to an external stimulus such as prolonged cold exposure.[11–13] Under cold conditions there is an activation of thermogenic mechanisms that can be divided into two distinct forms: shivering and non-shivering thermogenesis.[11]

Shivering thermogenesis relies on small repetitive muscle contractions that generate heat via three major ATP-consuming processes. First, the myosin ATPase hydrolyzes ATP in order to reverse the power stroke during cross-bridge cycling, specifically activating the myosin head for a new interaction with actin.[14] Second, the sarco(endo)plasmic reticulum Ca^{2+} ATPase (SERCA) pump harnesses the energy released from ATP hydrolysis to bring Ca^{2+} from the cytosol back into the sarcoplasmic reticulum (SR), eliciting muscle relaxation.[15] Finally, the Na^+/K^+-ATPase within the plasma membrane restores the Na^+ (extracellular > intracellular) and K^+ (intracellular > extracellular) concentration gradients required for subsequent action potentials and membrane depolarization events by transporting three Na^+ out of the cell and bringing two K^+ in at the expense of ATP.[16] Thus, muscle contraction and relaxation require large amounts of energy in the form of ATP, inherently releasing heat and providing an acute mode of thermoregulation. However, when exposed to cold temperatures for a prolonged period, eventually shivering thermogenesis stops and non-shivering thermogenic mechanisms become the dominant process for the maintenance of core temperature. Adipose tissue and skeletal muscle are two main sites of non-shivering thermogenesis and, in addition to cold exposure, can become activated in response to caloric excess to promote the combustion of the metabolic substrates being consumed. Briefly, beige adipose and brown adipose tissue are characterized by an abundance of mitochondria and high uncoupling protein 1 (UCP1) content, which acts to uncouple the proton gradient from ATP synthesis, ultimately dissipating the stored energy stored in the form of heat. Within the skeletal muscle, the SERCA pump is a major energy (ATP) consumer and is a primary mediator of muscle-based thermogenesis.

12.3 SERCA

12.3.1 Excitation-Contraction Coupling: The Role of Ca^{2+}

The SERCA pump is a vital physiological component of all living cells that regulates intracellular Ca^{2+} ($[Ca^{2+}]_i$) levels.[17,18] It is a P-type ATPase, meaning that it has a highly conserved phosphorylation site at its Asp^{351} residue, which is transiently phosphorylated upon ATP hydrolysis, causing a conformational change in the pump.[19,20] SERCA is mainly responsible for pumping Ca^{2+} from the cytosol against its concentration gradient into the SR at the expense of ATP (e.g., transport work). In skeletal muscle, SERCA's actions are vital in eliciting muscle relaxation, ensuring sufficient Ca^{2+} load for subsequent contractions, and maintaining low resting $[Ca^{2+}]_i$ levels.

To initiate muscle contraction, an action potential propagates along the plasma membrane and down the transverse tubules, activating the dihydropyridine receptors (DHPRs, Figure 12.1).[21–23] DHPRs consist of four subunits: α_{1S}, α_2/δ, β_1, and γ.[24] The α_1 subunit is the integral membrane protein and forms the voltage-sensing region with four hydrophobic domains that assemble and also form the central cation pore region. Each domain of the α_1 subunit has six transmembrane alpha-helices (denoted S1–S6), and it is the S4 helix of each domain that forms the voltage sensor, given its strong positive charge. In response to membrane depolarization, the S4 helices are displaced by the repulsive positive charge, which activates the DHPR and opens the pore.[22,24] In skeletal muscle, there is a small and slow Ca^{2+} influx corresponding with a conformational change in the DHPR.[25] However, physical protein–protein interactions of the DHPR with ryanodine receptors (RyRs) via the cytosolic region between domains II and III of the α_1 subunits[24] are what causes RyR opening, not the Ca^{2+} influx through the DHPRs.[25] The function of the small Ca^{2+} current in skeletal muscle remains to be elucidated. In contrast, cardiac muscle DHPR activation causes a large and rapid Ca^{2+} influx that activates the Ca^{2+}-sensitive RyRs.[23,25] The functional significance of the differences between skeletal and cardiac muscle with regards to Ca^{2+}-induced Ca^{2+} release, despite homologous proteins, remains unknown. It has been suggested that this mechanism may allow for tighter regulation of the strength and speed of the heartbeat.[26]

RyRs are large ion channels that are localized within the SR membrane.[21,27] There are three different isoforms: RyR 1–3, with RyR1 being widely expressed in skeletal muscle; and the Ca^{2+}-sensitive RyR2 being predominant in cardiac muscle.[25,27] When the voltage-sensitive DHPR pulls open the homotetramer RyR, there is a large efflux of Ca^{2+} from the SR, thus drastically increasing $[Ca^{2+}]_i$ (Figure 12.1).[21,28] The rise in free $[Ca^{2+}]_i$ allows for Ca^{2+} binding to troponin-C, displacing tropomyosin and thus exposing the myosin-binding site on actin.[14] The myosin head is then able to form a cross-bridge, which results in force production and cross-bridge cycling.[14] After muscle activation, and in order to elicit relaxation, the free $[Ca^{2+}]_i$ must be removed from the cytosol. SERCA is a vital component in muscle relaxation in that it removes the free $[Ca^{2+}]_i$ by pumping it back into the SR (Figure 12.1), which inherently ensures a sufficient Ca^{2+} load in the SR for subsequent muscle contractions.[17,18,20]

RyRs are also major sources of resting Ca^{2+} leak out of the SR, generating futile Ca^{2+} cycling with SERCA, which partly explains the contribution of both SERCA (discussed further in Section 12.3.5) and skeletal muscle to daily BMR.[29] A recently published study by Meizoso-Huesca et al.[29] demonstrated that changes in RyR1 leak in isolated skeletal muscle fibers results in direct changes in heat production of the fiber solely through SERCA Ca^{2+} pumping. The amount of Ca^{2+} leak out of RyR is mediated by various factors including interactions with calstabin, which stabilizes the closed state and reduces Ca^{2+} release, or phosphorylation of one of three serine sites, which has been shown to release calstabin and thus increase Ca^{2+} release.[30] Phosphorylation of RyR has also been shown to alter sensitivity to ATP, Mg^{2+}, and Ca^{2+}, ultimately increasing channel activity and the probability of opening.[30] Furthermore, RyRs are sensitive to reactive oxygen/nitrogen species (ROS/RNS), resulting in increased Ca^{2+} leak.[30,31] Thus, the local redox environment in both healthy and disease states can alter the rate of futile Ca^{2+} cycling through changes in RyR leak.

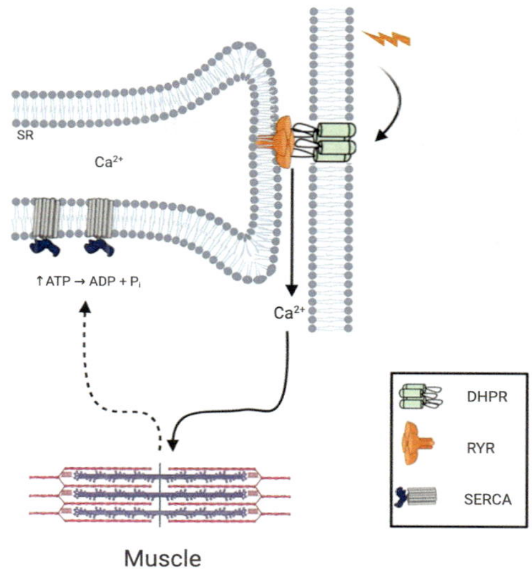

FIGURE 12.1 Excitation-contraction coupling in skeletal muscle. An action potential, initiated from a motor neuron, will propagate down the transverse tubules activating the voltage sensitive DHPRs. The DHPR pores will open in response to depolarization; however, in skeletal muscle it is a direct interaction and conformational change with RyR that causes activation of the RyR channel. The RyRs release large amounts (~1–10uM) of Ca^{2+} into the SR, resulting in a >1000-fold increase in myoplasmic Ca^{2+}, which is able to initiate force production in the sarcomeres. To elicit muscle relaxation, SERCA will pump the free Ca^{2+} back into the SR at the expense of ATP hydrolysis. Figure made with BioRender.

12.3.2 SERCA Protein Structure

The SERCA pump contains cytosolic and transmembrane domains that are critical for its catalytic cycle (Figure 12.2a). The cytoplasmic head consists of three domains. The P (phosphorylation) domain, which contains the Asp351 residue that is autophosphorylated during the reaction cycle, induces a reversible conformational change (Figure 12.2b).[19,32] The N (nucleotide binding domain) domain forms an ATP-binding pocket in which Phe487, a critical residue for ATP binding,[33] and two lysine residues (Lys515 and Lys492) produce a positive charge in this pocket relative to the phosphorylation site.[34] Lastly, the A domain (actuator domain) is a smaller, flexible domain that appears to be a gating mechanism required to allow for the conformational changes that occur with Ca^{2+} binding, transition, and release,[33,35,36] given that the removal of any residues in domain A results in complete inhibition of any conformational changes to the pump (Figure 12.2a).[36]

The transmembrane domain of SERCA contains 10 transmembrane α -helices (M1–10, Figure 12.2a).[34] Using x-ray crystallography, Toyoshima et al.[34] visualized the structure of the SERCA pump at a 2.6 Å (or 2.6 x 10^{-10} m) resolution, which ultimately revealed a Ca^{2+}-binding pocket located between M4–6 and M8 transmembrane helices.[34] SERCA contains two Ca^{2+}-binding sites (sites I and II). Site I is located between M5 and M6, with M8 acting distally, whereas site II can be found within M4 where the helix is unwound between Ile306 and Gly310.[34] Ca^{2+} appears to enter the pump, reaching its transmembrane binding sites, through the area surrounding M2, M4, and M6 whereby a row of exposed oxygen atoms create a hydrophilic pathway to the binding sites.[34] This pathway constricts as it nears sites I and II, displacing the water molecules from the divalent cation.[34]

Finally, it is important to briefly discuss the loop that connects M6 and M7 (L67; Phe809–Gly831, shown in red in Figure 12.2). This loop has been shown to directly or indirectly interact with most key parts of SERCA and thus appears to coordinate the different parts of the enzyme.[34] L67 runs below domain P and is able to mediate the interactions between the transmembrane domain and the

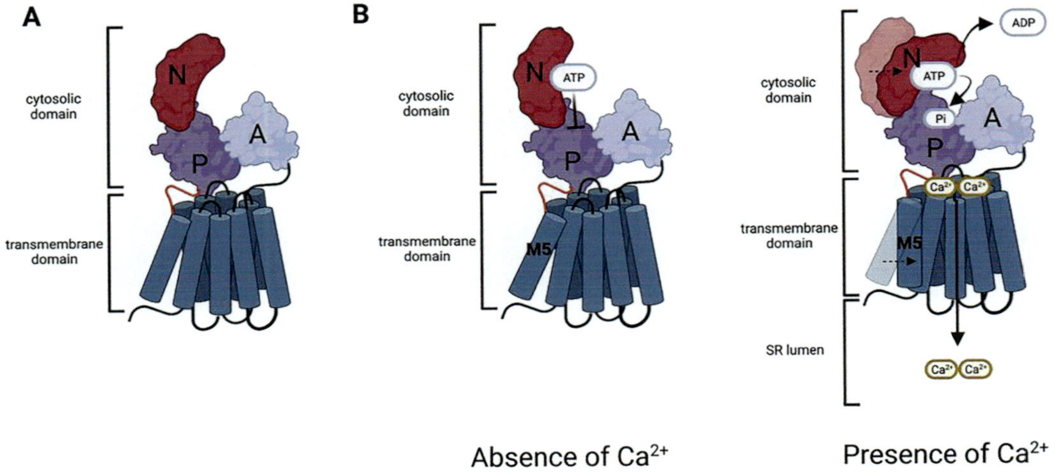

FIGURE 12.2 Basic representation of SERCA structure and conformational changes associated with Ca^{2+} binding. (a) The SERCA pump contains three cytosolic domains, the nucleotide binding domain (N), the phosphorylation domain (P), and the actuator domain (A), as well as 10 transmembrane helices. The L67 loop is highlighted in red. (b) In the absence of Ca^{2+}, ATP is able to bind to the N domain but is unable to reach the P domain due to restricted movement of M5 by L67. When Ca^{2+} binds, it induces a "straightening" of M5 with this conformational change being translated to the cytosolic domain by L67. This allows for interactions between the N and P domains, autophosphorylation, and Ca^{2+} transport into the SR. Figure made with BioRender.

phosphorylation domain.[34] More specifically, given that L67 is directly associated with M6, any changes to M6 caused by Ca^{2+} binding will be transmitted to domain P.[34] For example, in the absence of Ca^{2+}, the three cytoplasmic domains are in a closed position and are fairly immobile[34,37,38] (Figure 12.2b). In this state, ATP is able to bind to domain N but is not able to reach Asp351 on domain P (Figure 12.2B), mainly given that L67 is restricting M5 movement.[37] However, Ca^{2+} binding induces a straightening of M5, breaking the closed configuration of the cytoplasmic domains and allowing for domain N to interact with domain P through an ATP crosslink (Figure 12.2b).[37] Taken together, it becomes clear that SERCA ATPase activity is Ca^{2+} dependent, which is physiologically relevant, as the pump will not hydrolyze ATP without the potential of transporting Ca^{2+} into the SR. Phosphate transfer to Asp351 induces a rotation in M6 (translated through L67), which alters the Ca^{2+}-binding sites, reducing their affinity for Ca^{2+}, while simultaneously opening the luminal gate to release Ca^{2+} into the SR lumen.[37] Furthermore, mutational analyses have shown that even a single mutation within L67 drastically impairs ATPase activity,[39] demonstrating the functional significance of this loop.

12.3.3 SERCA Isoforms

In vertebrates, there are up to 11 identified isoforms of SERCA produced via alternative splicing of three genes (*ATP2A1, ATP2A2, ATP2A3*).[20] Of these, eight isoforms have been thoroughly studied and well characterized throughout the literature. There are two SERCA1 isoforms, SERCA1a and 1b, where SERCA1a is expressed in adult skeletal muscle primarily in fast-twitch muscles,[20] and SERCA1b in fetal/developing muscle.[40,41] There are three SERCA2 isoforms. SERCA2a is found mainly in slow-twitch and cardiac muscle but can be found in all muscle cells, and SERCA2b is ubiquitously expressed in all cell types, both muscle and non-muscle.[20] Finally, SERCA2c has been reported in cardiac muscle,[20,42] though SERCA2a is the dominant isoform in this tissue. SERCA3 encodes for six isoforms, 3a-3f, where SERCA3a-c are expressed mainly in non-muscle

TABLE 12.1
Summary of SERCA Isoform Characteristics

Isoform	Primary Expression	Ca^{2+} Affinity (of those characterized)	Velocity (of those characterized)	Leaky? (of those characterized)
SERCA1	Adult fast-twitch skeletal muscle	intermediate	fastest	++
SERCA1a	Fetal/developing muscle			
SERCA1b				
SERCA2	Slow-twitch skeletal, cardiac,	intermediate	intermediate	-
SERCA2a	and smooth muscle	highest	slowest	
SERCA2b	Ubiquitously expressed			
SERCA2c	Cardiac muscle			
SERCA3	Non-muscle cells	lowest	slowest	
SERCA3a	Non-muscle cells			
SERCA3b	Non-muscle cells			
SERCA3c	Cardiac muscle			
SERCA3d	Unknown			
SERCA3e	Cardiac muscle			
SERCA3f				

cells such as platelets and epithelial cells.[20,43] SERCA3a, 3d, and 3f have also been found in cardiac muscle,[44] but SERCA3e, at the protein level, has yet to be identified.

Functionally, the various SERCA isoforms differ mainly in their affinity for Ca^{2+} as well as pump velocity and leakiness. Of the isoforms that have been well characterized, SERCA2b has the highest affinity for Ca^{2+} followed by SERCA1a and SERCA2a of equivalent affinity, and finally SERCA3a having the lowest affinity for Ca^{2+}.[20] With regards to pump velocity, SERCA1a is the fastest, followed by SERCA2a, with SERCA2b and SERCA3a being the slowest.[20] Finally, SERCA1 is the only known isoform to show increased Ca^{2+} leak in response to a Ca^{2+} gradient being formed across the SR membrane.[45,46] In other words, once enough Ca^{2+} has been brought into the SR, a Ca^{2+} gradient forms (inside SR > cytosol) triggering SERCA1 to hydrolyze ATP without the translocation of Ca^{2+}[45,46] (described further in the section that follows). This unique characteristic allows for SERCA1 to increase heat production from ATP hydrolysis from approximately 10kcal/mol up to 20–30kcal/mol.[45,46] These functional differences, summarized in Table 12.1, contribute to the known differences in muscle relaxation rates, where SERCA1a is predominantly expressed in fast-twitch skeletal muscle and SERCA2a is predominantly expressed in slow-twitch muscle. However, it should be noted that the primary determinant of Ca^{2+} uptake velocity in muscle is SERCA density rather than SERCA isoform.[47]

12.3.4 Catalytic Cycle and Coupling Ratio

SERCA is responsible for maintaining low $[Ca^{2+}]_i$, not only to elicit muscle relaxation but also to avoid the cytotoxic effects of elevated $[Ca^{2+}]_i$, such as protein degradation and/or cell death leading to muscle weakness and wasting.[48,49] As briefly mentioned in Section 12.3.2, SERCA must undergo several conformational changes during the catalytic cycle to succeed in bringing Ca^{2+} into the SR. There are two main conformational states in which SERCA can be found: E_1 and E_2.[13,20,37,46,50] The E_1 state exhibits a higher affinity for Ca^{2+} than E_2, and because the E_2 Ca^{2+}-binding sites are facing the lumen, this facilitates Ca^{2+} release into the SR.[13,37,46] Following muscle contraction, $[Ca^{2+}]_i$ is high and Ca^{2+} cooperatively binds to site I and then II. SERCA then becomes autophosphorylated at the Asp351 site, generating a high-energy state, E_1-P (Figure 12.3, reactions 1 and 2).[13,37] The phosphoryl transfer from ATP to Asp351 allows for dissociation of ADP, triggering a conformational change into the low-energy E_2-P state.[13,37] This conformational change alters the Ca^{2+} affinity of the two Ca^{2+}-binding sites and opens the luminal gate, releasing Ca^{2+} into the SR (Figure 12.3, reactions 3 and 4).[37] Following Ca^{2+} release, a replacement of P_i for H_2O occurs on Asp351 resulting in a transition back to E_1, and the catalytic cycle is complete (Figure 12.3, reactions 5 and 6) [13,37,46].

As Toyoshima et al. [34] determined, based on SERCA's structure, its optimal stoichiometry is the transport of 2 Ca^{2+} per 1 ATP hydrolyzed; however, under physiological (and non-physiological) conditions this ratio is often reduced due to alterations in SERCA's catalytic cycle. In conditions of high [Ca^{2+}] in the SR lumen, Ca^{2+} may bind to the E_2 state and get carried back into the cytosol during $E_2 \rightarrow E_1$ transformation (Figure 12.3, reactions 7–9) and is called passive leak.[46,50] A second condition of altered coupling ratio again occurs with high luminal [Ca^{2+}] and causes a slowed transition of E_1-P to E_2-P, and P_i will be released prematurely with all energy released as heat (100% uncoupled, Figure 12.3, reaction 10), which is seen only with the SERCA1 isoform.[45,46,50] Finally, slippage occurs when Ca^{2+} is released prematurely from the E_1-P state, still on the cytosolic side, during the transition from E_1-P to E_2-P and after ATP hydrolysis (Figure 12.3, reaction 11).[50] These three uncoupling events are likely to occur at rest (high luminal [Ca^{2+}]) or when sarcolipin (SLN) is

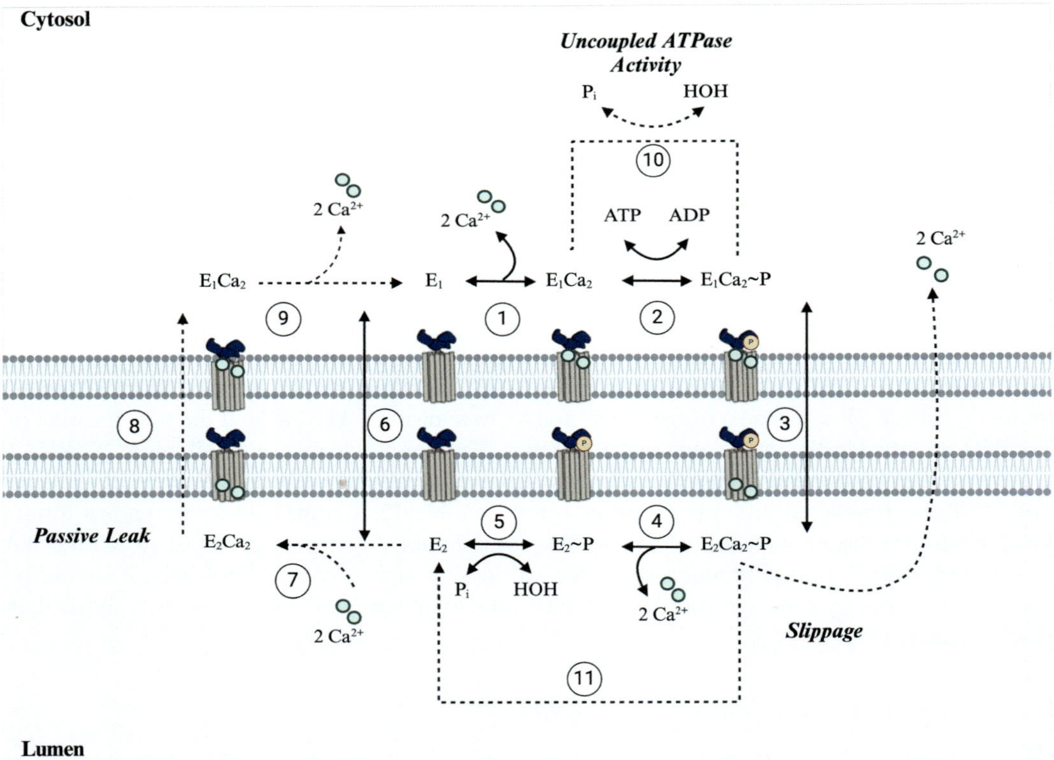

FIGURE 12.3 SERCA pump catalytic cycle and uncoupling conditions. Under optimal conditions, SERCA binds two Ca^{2+} ions and one ATP molecule (reactions 1, 2). Coinciding with ATP hydrolysis (reaction 2), both Ca^{2+} are transported into the SR lumen (reactions 3, 4) and SERCA returns to its high Ca^{2+} affinity state (reactions 5, 6). Uncoupling of the pump occurs in conditions of high luminal [Ca^{2+}] where Ca^{2+} can be passively leaked back into the cytosol (reactions 7, 8). Additionally, Ca^{2+} transport can be slowed, and premature P_i release can occur in the SERCA1 isoform[45,46] (reaction 10). Finally, slippage of Ca^{2+} from the binding sites can occur when Ca^{2+} is released back into the cytosol during transport, after ATP has been hydrolyzed (reaction 11). Figure created using Biorender.

bound, which can bind to SERCA at all points in its catalytic cycle (discussed in Section 12.4.2). Regardless, all alterations from the catalytic cycle will result in greater energetic demand through ATP hydrolysis to maintain the Ca^{2+} gradient, thus increasing energy expenditure and altering the metabolic rate.

12.3.5 SERCA's Contribution to Energy Expenditure

Muscle-based non-shivering thermogenesis is largely mediated through the SERCA pump, which has been determined to largely contribute to energy expenditure both at rest and during striated muscle contraction. Skeletal muscle accounts for approximately 40% of body weight and 20–30% of BMR.[1,52] Importantly, during muscle contraction, up to 80% of ATP consumption is devoted to myoplasmic Ca^{2+} removal,[53] demonstrating that SERCA is undoubtedly a major energy consumer during physical activity. In addition to this, a study by Smith et al.[52] measured SERCA's contribution to BMR. Specifically, oxygen consumption was measured through indirect calorimetry in isolated and intact murine soleus and extensor digitorum longus (EDL) muscles using a Clarke-type PO_2 electrode with and without a high concentration (10 mM) of $MgCl_2$.[52,54] Mg^{+} directly inhibits RyR channels by competing for the inhibitory cytoplasmic Ca^{2+}-binding domain with very high affinity,[55] preventing Ca^{2+} release from the SR and therefore attenuating SERCA mediated Ca^{2+} reuptake from the cytosol.[52] Comparing oxygen consumption measurements with and without $MgCl_2$ provided an indirect measurement of SERCA-specific contribution to skeletal muscle metabolic rate. It was determined that SERCA-mediated ATP hydrolysis contributes up to 50% of skeletal muscle BMR, with no differences found between muscle types.[52] Specifically, it was found that the ATP turnover rate by SERCA in resting EDL and soleus was 73 nmol/g/s and 82 nmol/g/s, respectively.[52] Given that BMR and physical activity account for 60% and 30% of total daily energy expenditure, respectively, and that skeletal muscle accounts for 30% of BMR and 90% of physical activity, Smith et al. concluded that if SERCA accounts for 50% of resting and active energy expenditure in skeletal muscle, then SERCA within skeletal muscle accounts for 20–22% of total daily energy expenditure.[52] Therefore, the SERCA pumps are recognized as a metabolic hub within the skeletal muscle, and alterations to SERCA's transport efficiency may influence total daily energy expenditure.

12.4 REGULATORS OF SERCA FUNCTION

12.4.1 Phospholamban

The SERCA pumps are mainly regulated by two proteins, sarcolipin (SLN) and phospholamban (PLN), which are both expressed in skeletal and cardiac muscle.[56,57] The latter is a highly conserved, 52-amino-acid protein that, in its unphosphorylated active state, allosterically binds to SERCA and decreases its affinity for Ca^{2+}.[58] PLN exists in an active monomeric form as well as inactive pentameric form. Monomeric PLN can bind to a groove formed between the transmembrane helices M2, M4, M6, and M9.[58–60] Binding of PLN and Ca^{2+} to the SERCA pump has been determined to be mutually exclusive, as PLN can only bind to SERCA in its E_2 conformational state (i.e., its Ca^{2+}-free state).[59,60] Thus, PLN may only regulate SERCA under conditions of low $[Ca^{2+}]_i$. In addition to high levels of $[Ca^{2+}]_i$, phosphorylation of PLN can also cause it to dissociate from SERCA, thereby restoring the pump's affinity for Ca^{2+}.[58,61,62] PLN phosphorylation occurs in two distinct pathways and at two distinct sites. The first phosphorylation pathway involves stimulation of the β-adrenergic system causing an increase in production of cyclic AMP by adenylyl cyclase, which in turn activates protein kinase A that phosphorylates PLN at Ser16.[63] Phosphorylation at Thr17 can occur independently of Ser16 through the Ca^{2+}/calmodulin-dependent protein kinase II (CaMKII) phosphorylation pathway, which is activated in the presence of high levels of intracellular Ca^{2+}.[64] It has been shown that dual phosphorylation of PLN it is not necessary to relieve inhibition of SERCA, nor does it improve pump function any further than phosphorylation at a single site.[65] Interestingly, it has been shown that Ser16 phosphorylation accumulates faster and correlates more closely with contraction amplitude and Ca^{2+} transient decline than Thr17 phosphorylation, indicating it may play a more important role in regulating Ca^{2+} handling.[65] Teleologically, this is logical given the evolutionary purpose of the β-adrenergic system in promoting organism survival and allowing for optimal Ca^{2+} handling and muscle performance to either face or escape a threat.

PLN and its role in regulating SERCA are a significant component of muscle function, and maintaining a proper PLN:SERCA ratio is vital fir maintaining Ca^{2+} homeostasis in muscle. Overregulation of SERCA can lead to low SR Ca^{2+} levels, affecting muscle contractility, and cytotoxic increases in $[Ca^{2+}]_i$, potentially leading to cell death. In fact, many genetic mutations to the *PLN* gene and/or alterations in the ratio can lead to severe muscle abnormalities, such as centronuclear myopathy and dilated cardiomyopathy.[62,66,67] More specifically, transgenic overexpression of PLN in murine cardiac tissue results in altered contractile kinetics,[68] prolonged Ca^{2+} transients, and reduced rates of fractional shortening,[69] all of which were shown to be reversed with the addition of the β-adrenergic receptor agonist, isoproterenol.[68,69] Similarly, the overexpression of PLN in the slow-twitch soleus muscle results in prolonged contraction and relaxation time, smaller muscle size and cross-sectional area, as well as an increased proportion of oxidative muscle fibers compared to wild-type mice.[70] Further investigation into the effects of PLN overexpression in murine skeletal muscle demonstrated significant SERCA dysfunction and overlap with human centronuclear myopathy with muscles having central nuclei, progressive fibrosis, and muscle weakness.[67] Indeed, assessing muscle biopsies from centronuclear myopathy patients showed increases in PLN content and reductions in SERCA activity.[67]

With regards to metabolism and adaptive thermogenesis, PLN does not appear to play a significant role, with mice lacking PLN showing neither differences in SERCA pumping efficiency[72] nor increased susceptibility to diet-induced obesity.[73] Thus, despite apparent similarities in protein structure and function between PLN and SLN, the two are functionally diverse and play unique roles in skeletal muscle health and disease.

12.4.2 Sarcolipin

SLN, like PLN, functions as a negative regulator of the SERCA pump. Similar to PLN, SLN physically interacts with SERCA by binding to a groove within the M2, M6, and M9 helices, and reduces its affinity for Ca^{2+}[74–76]. However, unlike PLN, SLN can bind to SERCA throughout its catalytic cycle (i.e., both E_1 and E_2 states) and is thus not mutually exclusive with Ca^{2+} binding.[72] This distinction gives SLN the unique ability to regulate the SERCA pump under high and low $[Ca^{2+}]_i$ conditions. Furthermore, this allows SLN to promote SERCA uncoupling. Unlike PLN, SLN can uncouple SERCA-mediated Ca^{2+} transport from ATP hydrolysis by causing premature Ca^{2+} slippage off the pump.[51] More specifically, the uncoupling action of SLN is mediated through its N-terminal tail (residues ERSTQ), which is lacking in PLN.[51,77,78] Given that SLN allows Ca^{2+} binding to the pump, it also allows for ATP hydrolysis (which is dependent on Ca^{2+} binding). It is the N-terminal tail of SLN that then prevents Ca^{2+} movement across the SR membrane following ATP hydrolysis, releasing it back into the cytosol. By uncoupling the SERCA pump, SLN increases the energy demands of SR Ca^{2+} pumping, resulting in an increased release of heat.[79] This makes SLN in skeletal muscle an important player in muscle-based non-shivering thermogenesis, and several studies have now been conducted in response to cold exposure and caloric excess.

SLN expression in oxidative skeletal muscles, such as the soleus, has been shown to be increased in response to prolonged cold exposure and excessive caloric intake, likely to increase heat production and promote the combustion of excess metabolic substrates and to raise basal energy expenditure, respectively.[80–82] Genetic ablation of SLN reduces thermogenic regulation in mice making them less able to maintain core temperature in response to cold exposure.[83] Furthermore, SLN knockout (Sln^{KO}) mice fed a high-fat diet gain more weight and are more glucose intolerant compared with their wild-type littermates—a result of enhanced SERCA transport efficiency and a reduction in daily energy expenditure.[81,83] Translating this to humans, a recent study found that knocking down SLN in human skeletal muscle cells reduced basal respiration and impaired substrate utilization resulting in increased lipid accumulation.[82] SLN also regulates skeletal muscle oxidative capacity whereby the deletion of SLN from skeletal muscle impaired oxidative capacity and the overexpression of SLN in glycolytic skeletal muscle promoted oxidative metabolism.[84] SLN is therefore an important regulator of skeletal muscle metabolism and oxidative capacity.

Given its importance for metabolic health, others have suggested that increasing SLN may be beneficial to muscle and whole-body health. However, some of the current literature suggests that this may be a double-edged sword, with an increase in SLN causing disturbances to Ca^{2+} regulation and muscle health. This has been shown primarily with muscular dystrophy. In conditions of muscle atrophy, such as disuse, myopathies, muscular dystrophies, and spaceflight, the upregulation of SLN is a consistent finding.[85–88] However, whether these alterations in SLN expression are beneficial or detrimental is still under considerable debate. Some studies have shown that the upregulation of SLN in muscular dystrophy exacerbates disease pathology,[89] whereby others have shown that its upregulation is critical for activating compensatory and disease-mitigating pathways.[90,91] Consistent with the latter, the overexpression of SLN in murine skeletal muscle resulted in improved exercise tolerance and fatigue resistance with no signs of muscle dysfunction.[92] This is in stark contrast with studies from the PLN-overexpressing mouse model, where signs of centronuclear myopathy were found. Taken together, increasing SLN-mediated SERCA uncoupling in muscle may be an attractive therapy for metabolic disorders such as obesity and type 2 diabetes, but whether this strategy will negatively affect muscle and thus whole-body health remains unknown.[93–95]

12.4.3 Other SERCA Regulators

In light of the SERCA pump being a viable therapeutic target for muscle disease and obesity, many researchers have sought to determine whether other regulators, aside from PLN and SLN, may exist. Several new regulators have been uncovered, one of which being a micropeptide called myoregulin (MLN), a 46-amino-acid skeletal muscle–specific protein that, similar to PLN, binds to and regulates SERCA1, SERCA2a, and SERCA2b through decreasing its affinity for Ca^{2+}.[96,97] Unlike PLN, which is expressed solely in cardiac and slow-twitch skeletal muscle, MLN is expressed in all skeletal muscle, regardless of slow-twitch vs. fast-twitch muscle types.[96,98] In addition to MLN in skeletal muscle, Olson and colleagues have also identified several more micropeptides with similar regulatory mechanisms on SERCA's affinity for Ca^{2+}, including endoregulin (ELN), expressed in endo/epithelial tissues, and another, regulin (ALN), which is expressed ubiquitously.[97] A 17kDa protein called small ankyrin 1 (sAnk1) has recently been shown to interact with SERCA1 through its transmembrane domain and is able to reduce Ca^{2+} affinity of SERCA1, but not as much as SLN, and has no effect on SERCA1 ATPase activity.[99] Interestingly, it has been shown that SLN and sAnk1 are able to interact whereby SLN is able to promote sAnk1 interactions with SERCA1, yet sAnk1 relieves SLN's inhibition of SERCA1,[100] similar to the interactions between SLN and PLN. It remains unclear, however, whether sAnk1 reduces SLN's ability to uncouple SERCA, and further investigation on this reciprocal relationship is still required. One other addition to the SERCA regulator family includes dwarf open reading frame (DWORF), which is expressed in cardiac muscle and slow skeletal muscle.[97,101,102] The unique feature of DWORF is that in binding to SERCA, it does not inhibit the pump, but displaces PLN and enhances its function, even in disease states.[97,101,102] DWORF is the only known micropeptide to enhance rather than inhibit SERCA function.[103,104]

Importantly, among these newly identified SERCA regulators, none have been shown to have the capability of uncoupling Ca^{2+} transport from ATP hydrolysis that SLN possesses. SLN expression in rodents and humans is mainly in the soleus and atria,[105] and despite the recent cell work in human skeletal muscle previously mentioned,[82] a recent human study has demonstrated that SLN expression in skeletal muscle does not appear to regulate fat mass in men, nor does it correlate with resting energy expenditure.[106] Thus, finding a novel SERCA uncoupler that is more ubiquitously expressed

TABLE 12.2
Summary of the Theoretical Effects of Known SERCA Regulators on SERCA Function

Regulator	Expression	Effect on SERCA Activity	Effect on SERCA Ca^{2+} Uptake
PLN	Cardiac and slow skeletal muscle	Neutral	Negative
SLN	Atria and soleus	Positive	Negative
MLN	Skeletal muscle	Neutral	Negative
ELN	Endo/epithelial tissues	Neutral	Negative
ALN	Ubiquitously expressed	Neutral	Negative
sAnk1	Striated muscle	Neutral	Negative
DWORF	Cardiac and slow skeletal muscle	Neutral	Neutral
NNAT	Brain, Adipose, skeletal muscle, hypothalamus, pancreatic β-cells	Positive	Negative

could have large implications. Recently, work investigating a small protein called neuronatin (NNAT) has demonstrated that it not only shares significant sequence homology with PLN and SLN[107,108] but appears to regulate Ca^{2+} levels in various tissues such as neurons,[109–111] adipocytes,[112] and pancreatic β-cells.[113] It is therefore postulated that NNAT is a novel regulator of SERCA function. Indeed, *in vitro* work has demonstrated NNAT to reduce SERCA's coupling ratio,[108] and, like *Sln*[KO] mice, NNAT knockout mice have shown increased susceptibility to diet-induced obesity and glucose intolerance.[114] Thus, NNAT may be a novel SERCA regulator and uncoupler, though further investigation into its role in skeletal muscle is warranted.

12.4.4 Sarcoplasmic Reticulum Lipid Composition

The SR is a membrane-bound organelle derived from the endoplasmic reticulum that is dedicated to Ca^{2+} storage in muscle cells.[115] It is organized into longitudinal tubules that surround each myofibril, ensuring sufficient storage space for intracellular Ca^{2+} and allowing for efficient contraction and relaxation cycles.[115] Importantly, various properties of the SR lipid bilayer such as bilayer thickness, lipid chain structure, and ratio of different fatty acid types can all influence SERCA function and possibly energy expenditure.[116–121] Two foundational studies by Froud et al.[118] and Cornea and Thomas[119] reconstituted SERCA into SRs containing fatty acyl chains of varying lengths and measured ATPase activity at maximally activating Ca^{2+} concentrations. Both studies demonstrated optimal SERCA ATP hydrolysis activity at chain lengths of 18 carbons with an inverted-U pattern whereas chain lengths shorter or longer than 18 carbons resulted in reductions in ATP hydrolysis activity.[118,119] However, a follow-up study using a range of physiological Ca^{2+} concentrations rather than a supraphysiological maximum showed that optimal ATPase activity of the SERCA pump occurred with chain lengths of 22 carbons.[116] Further, Gustavsson et al.[116] investigated the effect of carbon chain length on SERCA's affinity for Ca^{2+} and found an optimal chain length of 20 carbons. Given that naturally occurring skeletal muscle SR membranes have approximately 18-carbon-long chains,[122] it can be hypothesized that the SR is optimized for SERCA Ca^{2+} affinity rather than maximal ATP hydrolysis.[116] In addition to carbon chain length, the headgroups of phospholipids making up the SR lipid bilayer can affect SERCA function. For example, phosphatidylethanolamine (PE) makes up ~16% of skeletal muscle SR,[122] and increasing PE content in reconstituted SR membranes progressively increases SERCA ATPase activity.[116,123] Similarly, the less prominent phosphatidylserine (PS) and phosphatidylinositol (PI) also show activating effects on SERCA ATPase activity and Ca^{2+} uptake,[124] indicating that SERCA function is sensitive to both phospholipid length and head group.

Apart from membrane lipids and properties affecting the SERCA pump directly, they can also influence membrane permeability. For example, increases in docosahexenoic acid (DHA) in phospholipids and membrane unsaturation have been shown to increase permeability to cations, such as Na^+ and K^+, resulting in futile cycling by the Na^+/K^+ pump and increasing mitochondrial proton leak.[125,126] With regards to Ca^{2+}, a study by Fajardo et al.[127] demonstrated that the soleus muscle, which has high SR membrane permeability,[128] contains higher DHA content and greater SR unsaturation than the less leaky red and white gastrocnemius muscle. This suggests that DHA and unsaturation cultivate a leaky SR membrane, ultimately increasing SERCA ATPase activity to counteract this Ca^{2+} leak. This futile cycling can be beneficial for metabolic conditions such as obesity, as it would reduce SERCA's coupling ratio. In fact, a study by Fajardo et al.[117] demonstrated that following a dietary intake of DHA, it was incorporated into the SR membrane, promoting Ca^{2+} leak and raising energy expenditure via increased SERCA activity in rats. Other studies aimed at altering SR membrane lipid composition have also shown changes to SERCA's energy expenditure and whole-body susceptibility to diet-induced obesity[129] and muscle insulin sensitivity.[129,130] In summary, the composition of the SR membrane can dramatically affect SERCA function, either directly or indirectly via changes in Ca^{2+} leak, which may play a role in muscle health and disease.

12.5 EXERCISE AND SERCA FUNCTION

It is well understood that regular exercise is an important contributor to overall health that reduces the risk of obesity and its comorbidities surrounding metabolic complications in large part by increasing daily energy expenditure. Indeed, the biomedical perspective of obesity posits that it is a disease where caloric intake (energy in) exceeds energy expenditure. Physical activity or exercise accounts for 20–30% of daily energy output; therefore, increasing daily physical activity can counter the energy imbalance observed with obesity. Exercise is critical to maintaining healthy cardiac and skeletal muscle function, and the underlying molecular mechanisms involved are beginning to be uncovered. Exerkines, defined as signaling molecules released in response to acute or chronic exercise bouts, can be released through several different organs and demonstrate paracrine, endocrine, and autocrine effects. Myokines, released from the skeletal muscle in response to exercise-induced contractions, are thought to play a role in the beneficial whole-body effects of exercise. Myokine release at the skeletal muscle has been shown to initiate adipose-muscle crosstalk in response to exercise. Irisin and BDNF, both myokines released in response to exercise, have been shown to induce the beiging of white adipose tissue, thereby activating adaptive thermogenic mechanisms within the adipose. Whether exercise can stimulate SLN expression in skeletal muscle to promote SERCA uncoupling based thermogenesis remains unknown. It is understood that regular exercise can elevate PGC1a expression and activity. Further, increases in PGC1a have been shown to elevate the expression of SLN, suggesting the possibility of exercise influencing SERCA uncoupling. However, future studies are required to uncover this connection.

Exercise-induced changes in SERCA's regulation, function, and expression can also be seen in the context of muscle health and disease states. Skeletal muscle plasticity and structural remodeling of muscle fibers can be induced through exercise in response to metabolic and functional demands.[131] Exercise-stimulated changes in fiber type plasticity are often determined by elevated free $[Ca^{2+}]_i$ levels. SERCA isoform expression has been shown to be nearly fivefold greater in type II fibers compared to type I fibers, suggesting differences in capacity for Ca^{2+} sequestration back into the SR and slower rate of muscle relaxation. Likewise, slow-oxidative fibers report the greatest expression of SERCA2 where fast-glycolytic fibers possess SERCA1 isoforms. Indeed, calcineurin, a Ca^{2+}-dependent serine/threonine phosphatase, is involved in the promotion of the oxidative phenotype, whereas inhibition of calcineurin induces a slow-to-fast fiber type shift. Calcineurin can be activated in a Ca^{2+}-dependent manner, through increases in $[Ca^{2+}]_i$ levels following muscle contraction, allowing for dephosphorylation and nuclear translocation of the transcription factor NFATc.[132,133] In response to exercise, activation of calcineurin induces metabolic alterations in skeletal muscle, promoting mitochondrial biogenesis to match heightened energy demands.[134,135] During the transition from rest to exercise, energetic demand shifts upward 100-fold, and much of this energetic utilization is accounted for by the

SERCA pumps during muscle contraction.[136] Endurance exercise, defined by repeated, sustained low-intensity contractions, results in a fast-to-slow fiber type switch. It has been shown that 10 weeks of endurance exercise training in humans reduced SERCA1a expression with no changes to SERCA2a content in the vastus lateralis.[137] Further, in male rats, exercise training at moderate to high intensity increased SERCA2 mRNA content in the gastrocnemius, though there were no changes in either isoform found in the soleus.[138] Metabolic flexibility is a property of healthy skeletal muscle whereby substrate preference shifts in response to metabolic needs. Promotion of the oxidative phenotype observed with endurance exercise also promotes a shift in substrate utilization where fatty acid oxidation is the preferred source of fuel to sustain energy demands over carbohydrates.[139] Endurance exercise adaptations to skeletal muscle promote the oxidative capacity of fats through increased fatty acid flux, mitochondrial content, and oxidative enzymes. Likewise, the transition to type I fibers in response to endurance exercise promotes enhanced fatty acid oxidative capacity, as type I fibers have greater intramyocellular triacylglycerol storing capabilities.[140] Further, type I oxidative fibers possess higher quantities of the uncoupling protein SLN.[141] Indeed, $SLN^{O/E}$ mice show enhanced fatigue resistance in response to endurance exercise. In wild-type mice, SLN content has been shown to increase in response to diet-induced obesity thereby increasing energy demands on SERCA Ca^{2+} pumping under active states.[142] In conditions such as obesity and overweight, metabolic flexibility is impaired. Taken together, exercise-induced adaptations in skeletal muscle such as SERCA Ca^{2+} handling and fiber type plasticity play a pivotal role in combatting the energy imbalance associated with obesity and other conditions such as aging, disuse, and atrophy.

12.6 SERCA FUNCTION IN AGING AND DISUSE

12.6.1 Sarcopenia

Aging is associated with alterations to muscle quantity and quality, which can impact physical performance.[143] These changes and functional deficits are referred to as sarcopenia, the age-related loss of muscle mass and strength.[144] Progressive declines in muscle mass begin at approximately 40 years of age and are estimated at 8% per decade until the age of 70, after which the loss increases to 15% per decade.[145–148] Sarcopenia is characterized by preferential atrophy of type II fibers, leading to a greater proportion of type I fibers, which results in slower contraction and relaxation times.[149] Reduced relaxation times in aged muscle may be attributed to reduced SERCA activity caused by oxidative stress. The etiology of sarcopenia is multifactorial; however, oxidative stress is suggested to play an important role in the initiation and progression of muscle atrophy during aging.[150] There is an imbalance between the generation of RONS and the ability of antioxidant defense systems to remove RONS with biological aging, leading to oxidative stress.[151] RONS target proteins for post-translational modifications, leading to altered enzyme activity or loss of function, which may be responsible for age-related dysfunction.[152–154] Structurally, the SERCA pumps are highly susceptible to post-translational modifications by RONS, as they contain vulnerable cysteine, lysine, and tyrosine residues.[154–157] Further, SERCA has a relatively long half-life in the cell (~ 14–15 days) and is turned over even more slowly with age,[156] leading to an accumulation of modified SERCA protein in aged muscle.[156] The SERCA2a isoform progressively accumulates a significant amount of nitrotyrosine with age,[156] which may contribute to age-dependent reductions in the ATPase activity and Ca^{2+} uptake rate of the SERCA pump in slow-twitch muscles.[156,158] In turn, reduced SERCA activity can negatively impact muscle form and function, leading to atrophy and weakness. Previous studies have found impaired SERCA function in skeletal muscles of mice with a deletion of cellular antioxidant enzyme CuZn superoxide dismutase ($Sod1^{-/-}$), a model that is highly susceptible to oxidative stress.[159] This mouse model has many features of rapid and accelerated sarcopenia, including reduced specific force production.[159] Further, in naturally aged muscles of 26-month-old mice, there is evidence of elevated RONS production and reduced SERCA activity,[160] suggesting that oxidative stress impairs SERCA function.

There is limited literature on sarcopenia and muscle-based adaptive thermogenesis; however, it is important to discuss sarcopenic obesity, which is the coexistence of low lean muscle mass and excess adiposity with age.[161,162] Given that skeletal muscle is a major site of metabolic activity, the loss of muscle mass contributes, in part, to decreased resting energy expenditure and physical activity levels.[162,163] Consequently, low muscle mass, physical activity levels, and energy expenditure can lead to weight gain characterized predominately as adipose tissue.[164] Increasing SERCA uncoupling could be advantageous for aging and could potentially lower excess fat mass. Additionally, increased fat mass and obesity further compound the age-related loss of muscle by induing low-grade inflammation and impairing the regenerative capacity of skeletal muscle.[162] Thus, promoting SERCA uncoupling in aged can lower fat mass and protect against sarcopenia. On the other hand, promoting SERCA uncoupling can lead to elevated free $[Ca^{2+}]_i$ levels, which may have pathological consequences on excitation-contraction coupling and mitochondrial function.[166,167] Future studies investigating the mechanisms of muscle-based thermogenesis are necessary to confirm whether SERCA uncoupling can protect against sarcopenia and the effects of aging.

12.6.2 Spaceflight

The basis of skeletal muscle atrophy is quite straightforward; if you don't use it, you lose it. This is made evident in cases of disuse atrophy, such as aging or injury, as well as with spaceflight (often considered an aging analog), where significant muscle weakness and wasting are observed. Microgravity exposure is known to pose several hazards to astronaut health[165–167] that could impair their ability to perform duties during missions and upon return to Earth. Mammals have evolved under the constant influence of the gravitational pull, and thus our cardiovascular and musculoskeletal systems—especially those involved in maintaining posture—are radically affected in a microgravity environment. For example, without the influence of gravity, the cardiovascular system does not need to "work" as hard to pump blood to peripheral tissues, resulting in fluid shifts, reductions in blood pressure, and cardiac muscle atrophy.[168] Spaceflight also unloads our bones, resulting in reductions in bone mineral content and bone mineral density as well as decrements in bone structure, demonstrated in mice[169] and humans.[170] Finally, in rodents, the loss of gravity induces fiber type shifts toward the fast-glycolytic phenotype, muscle weakness and atrophy with postural muscles, such as the soleus, typically being the most affected.[171–175] Similar changes to skeletal muscle have also been reported in humans following just 17 days of spaceflight.[176] Importantly, changes to the musculoskeletal system become more pronounced with longer durations,[177,178] meaning that finding the molecular basis behind these changes as well as potential countermeasures is imperative to allow for longer-duration human missions.

An interesting finding regarding muscle unloading that may provide insight into the mechanisms underlying atrophy and weakness is that skeletal muscle shows a disproportionate loss in muscle mass and strength whereby the declines in strength occur earlier and faster than the declines in muscle mass.[179] This suggests that the muscle weakness seen in unloading is not due to atrophy alone, suggesting detriments to muscle quality such as dysfunctional Ca^{2+} handling,[179,180] though other mechanisms may be at play (i.e., mitochondrial impairments). This is important, as SERCA functions not only to elicit muscle relaxation but also to maintain low intracellular Ca^{2+}. SERCA dysfunction, therefore, can raise Ca^{2+} levels in the cytosol leading to cytotoxic events including, but not limited to, elevated RONS production, protein degradation, and cell death.[181–184] Consistent with this hypothesis, a study using hind-limb suspension, the best Earth-based spaceflight analog, demonstrated increases in intracellular Ca^{2+} in murine soleus muscle.[180] In humans, 10 days of bed rest resulted in impaired Ca^{2+} release in both slow and fast single muscle fibers.[179] In humans, 10 days of bed rest resulted in impaired Ca^{2+} release in both slow and fast single muscle fibers.[179] Increased SERCA1a *mRNA* and protein content[86] are observed in the soleus muscle following spaceflight as well as hindlimb unloading,[185,186] consistent with a fiber-type shift toward a glycolytic phenotype. SERCA2a content has been reported as unchanged[185,186] or increased,[86] but only one study thus far has characterized SERCA function following spaceflight in murine skeletal muscle. In it, the authors found drastically impaired SERCA Ca^{2+} uptake in the spaceflown soleus along with reductions in PLN and NNAT but increases in SLN.[86] On the contrary, in the glycolytic tibialis anterior (TA) muscle, SERCA Ca^{2+} uptake was increased following spaceflight despite signs of atrophy and a fiber type shift.[86] Upon further investigation it became clear that the soleus had elevated levels of RONS that was absent in the TA,[86] providing a potential explanation into the SERCA dysfunction in the soleus, but not in the TA following spaceflight. SERCA is highly susceptible to RONS modifications that can impair its function, indicating that elevated RONS can contribute to impairments in Ca^{2+} regulation. Importantly, elevated intracellular Ca^{2+} can also influence RONS production, leading to a cycle of damage in skeletal muscle highlighting both SERCA function and RONS production as possible targets for spaceflight-related muscle weakness and atrophy. Indeed, SERCA activation has been shown to reduce RONS production and improve muscle health in aged mice as well as a mouse model of oxidative stress.[183] That being said, research remains limited, and targeting SERCA dysfunction as a therapeutic for spaceflight-related muscle atrophy and weakness remains in early stages.

The role of muscle-based thermogenesis in spaceflight remains unexplored. The increases in soleus SLN content following spaceflight suggest a role of SERCA uncoupling in spaceflight atrophy and weakness; however, whether this is protective or pathological to the muscle remains to be confirmed. Previous work using a ground-based tenotomy model, where the soleus and gastrocnemius tendons are severed, unloading the soleus and overloading the plantaris, has hinted at SLN playing a protective effect. In this study it was found that in the absence of SLN, through the use of *Sln*$^{-/-}$ mice, both the overloaded plantaris and the unloaded soleus showed smaller myofiber cross-sectional area and greater proportions of fast-glycolytic fibers, attributed to reductions in calcineurin signaling.[88] Thus, muscle-based thermogenesis via SERCA uncoupling may play an important role in counteracting spaceflight-induced atrophy through the promotion of calcineurin signaling and muscle hypertrophy pathways. Further work is necessary to confirm and understand the mechanisms of muscle-based thermogenesis and spaceflight-induced atrophy as well as to investigate potential contributions of NNAT and PLN.

12.6.3 Muscular Dystrophy

Muscular dystrophies (MDs) are a group of degenerative muscle disorders characterized by progressive muscle wasting and weakness.[187] It is caused by the absence of functional dystrophin—a structural protein that provides membrane stability to muscle throughout strenuous contraction-relaxation cycles.[188] Duchenne muscular dystrophy (DMD) is the most common form of MD and has been studied extensively using the preclinical *mdx* mouse model.[189] The loss of dystrophin impacts metabolic function, namely that *mdx* mice exhibit a lean phenotype and possess less body fat mass compared with WT mice.[190] It is important to note that *mdx* mice present with a hypermetabolic phenotype,[191,192] expending a significant amount of energy on a daily basis to support the costs of muscle degeneration and regeneration. Additionally, *mdx* mice have lower core body temperature compared with wild-type mice[190] and thus likely need to expend energy in order to generate heat.

SLN, an uncoupler of the SERCA pump, is upregulated in skeletal muscles of *mdx* and *mdx/utr*$^{-/-}$ (DKO) mice.[87] DKO mice are deficient in dystrophin and utrophin, which is a dystrophin homolog that is upregulated in the *mdx* mouse to provide compensatory membrane stability.[87] Increased levels of SLN can uncouple the SERCA pump and increase ATP utilization,[193] which may be an attempt to increase body temperature. Further, since SLN is known to uncouple the SERCA pump and increase the energetic requirements of Ca^{2+} transport, increased SLN expression in *mdx* muscles may, in part, explain the hypermetabolic phenotype.[51,191,194] However, there is conflicting evidence on whether or not increases in SLN are beneficial.[91,195–197] Schneider et al. [87] were the first to report elevated SLN expression in muscles of *mdx* and DKO mice. Additionally, they demonstrated reduced SERCA activity in various muscle types and suggested that elevated SLN content and impaired SERCA activity worsen dystrophic pathology.[87] On the contrary, Fajardo et al.[91] demonstrated that deletion of SLN in *mdx* mice impaired calcineurin signaling, which led to worsened dystrophic pathology compared with *mdx* control mice. As previously described, calcineurin is a Ca^{2+}/calmodulin-dependent serine/threonine phosphatase that promotes the expression of genes associated with the oxidative fiber type, such as utrophin.[131] This is important, as type II fibers are more suspectable to atrophy in DMD, and promoting the oxidative phenotype is found to alleviate the dystrophic phenotype.[149,198] Consistent with Fajardo et al.,[91] Tanihata et al.[197] reported significant reductions in muscle mass in *mdx*/SLN-knockout mice, which may be due to impaired muscle regeneration, since calcineurin is a known activator of muscle regeneration.[195] Voit et al.[196] found that reducing SLN expression in DKO mice using adeno-associated virus 9–mediated RNA interference attenuated dystrophic pathology in skeletal muscles as demonstrated by lower necrosis and fibrosis. Moreover, reducing SLN expression increased the life span of DKO mice.[196] However, total deletion of SLN in DKO mice did not improve disease phenotype in the diaphragm, as force production and fibrosis were similar to control mice.[196] Thus, SLN overexpression may have compensatory effects on muscle function in *mdx* and DKO mice, and total SLN deletion may exacerbate muscle atrophy and weakness in dystrophic muscles.

12.7 THE ROLE OF BIOLOGICAL SEX

Skeletal muscle exhibits sexual dimorphic features in which total lean muscle mass, size of individual fibers, fiber type composition, energy metabolism, contractile strength, and fatigability are different between males and females.[199] It is well established that males have greater force production, lean body mass, and CSA of individual fibers than do females.[200] Males are also known to have a greater proportion of fast-twitch fibers than females. As a result, force generation and relaxation velocity are faster in males than in females, but female muscles are more fatigue resistant and recover faster than male muscles.[201] Further, females have a lower respiratory exchange ratio than males, which is indicative of a greater reliance on fat oxidation for fuel as opposed to carbohydrate and glycogen utilization.[202,203] These sex-based differences are likely due to the expression of estrogen receptors in skeletal muscle, suggesting that estrogen has a direct effect on skeletal muscle function and metabolism. A recent study found that ovariectomized rats, a preclinical model of menopause in which ovaries are surgically removed, have reduced SERCA and sarcolipin content compared with sham control rats,[204] suggesting that estrogen regulates SERCA uncoupling, and estrogen deficiency may impair muscle-based adaptive thermogenesis. However, despite all of these sex-based differences, there is limited literature on muscle-based adaptive thermogenesis and skeletal muscle SERCA function in females.

REFERENCES

1. Zurlo, F., Larson, K., Bogardus, C. & Ravussin, E. Skeletal muscle metabolism is a major determinant of resting energy expenditure. *J Clin Invest* **86**, 1423–1427 (1990).
2. Kalyani, R.R., Corriere, M. & Ferrucci, L. Age-related and disease-related muscle loss: the effect of diabetes, obesity, and other diseases. *Lancet Diabetes Endocrinol* **2**, 819–829 (2014).
3. Picard, M., Hepple, R.T. & Burelle, Y. Mitochondrial functional specialization in glycolytic and oxidative muscle fibers: tailoring the organelle for optimal function. *American Journal of Physiology-Cell Physiology* **302**, C629–C641 (2012).
4. Bourdeau Julien, I., Sephton, C.F. & Dutchak, P.A. Metabolic Networks Influencing Skeletal Muscle Fiber Composition. *Front Cell Dev Biol* **6**, 125 (2018).
5. Severinsen, M.C.K. & Pedersen, B.K. Muscle-Organ Crosstalk: The Emerging Roles of Myokines. *Endocr Rev* **41**, 594–609 (2020).
6. Voet, D., Voet, J.G. & Pratt, C.W. *Fundamentals of Biochemistry*, (John Wiley & Sons., 2016).
7. McArdle, W.D., Katch, F.I. & Katch, V.L. *Exercise Physiology: Energy, Nutrition, and Human Performance*, (Wolters Kluwer Health, 2015).

8. Törnroth-Horsefield, S. & Neutze, R. Opening and closing the metabolite gate. *Proc Natl Acad Sci U S A* **105**, 19565–19566 (2008).
9. Trexler, E.T., Smith-Ryan, A.E. & Norton, L.E. Metabolic adaptation to weight loss: implications for the athlete. *Journal of the International Society of Sports Nutrition* **11**, 7 (2014).
10. Rolfe, D.F. & Brown, G.C. Cellular Energy Utilization and Molecular Origin of Standard Metabolic Rate in Mammals. *Physiological Reviews* **77**, 731–758 (1997).
11. Chouchani, E.T., Kazak, L. & Spiegelman, B.M. New Advances in Adaptive Thermogenesis: UCP1 and Beyond. *Cell Metab* **29**, 27–37 (2019).
12. Ikeda, K., et al. UCP1-independent signaling involving SERCA2b-mediated calcium cycling regulates beige fat thermogenesis and systemic glucose homeostasis. *Nat Med* **23**, 1454–1465 (2017).
13. Gamu, D., Bombardier, E., Smith, I.C., Fajardo, V.A. & Tupling, A.R. Sarcolipin Provides a Novel Muscle-Based Mechanism for Adaptive Thermogenesis. *Exercise and Sport Sciences Reviews* **42**, 136–142 (2014).
14. Landesberg, A. & Sideman, S. Coupling calcium binding to troponin C and cross-bridge cycling in skinned cardiac cells. *Am J Physiol* **266**, H1260–1271 (1994).
15. Periasamy, M., et al. Role of SERCA Pump in Muscle Thermogenesis and Metabolism. *Compr Physiol* **7**, 879–890 (2017).
16. DiFranco, M., Hakimjavadi, H., Lingrel, J.B. & Heiny, J.A. Na,K-ATPase α2 activity in mammalian skeletal muscle T-tubules is acutely stimulated by extracellular K+. *J Gen Physiol* **146**, 281–294 (2015).
17. Periasamy, M. & Huke, S. SERCA pump level is a critical determinant of Ca(2+)homeostasis and cardiac contractility. *J Mol Cell Cardiol* **33**, 1053–1063 (2001).
18. Misquitta, C.M., Mack, D.P. & Grover, A.K. Sarco/endoplasmic reticulum Ca2+ (SERCA)-pumps: link to heart beats and calcium waves. *Cell Calcium* **25**, 277–290 (1999).
19. Palmgren, M.G. & Nissen, P. P-type ATPases. *Annu Rev Biophys* **40**, 243–266 (2011).
20. Periasamy, M. & Kalyanasundaram, A. SERCA pump isoforms: their role in calcium transport and disease. *Muscle Nerve* **35**, 430–442 (2007).
21. Protasi, F. Structural interaction between RYRs and DHPRs in calcium release units of cardiac and skeletal muscle cells. *Front Biosci* **7**, d650–658 (2002).
22. Calderón, J.C., Bolaños, P. & Caputo, C. The excitation-contraction coupling mechanism in skeletal muscle. *Biophys Rev* **6**, 133–160 (2014).
23. Lamb, G.D. Excitation-contraction coupling in skeletal muscle: comparisons with cardiac muscle. *Clin Exp Pharmacol Physiol* **27**, 216–224 (2000).
24. Samsó, M. 3D Structure of the Dihydropyridine Receptor of Skeletal Muscle. *Eur J Transl Myol* **25**, 4840 (2015).
25. Schrötter, K., Dayal, A. & Grabner, M. The mammalian skeletal muscle DHPR has larger Ca(2+) conductance and is phylogenetically ancient to the early ray-finned fish sterlet (Acipenser ruthenus). *Cell Calcium* **61**, 22–31 (2017).
26. Shiels, H.A. *Design and Physiology of the Heart: Cardiac Excitation-Contraction Coupling: Routes of Cellular Calcium Flux*, (Academic Press, 2011).
27. Van Petegem, F. Ryanodine receptors: structure and function. *J Biol Chem* **287**, 31624–31632 (2012).
28. Lanner, J.T., Georgiou, D.K., Joshi, A.D. & Hamilton, S.L. Ryanodine receptors: structure, expression, molecular details, and function in calcium release. *Cold Spring Harb Perspect Biol* **2**, a003996 (2010).
29. Meizoso-Huesca, A., Pearce, L., Barclay, C.J. & Launikonis, B.S. Ca(2+) leak through ryanodine receptor 1 regulates thermogenesis in resting skeletal muscle. *Proc Natl Acad Sci U S A* **119**(2022).
30. Meissner, G. Regulation of Ryanodine Receptor Ion Channels Through Posttranslational Modifications. *Curr Top Membr* **66**, 91–113 (2010).
31. Bellinger, A.M., et al. Hypernitrosylated ryanodine receptor calcium release channels are leaky in dystrophic muscle. *Nat Med* **15**, 325–330 (2009).
32. MacLennan, D.H., Rice, W.J. & Green, N.M. The mechanism of Ca2+ transport by sarco(endo)plasmic reticulum Ca2+-ATPases. *J Biol Chem* **272**, 28815–28818 (1997).
33. McIntosh, D.B., Woolley, D.G., Vilsen, B. & Andersen, J.P. Mutagenesis of segment 487Phe-Ser-Arg-Asp-Arg-Lys492 of sarcoplasmic reticulum Ca2+-ATPase produces pumps defective in ATP binding. *J Biol Chem* **271**, 25778–25789 (1996).
34. Toyoshima, C., Nakasako, M., Nomura, H. & Ogawa, H. Crystal structure of the calcium pump of sarcoplasmic reticulum at 2.6 Å resolution. *Nature* **405**, 647–655 (2000).

35. Nagarajan, A., Andersen, J.P. & Woolf, T.B. The role of domain: domain interactions versus domain: water interactions in the coarse-grained simulations of the E1P to E2P transitions in Ca-ATPase (SERCA). *Proteins* **80**, 1929–1947 (2012).
36. Daiho, T., et al. Deletions of any single residues in Glu40-Ser48 loop connecting a domain and the first transmembrane helix of sarcoplasmic reticulum Ca(2+)-ATPase result in almost complete inhibition of conformational transition and hydrolysis of phosphoenzyme intermediate. *J Biol Chem* **278**, 39197–39204 (2003).
37. Toyoshima, C. Ion pumping by calcium ATPase of sarcoplasmic reticulum. in *Regulatory mechanisms of striated muscle contraction* 295–303 (Springer, 2007).
38. Huang, S. & Squier, T.C. Enhanced rotational dynamics of the phosphorylation domain of the Ca-ATPase upon calcium activation. *Biochemistry* **37**, 18064–18073 (1998).
39. Zhang, Z., Lewis, D., Sumbilla, C., Inesi, G. & Toyoshima, C. The role of the M6-M7 loop (L67) in stabilization of the phosphorylation and Ca(2+) binding domains of the sarcoplasmic reticulum Ca(2+)-ATPase (SERCA). *J Biol Chem* **276**, 15232–15239 (2001).
40. Zador, E., Vangheluwe, P. & Wuytack, F. The expression of the neonatal sarcoplasmic reticulum Ca2+ pump (SERCA1b) hints to a role in muscle growth and development. *Cell Calcium* **41**, 379–388 (2007).
41. Zhao, Y., et al. Functional analysis of SERCA1b, a highly expressed SERCA1 variant in myotonic dystrophy type 1 muscle. *Biochim Biophys Acta* **1852**, 2042–2047 (2015).
42. Dally, S., et al. Ca2+-ATPases in non-failing and failing heart: evidence for a novel cardiac sarco/endoplasmic reticulum Ca2+-ATPase 2 isoform (SERCA2c). *Biochem J* **395**, 249–258 (2006).
43. Gamu, D., Juracic, E.S., Hall, K.J. & Tupling, A.R. The sarcoplasmic reticulum and SERCA: a nexus for muscular adaptive thermogenesis. *Appl Physiol Nutr Metab* **45**, 1–10 (2020).
44. Dally, S., et al. Compartmentalized expression of three novel sarco/endoplasmic reticulum Ca2+ATPase 3 isoforms including the switch to ER stress, SERCA3f, in non-failing and failing human heart. *Cell Calcium* **45**, 144–154 (2009).
45. de Meis, L., Arruda, A.P. & Carvalho, D.P. Role of sarco/endoplasmic reticulum Ca(2+)-ATPase in thermogenesis. *Biosci Rep* **25**, 181–190 (2005).
46. de Meis, L. Uncoupled ATPase activity and heat production by the sarcoplasmic reticulum Ca2+-ATPase. Regulation by ADP. *J Biol Chem* **276**, 25078–25087 (2001).
47. Tupling, A.R. The sarcoplasmic reticulum in muscle fatigue and disease: role of the sarco(endo)plasmic reticulum Ca2+-ATPase. *Can J Appl Physiol* **29**, 308–329 (2004).
48. Ermak, G. & Davies, K.J. Calcium and oxidative stress: from cell signaling to cell death. *Mol Immunol* **38**, 713–721 (2002).
49. Zhivotovsky, B. & Orrenius, S. Calcium and cell death mechanisms: a perspective from the cell death community. *Cell Calcium* **50**, 211–221 (2011).
50. Berman, M.C. Slippage and uncoupling in P-type cation pumps; implications for energy transduction mechanisms and regulation of metabolism. *Biochim Biophys Acta* **1513**, 95–121 (2001).
51. Bombardier, E., Smith, I.C., Vigna, C., Fajardo, V.A. & Tupling, A.R. Ablation of sarcolipin decreases the energy requirements for Ca2+ transport by sarco(endo)plasmic reticulum Ca2+-ATPases in resting skeletal muscle. *FEBS Lett* **587**, 1687–1692 (2013).
52. Smith, I.C., Bombardier, E., Vigna, C. & Tupling, A.R. ATP Consumption by Sarcoplasmic Reticulum Ca2+ Pumps Accounts for 40–50% of Resting Metabolic Rate in Mouse Fast and Slow Twitch Skeletal Muscle. *PLoS One*, Vol. **8** (2013).
53. Allen, D.G., Lamb, G.D. & Westerblad, H. Skeletal muscle fatigue: cellular mechanisms. *Physiol Rev* **88**, 287–332 (2008).
54. Speakman, J.R. Measuring energy metabolism in the mouse - theoretical, practical, and analytical considerations. *Front Physiol* **4**, 34 (2013).
55. Laver, D.R. Regulation of the RyR channel gating by Ca(2+) and Mg(2). *Biophys Rev* **10**, 1087–1095 (2018).
56. Periasamy, M., et al. Regulation of sarcoplasmic reticulum Ca2+ ATPase pump expression and its relevance to cardiac muscle physiology and pathology. *Cardiovascular Research* **77**, 265–273 (2007).
57. Primeau, J.O., Armanious, G.P., Fisher, M.E. & Young, H.S. The SarcoEndoplasmic Reticulum Calcium ATPase. *Subcell Biochem* **87**, 229–258 (2018).
58. Gustavsson, M., et al. Allosteric regulation of SERCA by phosphorylation-mediated conformational shift of phospholamban. *Proc Natl Acad Sci U S A* **110**, 17338–17343 (2013).

59. Akin, B.L., Hurley, T.D., Chen, Z. & Jones, L.R. The structural basis for phospholamban inhibition of the calcium pump in sarcoplasmic reticulum. *J Biol Chem* **288**, 30181–30191 (2013).
60. Jones, L.R., Cornea, R.L. & Chen, Z. Close proximity between residue 30 of phospholamban and cysteine 318 of the cardiac Ca2+ pump revealed by intermolecular thiol cross-linking. *J Biol Chem* **277**, 28319–28329 (2002).
61. Frank, K. & Kranias, E.G. Phospholamban and cardiac contractility. *Ann Med* **32**, 572–578 (2000).
62. MacLennan, D.H. & Kranias, E.G. Phospholamban: a crucial regulator of cardiac contractility. *Nat Rev Mol Cell Biol* **4**, 566–577 (2003).
63. Mattiazzi, A. & Kranias, E.G. The role of CaMKII regulation of phospholamban activity in heart disease. *Front Pharmacol* **5**, 5 (2014).
64. Mattiazzi, A., Mundiña-Weilenmann, C., Guoxiang, C., Vittone, L. & Kranias, E. Role of phospholamban phosphorylation on Thr17 in cardiac physiological and pathological conditions. *Cardiovasc Res* **68**, 366–375 (2005).
65. Colyer, J. Phosphorylation states of phospholamban. *Ann N Y Acad Sci* **853**, 79–91 (1998).
66. Schmitt, J.P., et al. Dilated cardiomyopathy and heart failure caused by a mutation in phospholamban. *Science* **299**, 1410–1413 (2003).
67. Fajardo, V.A., et al. Phospholamban overexpression in mice causes a centronuclear myopathy-like phenotype. in *Dis Model Mech*, Vol. **8** 999–1009 (2015).
68. Neumann, J., et al. Targeted overexpression of phospholamban to mouse atrium depresses Ca2+ transport and contractility. *J Mol Cell Cardiol* **30**, 1991–2002 (1998).
69. Kadambi, V.J., et al. Cardiac-specific overexpression of phospholamban alters calcium kinetics and resultant cardiomyocyte mechanics in transgenic mice. *J Clin Invest* **97**, 533–539 (1996).
70. Song, Q., et al. Overexpression of phospholamban in slow-twitch skeletal muscle is associated with depressed contractile function and muscle remodeling. *FASEB J* **18**, 974–976 (2004).
71. Fajardo, V.A., et al. Phospholamban overexpression in mice causes a centronuclear myopathy-like phenotype. *Dis Model Mech* **8**, 999–1009 (2015).
72. Sahoo, S.K., Shaikh, S.A., Sopariwala, D.H., Bal, N.C. & Periasamy, M. Sarcolipin protein interaction with sarco(endo)plasmic reticulum Ca2+ ATPase (SERCA) is distinct from phospholamban protein, and only sarcolipin can promote uncoupling of the SERCA pump. *J Biol Chem* **288**, 6881–6889 (2013).
73. Gamu, D., et al. Phospholamban deficiency does not alter skeletal muscle SERCA pumping efficiency or predispose mice to diet-induced obesity. *Am J Physiol Endocrinol Metab* **316**, E432–e442 (2019).
74. Glaves, J.P., et al. Interaction of a Sarcolipin Pentamer and Monomer with the Sarcoplasmic Reticulum Calcium Pump, SERCA. *Biophys J* **118**, 518–531 (2020).
75. Winther, A.M., et al. The sarcolipin-bound calcium pump stabilizes calcium sites exposed to the cytoplasm. *Nature* **495**, 265–269 (2013).
76. Toyoshima, C., et al. Crystal structures of the calcium pump and sarcolipin in the Mg2+-bound E1 state. *Nature* **495**, 260–264 (2013).
77. Sahoo, S.K., et al. The N Terminus of Sarcolipin Plays an Important Role in Uncoupling Sarco-endoplasmic Reticulum Ca2+-ATPase (SERCA) ATP Hydrolysis from Ca2+ Transport. *J Biol Chem* **290**, 14057–14067 (2015).
78. Fernández-de Gortari, E., Aguayo-Ortiz, R., Autry, J.M. & Michel Espinoza-Fonseca, L. A hallmark of phospholamban functional divergence is located in the N-terminal phosphorylation domain. *Comput Struct Biotechnol J* **18**, 705–713 (2020).
79. Mall, S., et al. The presence of sarcolipin results in increased heat production by Ca(2+)-ATPase. *J Biol Chem* **281**, 36597–36602 (2006).
80. Ceddia, D.M.S.-K., Yass, S.-N., Ayesha, I., George, B. & Rolando, B. Cold acclimation causes fiber type-specific responses in glucose and fat metabolism in rat skeletal muscles. *Scientific Reports* **7**, 1–12 (2017).
81. Bombardier, E., et al. Sarcolipin trumps beta-adrenergic receptor signaling as the favored mechanism for muscle-based diet-induced thermogenesis. *FASEB J* **27**, 3871–3878 (2013).
82. Mengeste, A.M., et al. Knockdown of sarcolipin (SLN) impairs substrate utilization in human skeletal muscle cells. *Mol Biol Rep* **49**, 6005–6017 (2022).
83. Bal, N.C., et al. Sarcolipin is a newly identified regulator of muscle-based thermogenesis in mammals. *Nat Med* **18**, 1575–1579 (2012).
84. Maurya, S.K., et al. Sarcolipin Signaling Promotes Mitochondrial Biogenesis and Oxidative Metabolism in Skeletal Muscle. *Cell Rep* **24**, 2919–2931 (2018).

85. Tomiya, S., et al. Cast immobilization of hindlimb upregulates sarcolipin expression in atrophied skeletal muscles and increases thermogenesis in C57BL/6J mice. *Am J Physiol Regul Integr Comp Physiol* **317**, R649–r661 (2019).
86. Braun, J.L., Geromella, M.S., Hamstra, S.I., Messner, H.N. & Fajardo, V.A. Characterizing SERCA Function in Murine Skeletal Muscles after 35–37 Days of Spaceflight. *Int J Mol Sci* **22**(2021).
87. Schneider, J.S., et al. Increased sarcolipin expression and decreased sarco(endo)plasmic reticulum Ca2+ uptake in skeletal muscles of mouse models of Duchenne muscular dystrophy. *Journal of Muscle Research and Cell Motility* **34**, 349–356 (2013).
88. Fajardo, V.A., et al. Effects of sarcolipin deletion on skeletal muscle adaptive responses to functional overload and unload. *Am J Physiol Cell Physiol* **313**, C154–C161 (2017).
89. Niranjan, N., et al. Sarcolipin overexpression impairs myogenic differentiation in Duchenne muscular dystrophy. *American Journal of Physiology-Cell Physiology* **317**, C813–C824 (2019).
90. Wang, W.P., et al. Progerin in muscle leads to thermogenic and metabolic defects via impaired calcium homeostasis. *Aging Cell* **19**, e13090 (2020).
91. Fajardo, V.A., et al. Sarcolipin deletion in mdx mice impairs calcineurin signalling and worsens dystrophic pathology. *Hum Mol Genet* **27**, 4094–4102 (2018).
92. Sopariwala, D.H., et al. Sarcolipin overexpression improves muscle energetics and reduces fatigue. *J Appl Physiol (1985)* **118**, 1050–1058 (2015).
93. Asahi, M., Kurzydlowski, K., Tada, M. & MacLennan, D.H. Sarcolipin inhibits polymerization of phospholamban to induce superinhibition of sarco(endo)plasmic reticulum Ca2+-ATPases (SERCAs). *J Biol Chem* **277**, 26725–26728 (2002).
94. Asahi, M., et al. Sarcolipin regulates sarco(endo)plasmic reticulum Ca2+-ATPase (SERCA) by binding to transmembrane helices alone or in association with phospholamban. *Proc Natl Acad Sci U S A* **100**, 5040–5045 (2003).
95. Morita, T., et al. Interaction sites among phospholamban, sarcolipin, and the sarco(endo)plasmic reticulum Ca(2+)-ATPase. *Biochem Biophys Res Commun* **369**, 188–194 (2008).
96. Tonkin, J. & Rosenthal, N. One small step for muscle: a new micropeptide regulates performance. *Cell Metab* **21**, 515–516 (2015).
97. Singh, D.R., et al. Newly Discovered Micropeptide Regulators of SERCA Form Oligomers but Bind to the Pump as Monomers. *J Mol Biol* (2019).
98. Anderson, D.M., et al. A micropeptide encoded by a putative long noncoding RNA regulates muscle performance. *Cell* **160**, 595–606 (2015).
99. Desmond, P.F., Muriel, J., Markwardt, M.L., Rizzo, M.A. & Bloch, R.J. Identification of Small Ankyrin 1 as a Novel Sarco(endo)plasmic Reticulum Ca2+-ATPase 1 (SERCA1) Regulatory Protein in Skeletal Muscle. *J Biol Chem* **290**, 27854–27867 (2015).
100. Desmond, P.F., et al. Interactions between small ankyrin 1 and sarcolipin coordinately regulate activity of the sarco(endo)plasmic reticulum Ca(2+)-ATPase (SERCA1). *J Biol Chem* **292**, 10961–10972 (2017).
101. Makarewich, C.A., et al. The DWORF micropeptide enhances contractility and prevents heart failure in a mouse model of dilated cardiomyopathy. in *eLife*, Vol. **7** (2018).
102. Nelson, B.R., et al. A peptide encoded by a transcript annotated as long noncoding RNA enhances SERCA activity in muscle. *Science* **351**, 271–275 (2016).
103. Pauli, L.E.C.-Q., Sarah, H. & Andrea. Decoding sORF translation – from small proteins to gene regulation. *RNA Biology* **13**, 1051–1059 http://doi.org/10.1080/15476286.2016.1218589 (2016).
104. Timmons, V.G., Mark, E.C. & James, A. Loss of neuronatin promotes "browning" of primary mouse adipocytes while reducing Glut1-mediated glucose disposal. https://doi.org/10.1152/ajpendo.00463.2012 (2013).
105. Odermatt, A., et al. Characterization of the gene encoding human sarcolipin (SLN), a proteolipid associated with SERCA1: absence of structural mutations in five patients with Brody disease. *Genomics* **45**, 541–553 (1997).
106. Morales-Alamo, D., et al. Sarcolipin expression in human skeletal muscle: Influence of energy balance and exercise. *Scand J Med Sci Sports* **30**, 408–420 (2020).
107. Pitale, P.M., Howse, W. & Gorbatyuk, M. Neuronatin Protein in Health and Disease. *J Cell Physiol* **232**, 477–481 (2017).
108. Braun, J.L., et al. Neuronatin promotes SERCA uncoupling and its expression is altered in skeletal muscles of high-fat diet-fed mice. *FEBS Lett* **595**, 2756–2767 (2021).

109. Joseph, R., Dou, D. & Tsang, W. Neuronatin mRNA: alternatively spliced forms of a novel brain-specific mammalian developmental gene. *Brain Res* **690**, 92–98 (1995).
110. Lin, H.H., et al. Neuronatin promotes neural lineage in ESCs via Ca(2+) signaling. *Stem Cells* **28**, 1950–1960 (2010).
111. Oyang, E.L., Davidson, B.C., Lee, W. & Poon, M.M. Functional Characterization of the Dendritically Localized mRNA Neuronatin in Hippocampal Neurons. in *PLoS One*, Vol. **6** (2011).
112. Suh, Y.H., et al. Ectopic expression of Neuronatin potentiates adipogenesis through enhanced phosphorylation of cAMP-response element-binding protein in 3T3-L1 cells. *Biochem Biophys Res Commun* **337**, 481–489 (2005).
113. Joe, M.K., et al. Crucial roles of neuronatin in insulin secretion and high glucose-induced apoptosis in pancreatic beta-cells. *Cell Signal* **20**, 907–915 (2008).
114. Millership, S.J., et al. Neuronatin deletion causes postnatal growth restriction and adult obesity in 129S2/Sv mice. in *Mol Metab*, Vol. **18** 97–106 (2018).
115. Rossi, D., et al. The Sarcoplasmic Reticulum of Skeletal Muscle Cells: A Labyrinth of Membrane Contact Sites. *Biomolecules* **12**(2022).
116. Gustavsson, M., Traaseth, N.J. & Veglia, G. Activating and deactivating roles of lipid bilayers on the Ca(2+)-ATPase/phospholamban complex. *Biochemistry* **50**, 10367–10374 (2011).
117. Fajardo, V.A., et al. Dietary docosahexaenoic acid supplementation reduces SERCA Ca2+ transport efficiency in rat skeletal muscle. *Chem Phys Lipids* **187**, 56–61 (2015).
118. Froud, R.J., Earl, C.R., East, J.M. & Lee, A.G. Effects of lipid fatty acyl chain structure on the activity of the (Ca2+ + Mg2+)-ATPase. *Biochim Biophys Acta* **860**, 354–360 (1986).
119. Cornea, R.L. & Thomas, D.D. Effects of membrane thickness on the molecular dynamics and enzymatic activity of reconstituted Ca-ATPase. *Biochemistry* **33**, 2912–2920 (1994).
120. Caffrey, M. & Feigenson, G.W. Fluorescence quenching in model membranes. 3. Relationship between calcium adenosinetriphosphatase enzyme activity and the affinity of the protein for phosphatidylcholines with different acyl chain characteristics. *Biochemistry* **20**, 1949–1961 (1981).
121. Dalton, K.A., Pilot, J.D., Mall, S., East, J.M. & Lee, A.G. Anionic phospholipids decrease the rate of slippage on the Ca(2+)-ATPase of sarcoplasmic reticulum. *Biochem J* **342** (**Pt 2**), 431–438 (1999).
122. Bick, R.J., Buja, L.M., Van Winkle, W.B. & Taffet, G.E. Membrane asymmetry in isolated canine cardiac sarcoplasmic reticulum: comparison with skeletal muscle sarcoplasmic reticulum. *J Membr Biol* **164**, 169–175 (1998).
123. Hunter, G.W., Negash, S. & Squier, T.C. Phosphatidylethanolamine modulates Ca-ATPase function and dynamics. *Biochemistry* **38**, 1356–1364 (1999).
124. Szymańska, G., Kim, H.W. & Kranias, E.G. Reconstitution of the skeletal sarcoplasmic reticulum Ca2(+)-pump: influence of negatively charged phospholipids. *Biochim Biophys Acta* **1091**, 127–134 (1991).
125. Hulbert, A.J. & Else, P.L. Membranes as possible pacemakers of metabolism. *J Theor Biol* **199**, 257–274 (1999).
126. Hulbert, A.J., Turner, N., Storlien, L.H. & Else, P.L. Dietary fats and membrane function: implications for metabolism and disease. *Biol Rev Camb Philos Soc* **80**, 155–169 (2005).
127. Fajardo, V.A., et al. Sarcoplasmic Reticulum Phospholipid Fatty Acid Composition and Sarcolipin Content in Rat Skeletal Muscle. *J Membr Biol* **248**, 1089–1096 (2015).
128. Holloway, G.P., Green, H.J. & Tupling, A.R. Differential effects of repetitive activity on sarcoplasmic reticulum responses in rat muscles of different oxidative potential. *Am J Physiol Regul Integr Comp Physiol* **290**, R393–404 (2006).
129. Verkerke, A.R.P., et al. Phospholipid methylation regulates muscle metabolic rate through Ca(2+) transport efficiency. *Nat Metab* **1**, 876–885 (2019).
130. Funai, K., et al. Muscle lipogenesis balances insulin sensitivity and strength through calcium signaling. *J Clin Invest* **123**, 1229–1240 (2013).
131. Qaisar, R., Bhaskaran, S. & Van Remmen, H. Muscle fiber type diversification during exercise and regeneration. *Free Radical Biology and Medicine* **98**, 56–67 (2016).
132. Rotter, D., et al. Regulator of Calcineurin 1 helps coordinate whole-body metabolism and thermogenesis. *EMBO Rep* **19**(2018).
133. Villota-Narvaez, Y., Ramírez-Martínez, A., Garzón-Alvarado, D. & Röhrle, O. A dynamical model for the calcineurin-NFATc signaling pathway and muscle fiber shifting. *PAMM* **20**, e202000274 (2021).

134. Garcia-Roves, P.M., Huss, J. & Holloszy, J.O. Role of calcineurin in exercise-induced mitochondrial biogenesis. *American Journal of Physiology-Endocrinology and Metabolism* **290**, E1172–E1179 (2006).
135. Luo, M. & Joiner, M.-l.A. Stress response signaling pathways may lead to mitochondrial biogenesis. *Diabetes* **63**, 1831–1832 (2014).
136. Sahlin, K., Tonkonogi, M. & Söderlund, K. Energy supply and muscle fatigue in humans. *Acta Physiologica Scandinavica* **162**, 261–266 (1998).
137. Green, H., Ballantyne, C., MacDougall, J., Tarnopolsky, M. & Schertzer, J. Adaptations in human muscle sarcoplasmic reticulum to prolonged submaximal training. *Journal of applied physiology* **94**, 2034–2042 (2003).
138. Kubo, H., et al. Differential effects of exercise training on skeletal muscle SERCA gene expression. *Medicine & Science in Sports & Exercise* **35**, 27–31 (2003).
139. Overmyer, K.A., et al. Maximal oxidative capacity during exercise is associated with skeletal muscle fuel selection and dynamic changes in mitochondrial protein acetylation. *Cell metabolism* **21**, 468–478 (2015).
140. Van Loon, L.J., et al. Intramyocellular lipids form an important substrate source during moderate intensity exercise in endurance-trained males in a fasted state. *The Journal of physiology* **553**, 611–625 (2003).
141. Fajardo, V.A., et al. Co-Expression of SERCA Isoforms, Phospholamban and Sarcolipin in Human Skeletal Muscle Fibers. *PLoS One* **8**, e84304 (2013).
142. Gamu, D., Bombardier, E., Smith, I.C., Fajardo, V.A. & Tupling, A.R. Sarcolipin provides a novel muscle-based mechanism for adaptive thermogenesis. *Exerc Sport Sci Rev* **42**, 136–142 (2014).
143. Hunter, G.R., Singh, H., Carter, S.J., Bryan, D.R. & Fisher, G. Sarcopenia and Its Implications for Metabolic Health. *J Obes* (**2019**), 8031705 2019.
144. Walston, J.D. Sarcopenia in older adults. *Curr Opin Rheumatol* **24**, 623–627 (2012).
145. Wilkinson, D.J., Piasecki, M. & Atherton, P.J. The age-related loss of skeletal muscle mass and function: Measurement and physiology of muscle fibre atrophy and muscle fibre loss in humans. *Ageing research reviews* **47**, 123–132 (2018).
146. Petermann-Rocha, F., et al. Global prevalence of sarcopenia and severe sarcopenia: a systematic review and meta-analysis. *Journal of cachexia, sarcopenia and muscle* **13**, 86–99 (2022).
147. Azzolino, D., et al. Musculoskeletal Changes Across the Lifespan: Nutrition and the Life-Course Approach to Prevention. *Frontiers in Medicine* **8**(2021).
148. Kim, T.N. & Choi, K.M. Sarcopenia: definition, epidemiology, and pathophysiology. *J Bone Metab* **20**, 1–10 (2013).
149. Talbot, J. & Maves, L. Skeletal muscle fiber type: using insights from muscle developmental biology to dissect targets for susceptibility and resistance to muscle disease. *Wiley Interdiscip Rev Dev Biol* **5**, 518–534 (2016).
150. Meng, S.J. & Yu, L.J. Oxidative stress, molecular inflammation and sarcopenia. *Int J Mol Sci* **11**, 1509–1526 (2010).
151. Liguori, I., et al. Oxidative stress, aging, and diseases. *Clin Interv Aging* **13**, 757–772 (2018).
152. Muller, F.L., et al. Denervation-induced skeletal muscle atrophy is associated with increased mitochondrial ROS production. *Am J Physiol Regul Integr Comp Physiol* **293**, R1159–1168 (2007).
153. Xu, H. & Van Remmen, H. The SarcoEndoplasmic Reticulum Calcium ATPase (SERCA) pump: a potential target for intervention in aging and skeletal muscle pathologies. *Skeletal Muscle* **11**, 25 (2021).
154. Viner, R.I., Ferrington, D.A., Huhmer, A.F., Bigelow, D.J. & Schoneich, C. Accumulation of nitrotyrosine on the SERCA2a isoform of SR Ca-ATPase of rat skeletal muscle during aging: a peroxynitrite-mediated process? *FEBS Lett* **379**, 286–290 (1996).
155. Braun, J.L., Hamstra, S.I., Messner, H.N. & Fajardo, V.A. SERCA2a tyrosine nitration coincides with impairments in maximal SERCA activity in left ventricles from tafazzin-deficient mice. *Physiol Rep* **7**, e14215 (2019).
156. Viner, R.I., Ferrington, D.A., Williams, T.D., Bigelow, D.J. & Schöneich, C. Protein modification during biological aging: selective tyrosine nitration of the SERCA2a isoform of the sarcoplasmic reticulum Ca2+-ATPase in skeletal muscle. *Biochem J* **340** (**Pt 3**), 657–669 (1999).
157. Viner, R.I., Williams, T.D. & Schöneich, C. Peroxynitrite modification of protein thiols: oxidation, nitrosylation, and S-glutathiolation of functionally important cysteine residue(s) in the sarcoplasmic reticulum Ca-ATPase. *Biochemistry* **38**, 12408–12415 (1999).
158. Narayanan, N., Jones, D.L., Xu, A. & Yu, J.C. Effects of aging on sarcoplasmic reticulum function and contraction duration in skeletal muscles of the rat. *Am J Physiol* **271**, C1032–1040 (1996).

159. Qaisar, R., et al. Oxidative stress-induced dysregulation of excitation-contraction coupling contributes to muscle weakness. *J Cachexia Sarcopenia Muscle* **9**, 1003–1017 (2018).
160. Qaisar, R., et al. Restoration of Sarcoplasmic Reticulum Ca(2+) ATPase (SERCA) Activity Prevents Age-Related Muscle Atrophy and Weakness in Mice. *Int J Mol Sci* **22**(2020).
161. Kalinkovich, A. & Livshits, G. Sarcopenic obesity or obese sarcopenia: A cross talk between age-associated adipose tissue and skeletal muscle inflammation as a main mechanism of the pathogenesis. *Ageing Res Rev* **35**, 200–221 (2017).
162. Batsis, J.A. & Villareal, D.T. Sarcopenic obesity in older adults: aetiology, epidemiology and treatment strategies. *Nat Rev Endocrinol* **14**, 513–537 (2018).
163. Roh, E. & Choi, K.M. Health Consequences of Sarcopenic Obesity: A Narrative Review. *Front Endocrinol (Lausanne)* **11**, 332 (2020).
164. Zamboni, M., Mazzali, G., Fantin, F., Rossi, A. & Di Francesco, V. Sarcopenic obesity: a new category of obesity in the elderly. *Nutr Metab Cardiovasc Dis* **18**, 388–395 (2008).
165. Shen, M. & Frishman, W.H. Effects of Spaceflight on Cardiovascular Physiology and Health. *Cardiol Rev* **27**, 122–126 (2019).
166. Van Ombergen, A., et al. The effect of spaceflight and microgravity on the human brain. *J Neurol* **264**, 18–22 (2017).
167. Juhl, O.J.T., et al. Update on the effects of microgravity on the musculoskeletal system. *NPJ Microgravity* **7**, 28 (2021).
168. Norsk, P., Asmar A., Damgaard, M., & Christensen, N.J. Fluid shifts, vasodilation and ambulatory blood pressure reduction during long duration spaceflight. *J Physiol* **593**, 573–584) (n.d.).
169. Baranowski, R.W., et al. Toward countering muscle and bone loss with spaceflight: GSK3 as a potential target. *iScience* **26**, 107047 (2023).
170. Coulombe, J.C., Senwar, B. & Ferguson, V.L. Spaceflight-Induced Bone Tissue Changes that Affect Bone Quality and Increase Fracture Risk. *Curr Osteoporos Rep* **18**, 1–12 (2020).
171. Fitts, R.H., Brimmer, C.J., Heywood-Cooksey, A. & Timmerman, R.J. Single muscle fiber enzyme shifts with hindlimb suspension and immobilization. *Am J Physiol* **256**, C1082–1091 (1989).
172. Schiaffino, S. & Reggiani, C. Fiber types in mammalian skeletal muscles. *Physiol Rev* **91**, 1447–1531 (2011).
173. Sandona, D., et al. Adaptation of mouse skeletal muscle to long-term microgravity in the MDS mission. *PLoS One* **7**, e33232 (2012).
174. Okada, R., et al. Transcriptome analysis of gravitational effects on mouse skeletal muscles under microgravity and artificial 1 g onboard environment. *Scientific Reports* **11**, 9168 (2021).
175. Tascher, G., et al. Proteome-wide Adaptations of Mouse Skeletal Muscles during a Full Month in Space. *J Proteome Res* **16**, 2623–2638 (2017).
176. Widrick, J.J., et al. Effect of a 17 day spaceflight on contractile properties of human soleus muscle fibres. *J Physiol* **516 (Pt 3**), 915–930 (1999).
177. Meck, J.V., Reyes, C.J., Perez, S.A., Goldberger, A.L. & Ziegler, M.G. Marked exacerbation of orthostatic intolerance after long- vs. short-duration spaceflight in veteran astronauts. *Psychosom Med* **63**, 865–873 (2001).
178. Perhonen, M.A., et al. Cardiac atrophy after bed rest and spaceflight. *J Appl Physiol (1985)* **91**, 645–653 (2001).
179. Monti, E., et al. Neuromuscular junction instability and altered intracellular calcium handling as early determinants of force loss during unloading in humans. *J Physiol* **599**, 3037–3061 (2021).
180. Ingalls, C.P., Wenke, J.C. & Armstrong, R.B. Time course changes in [Ca2+]i, force, and protein content in hindlimb-suspended mouse soleus muscles. *Aviat Space Environ Med* **72**, 471–476 (2001).
181. Hosfield, C.M., Elce, J.S., Davies, P.L. & Jia, Z. Crystal structure of calpain reveals the structural basis for Ca(2+)-dependent protease activity and a novel mode of enzyme activation. *Embo j* **18**, 6880–6889 (1999).
182. Eisner, V., Csordás, G. & Hajnóczky, G. Interactions between sarco-endoplasmic reticulum and mitochondria in cardiac and skeletal muscle - pivotal roles in Ca^{2+} and reactive oxygen species signaling. *J Cell Sci* **126**, 2965–2978 (2013).
183. Qaisar, R., et al. Restoration of SERCA ATPase prevents oxidative stress-related muscle atrophy and weakness. *Redox Biol* **20**, 68–74 (2019).

184. Enns, D.L. & Belcastro, A.N. Early activation and redistribution of calpain activity in skeletal muscle during hindlimb unweighting and reweighting. *Can J Physiol Pharmacol* **84**, 601–609 (2006).
185. Altaeva, E.G., Ogneva, I.V. & Shenkman, B. Dynamics of calcium ions accumulation and changes of isoforms of sarcoplasmic reticulum Ca-ATPase (SERCA) in soleus muscle fibers of rats and Mongolian gerbils under gravitational unloading of various duration. *Cell and Tissue Biology* **4**, 594–599 (2010).
186. Shenkman, B.S., Moukhina, A.M., Litvinova, K.S. & Nemirovskaya, T. Non-uniform shifts in MHC and SERCA isoform patterns in unloaded rat soleus. Effects of Ca-binding agent. *Conference: 9th European Symposium on Life Sciences in Space/26th Annual International Gravitational Physiology Meeting* (2005).
187. Lovering, R.M., Porter, N.C. & Bloch, R.J. The muscular dystrophies: from genes to therapies. *Phys Ther* **85**, 1372–1388 (2005).
188. Gao, Q.Q. & McNally, E.M. The Dystrophin Complex: Structure, Function, and Implications for Therapy. *Compr Physiol* **5**, 1223–1239 (2015).
189. Consalvi, S., et al. Preclinical Studies in the mdx Mouse Model of Duchenne Muscular Dystrophy with the Histone Deacetylase Inhibitor Givinostat. *Molecular Medicine* **19**, 79–87 (2013).
190. Strakova, J., et al. Integrative effects of dystrophin loss on metabolic function of the mdx mouse. *Sci Rep* **8**, 13624 (2018).
191. Cleverdon, R.E.G., et al. Sarco(endo)plasmic reticulum Ca(2+)-ATPase function is impaired in skeletal and cardiac muscles from young DBA/2J mdx mice. *iScience* **25**, 104972 (2022).
192. Strakova, J., et al. Integrative effects of dystrophin loss on metabolic function of the mdx mouse. *Scientific Reports* **8**, 13624 (2018).
193. Balakrishnan, R., Mareedu, S. & Babu, G.J. Reducing sarcolipin expression improves muscle metabolism in mdx mice. *Am J Physiol Cell Physiol* **322**, C260–c274 (2022).
194. Smith, W.S., Broadbridge, R., East, J.M. & Lee, A.G. Sarcolipin uncouples hydrolysis of ATP from accumulation of Ca2+ by the Ca2+-ATPase of skeletal-muscle sarcoplasmic reticulum. *Biochem J* **361**, 277–286 (2002).
195. Chambers, P.J., Juracic, E.S., Fajardo, V.A. & Tupling, A.R. Role of SERCA and sarcolipin in adaptive muscle remodeling. *Am J Physiol Cell Physiol* **322**, C382–c394 (2022).
196. Voit, A., et al. Reducing sarcolipin expression mitigates Duchenne muscular dystrophy and associated cardiomyopathy in mice. *Nat Commun* **8**, 1068 (2017).
197. Tanihata, J., et al. Truncated dystrophin ameliorates the dystrophic phenotype of mdx mice by reducing sarcolipin-mediated SERCA inhibition. *Biochem Biophys Res Commun* **505**, 51–59 (2018).
198. Marcella, B.M., et al. Treating muscle and brain alike: benefits of GSK3 inhibition in mdx mice. *bioRxiv*, 2022.2002. 2016.480726 (2022).
199. Christianto, A., et al. Sex differences in metabolic pathways are regulated by Pfkfb3 and Pdk4 expression in rodent muscle. *Communications Biology* **4**, 1264 (2021).
200. Haizlip, K.M., Harrison, B.C. & Leinwand, L.A. Sex-based differences in skeletal muscle kinetics and fiber-type composition. *Physiology (Bethesda)* **30**, 30–39 (2015).
201. Glenmark, B., et al. Difference in skeletal muscle function in males vs. females: role of estrogen receptor-β. *American Journal of Physiology-Endocrinology and Metabolism* (2004).
202. Maher, A.C., Akhtar, M. & Tarnopolsky, M.A. Men supplemented with 17β-estradiol have increased β-oxidation capacity in skeletal muscle. *Physiological Genomics* **42**, 342–347 (2010).
203. Brown, M. Skeletal muscle and bone: effect of sex steroids and aging. *Advances in Physiology Education* **32**, 120–126 (2008).
204. Gheit, R.E.A.E., et al. Irisin improves adiposity and exercise tolerance in a rat model of postmenopausal obesity through enhancing adipo-myocyte thermogenesis. *Journal of Physiology and Biochemistry* **78**, 897–913 (2022).

13 Recruitment of Brown and Beige Adipose Tissue Thermogenesis through Muscular Activity

Leslie A. Rowland
University of Massachusetts Chan Medical School, Worcester, Massachusetts, United States

13.1 INTRODUCTION

Upon exposure to a cold environment, multiple systems in the body are activated to defend core body temperature, including heat production, heat conservation, and energy mobilization mechanisms. These systems must be highly coordinated to effect whole-body thermogenic responses that ensure survival under harsh conditions. Of the heat production mechanisms, shivering, defined as involuntary skeletal muscle contractions, is the first line of defense against cold exposure. However, shivering is uncomfortable and can impede skeletal muscle from performing work effectively; therefore, during prolonged cold exposure, shivering thermogenesis is replaced by non-shivering thermogenesis (NST). In mammals, brown and beige adipose tissues are the predominant source of non-shivering heat production. Recruitment of brown and beige adipose tissues occurs by direct innervation and activation by the sympathetic nervous system as well as from humoral factors. This latter mechanism raises the possibility that other tissues involved in the thermogenic response could modulate brown and beige adipose NST. Since shivering is the first response to cold, it's been hypothesized that these skeletal muscle contractions induce NST recruitment and activation through the release of circulating factors. Thus, by this mechanism, the whole-body thermogenic response can be coordinated in part by skeletal muscle.

In rodents, classical interscapular BAT is the predominant site of NST and is supplemented by recruitable brown adipocytes in white adipose depots, termed *beige adipocytes*. In adult humans and larger mammals, this interscapular BAT depot is lacking, though BAT depots have been found in multiple locations in the human body [1, 2]. Characterization of human BAT has suggested these depots resemble rodent beige adipocytes and classical BAT [3, 4]. Nonetheless, regardless of species, brown and beige adipocytes are both cold-recruitable and thermogenic, and both will be discussed in this chapter [5]. A complete detailing of brown/beige fat activation and function is provided in Chapter 3 and thus will not be discussed here. Instead, this chapter will focus on mechanisms of adipose tissue thermogenesis activation specifically by striated muscle.

13.2 SHIVERING AND EXERCISE: SAME PROCESS WITH DIFFERENT GOALS

Due to the fairly recent discovery of functional and cold-activatable human BAT in 2009, data detailing the specific cold-induced activation of BAT by muscle-derived factors are relatively sparse [1, 2]. However, since the 2009 discovery, it has been uncovered that exercise can also stimulate the recruitment of beige adipocytes in rodent models, similar to cold exposure [6–10]. This is

DOI: 10.1201/9781003391470-16

noteworthy because both cold-induced shivering and exercise similarly engage repetitive muscle contractions. Shivering is the result of involuntary muscle contractions that produce heat without an associated work output, whereas exercise is the result of voluntary muscle contractions that include a major work output in addition to heat production. In fact, of all cold-activated thermogenic mechanisms, exercise produces the greatest amount of heat, and exercise-induced heat production can supplant both shivering and non-shivering thermogenesis in rats [11, 12]. Shivering thermogenesis, on the other hand, produces three to four times less heat than exercise, likely due to a variety of factors that are incompletely understood [11, 13]. However, exercise is not sustainable for extended periods of time, and consequently there is a greater reliance on shivering for muscle-based thermogenesis during cold stress. Nevertheless, exercise and shivering utilize similar mechanisms of action and elicit analogous physiologic responses. For instance, during shivering, two distinct muscle recruitment patterns are observed: continuous, low-intensity muscle contractions and short bursts of high-intensity contractions. The continuous, low-intensity contractions result from type I fiber (slow-twitch, fatigue resistant) recruitment, whereas, the high-intensity burst contractions result from type II fiber (fast-twitch, fatigable) recruitment [11]. Similarly, during exercise, type I or type II fibers or a combination are recruited, depending on the exercise modality [14]. With regard to physiologic responses, both exercise and cold-induced shivering increase muscle glucose uptake [15]. Mimicking shivering through electrical stimulation of unloaded muscles leads to increases in energy expenditure, heart rate, and carbon dioxide production, as are observed during exercise [16]. Whole-body vibration exercise, also thought to mimic shivering, increases known circulating markers of exercise [17]. Further evidence that exercise and shivering can be considered analogous stimuli is that prior exercise training can improve the shivering response to cold. In mice, endurance exercise training before a cold challenge improves cold tolerance, an effect that is attributed to enhanced muscle-based thermogenesis [18]. In another study, exercise training in the cold increased shivering thermogenesis in adult and aged mice, and in aged animals the increase in shivering mediated an improvement in cold tolerance [19]. These data indicate exercise training can prime or "train" the muscle for shivering thermogenesis and support the existence of a shared mechanism utilized by these processes.

Based on the data that show exercise and shivering result from muscular contractions that can induce analogous physiologic responses, operate with a common mechanism of action, and induce beige adipocyte recruitment, at least in rodents, this chapter will include muscle-derived factors from both stimuli, with the hypothesis that an exercise-stimulated factor may also be released during cold-induced shivering to elicit similar responses in adipose tissues.

13.2.1 Myokines

Myokine is the term for any secreted factor originating from skeletal muscle. Myokines can be secreted from muscle under various states (rest, exercise, etc.) and different stressors (obesity, diabetes, etc.). Myokines are recognized to include many types of signaling factors, including proteins (hormones), microRNAs, metabolites, and lipids, which are often contained and transported in extracellular vesicles (EVs) expelled from the muscle. Due to increasing understanding of EV biology and methodology, it's highly probable that the number of myokine discoveries will increase, as this is likely an underexplored aspect of muscle biology. To date, several exercise and cold-induced myokines have been uncovered and are discussed below.

13.2.2 Irisin

Irisin was discovered in 2012 as an exercise-induced myokine [7]. Irisin secretion results from the proteolytic cleavage of the transmembrane protein fibronectin type III domain-containing protein 5 (Fndc5) and release of the irisin polypeptide fragment. In mice and humans, exercise training increases plasma irisin levels, and gene expression of Fndc5/irisin increases proportionally in skeletal muscle. Later studies revealed cold exposure also increases circulating irisin levels in humans, and these levels directly correlate with the amount of shivering that occurs during the cold exposure, implying shivering is a stimulus for irisin secretion [20]. In mice, elevated circulating irisin levels lead to the induction of the thermogenic gene program in white adipose tissue and increase energy expenditure [7]. Treatment of human white adipocytes with irisin also leads to an induction of the brown fat gene program, increased thermogenesis, and increased oxygen consumption [20]. These findings were among the first of their kind to reveal a conserved mechanism by which cold-induced shivering stimulates the release of a factor (irisin) from muscle, which recruits thermogenesis in white adipose tissue.

13.2.3 FGF21

Fibroblast growth factor 21 (FGF21) is a hormone secreted by various tissues including liver, adipose tissue, and muscle. In mice and humans, FGF21 secretion from skeletal muscle is increased following exercise training or exercise mimetic [21, 22]. While plasma FGF21 levels are not changed upon cold exposure in humans, the amount of FGF21 in the circulation positively correlates with the thermogenic response, indicating FGF21 is associated with thermogenesis [20]. FGF21 is a well-known stimulant of adipose tissue thermogenesis, inducing both brown and beige adipocyte thermogenic function in mice and humans [20, 23, 24]. Interestingly, FGF21 treatment in combination with irisin had an additive effect on thermogenic gene expression, heat production, and oxygen consumption in human white adipocytes [20], suggesting multiple factors from muscle may work together to coordinate the full thermogenic response.

FGF21 is a bona fide exercise-stimulated myokine and activator of adipocyte thermogenesis. However, circulating FGF21 levels are not increased by cold exposure, and the effects of exercise-regulated FGF21 on adipose tissue remain unclear. *In vivo*, FGF21 appears to increase adipocyte thermogenesis via autocrine/paracrine signaling from adipose-derived FGF21 [24] rather than muscle. Thus, the role of muscle-derived FGF21 in the response to cold and exercise remains to be clarified.

13.2.4 Metrnl

Meteorin-like (Metrnl) is another myokine induced by exercise and cold exposure that can stimulate adipose tissue thermogenesis [8]. Metrnl expression is increased in muscle by eccentric exercise training in mice and combined resistance and endurance exercise training in humans. Circulating Metrnl levels are elevated post-exercise and after an acute exposure to cold in mice. Chronically increased circulating Metrnl levels by overexpression of Metrnl in the liver induces beiging of the subcutaneous and epididymal white adipose tissue depots with moderate effects on BAT. In line with the effects on thermogenesis, whole-body energy expenditure is elevated. In contrast, neutralization of circulating Metrnl levels using an anti-Metrnl antibody substantially blunts the induction of thermogenic gene expression in subcutaneous adipose tissue of mice exposed to cold. These effects of Metrnl on adipose tissue thermogenesis were shown to be indirect. Rather than targeting adipocytes, circulating Metrnl stimulates immune cell IL4/IL13 expression and alternative macrophage activation within adipose depots, which are required for the induction and maintenance of adipose tissue thermogenesis.

While exercise and cold exposure increase circulating Metrnl levels in mice and humans, the exact tissue source was not defined in these studies. Future studies will need to be conducted to determine whether muscle-derived Metrnl, specifically, can enter the circulation during cold exposure and stimulate adipose tissue thermogenesis.

13.2.5 GDF15

Growth and differentiation factor 15 (GDF15) was identified as an exercise-stimulated myokine using an *in vitro* exercise model [25]. Electrical pulse stimulation (EPS) of cultured human myotubes mimics the effects of exercise on lactate production, glycogen depletion, and myokine expression. Treatment of human cultured adipocytes with conditioned media from EPS-stimulated myotubes increased lipolysis, indicating the presence of a secreted factor from myotubes. Proteomics revealed GDF15 is released from stimulated myotubes, and expression of GDF15 is induced by EPS. These data were confirmed *in vivo*, where expression of GDF15 in skeletal muscle is increased following an exercise bout as well as circulating GDF15 [25, 26]. Treatment of human adipose tissue with GDF15 induces lipolysis, recapitulating the effect of the myotube-conditioned media. Thus, GDF15 is secreted by muscle and functions, in part, to stimulate adipose tissue lipolysis.

The effects of circulating GDF15 on adipose tissue thermogenesis have not been established, nor have the effects of cold exposure on circulating GDF15. However, adipocyte lipolysis and thermogenesis are often under control by the same regulatory factors and pathways, i.e., catecholamines signaling, indicating that a factor, such as GDF15, that induces lipolysis has the potential to concomitantly induce thermogenic gene expression. Independent of effects on the thermogenic gene program, stimulation of lipolysis in adipose tissue could enhance thermogenesis in brown/beige fat by providing fatty acids for fuel. Thus, GDF15 secretion from skeletal muscle during cold could be a signal to mobilize energy stores to fuel thermogenesis.

13.2.6 Musclin

In contrast to the aforementioned myokines, which positively modulate adipose tissue thermogenesis, musclin is a myokine that negatively affects thermogenesis. Identified in a transcriptomic screen of skeletal muscle from overweight individuals, musclin expression positively correlated with body mass index, and plasma musclin levels are elevated in overweight individuals [27]. Similarly, various models of mouse obesity show higher muscle and plasma musclin levels. In mice, muscle and plasma musclin levels are additionally regulated by environmental temperature, where thermoneutral (30°C) housing increases and cold exposure decreases musclin levels. Overexpression of musclin in skeletal muscle elevates circulating musclin and leads to impaired cold tolerance, which is attributed to a suppression of beige adipocyte recruitment and thermogenesis. Mechanistically, musclin suppresses beige adipocyte thermogenesis via inhibition of cAMP/PKA signaling, a major stimulus of the thermogenic program. Thus, a muscle-derived circulating factor can also negatively influence adipose tissue thermogenesis.

Musclin has also been reported to be secreted by muscle in response to exercise [28]. In these studies, musclin was shown to exert autocrine signaling effects on muscle to regulate mitochondrial biogenesis and endurance exercise adaptations. However, these studies were performed on whole-body musclin-knockout mice, potentially leading to developmental effects and non-muscle phenotypes. As exercise is negatively associated with body weight gain and can lead to beiging of white adipose tissue, the exercise-induced secretion of musclin seems counterintuitive. It's apparent that musclin secretion from muscle is highly context dependent and likely depends on the environment and muscle stimulus. Nonetheless, these data also raise the possibility that musclin release during shivering or cold exposure could be a signal to the adipose tissue that muscle-based thermogenesis is sufficient to maintain body temperature, thus shutting off the unnecessary activation of adipose tissue NST. Future studies are needed to define the precise stimulus for musclin release and its effect on adipose tissue.

13.2.7 BAIBA

In addition to proteins, metabolites can function as signaling intermediaries. β-aminoisobutyric acid (BAIBA) is a metabolite produced and released by skeletal muscle during exercise in mice and humans [9]. Treatment of human and mouse white adipocytes with BAIBA stimulates thermogenic gene expression, glucose uptake, and oxygen consumption. Elevation of circulating BAIBA levels by providing BAIBA in drinking water induces beiging of the subcutaneous white adipose tissue and increases energy expenditure, consistent with increased thermogenesis. BAIBA does not affect classical BAT in mice. As yet, the BAIBA response to cold exposure has not been elucidated; thus, BAIBA is an exercise-stimulated myokine that increases beiging of adipose tissue, but whether it plays a role in cold-induced thermogenesis is not known.

13.2.8 Lactate

Lactate is the end product of anaerobic glycolysis, which occurs in contracting skeletal muscle. During exercise, lactate is secreted from muscle and is considered an exercise-induced myokine. Exposure of adipose tissue to lactate induces the expression and secretion of the adipokine transforming growth factor-β2 (TGFβ2), which modulates some of the beneficial effects of exercise on glucose metabolism, indicating lactate can meaningfully affect adipose tissue function [29]. While circulating lactate is also increased during cold exposure, it's not clear whether this lactate regulates adipose tissue thermogenesis or other functions [30]. This is largely because the effects of lactate on beiging of white adipose tissue are unresolved. Many studies have indicated lactate can in fact increase adipocyte beiging, though these conclusions are confounded by experimental designs that did not include controls for osmolarity and counterions [31]. Future well-controlled studies will resolve the role of muscle-derived lactate on exercise and cold-induced beiging of adipose tissue.

Aside from its signaling function, lactate can be a major fuel source. In mice, under all housing temperatures, lactate is one of the top fuel sources consumed by BAT [32]. Therefore, muscle-derived lactate during cold exposure or exercise may simply fuel BAT thermogenesis.

13.3 NOREPINEPHRINE SPILLOVER

Catecholamines (norepinephrine and epinephrine) are perhaps the strongest stimulators of thermogenesis in adipose tissue, both activating thermogenic gene expression and fuel mobilization. During exercise, plasma catecholamines increase exponentially with work intensity [33]. There is evidence that active skeletal muscle is a major source of this circulating norepinephrine in particular, likely due to spillover from unbalanced sympathetic nervous stimulation of the muscle and uptake/extraction of catecholamines into the muscle [33, 34]. While catecholamines are not myokines per se, muscular activity can stimulate the sympathetic nervous system to release catecholamines, which can leak into the circulation and potentially affect adipose tissue thermogenesis. Whether shivering or cold exposure is a significant stimulus for muscle catecholamine release, and whether the muscular contribution to circulating catecholamines would be consequential to induce adipose tissue-based thermogenesis, is unknown.

13.4 CONCLUSIONS

The role of cold-induced myokines in the regulation of adipose tissue thermogenesis is an emerging field. A number of myokines released in response to cold or exercise with effects on adipose tissue thermogenesis have been identified and are summarized in Figure 13.1. Altogether, these myokines can elicit two main effects: (1) stimulation of the thermogenic gene program and/or activation of thermogenesis in adipose tissue; and (2) provision of fuel to supply the energy demands of whole-body thermogenesis. Despite these pioneering studies, many unanswered questions remain. One caveat to many of the myokines described in this chapter is that muscle-specific knockout models have not been characterized; therefore, it is difficult to ascertain whether the effects ascribed to a cold or exercise-induced myokine are exclusively due to muscle secretion or contributions from other tissues. Rodents are often the model of choice for identifying novel myokines; however, rodent and human physiological' responses to cold differ greatly, largely due to the sizable contribution of BAT in rodents versus a minor one in humans, often making it difficult to replicate findings and conditions in each species. Lastly, each of the myokines described here respond to different environmental cues, i.e., acute vs.

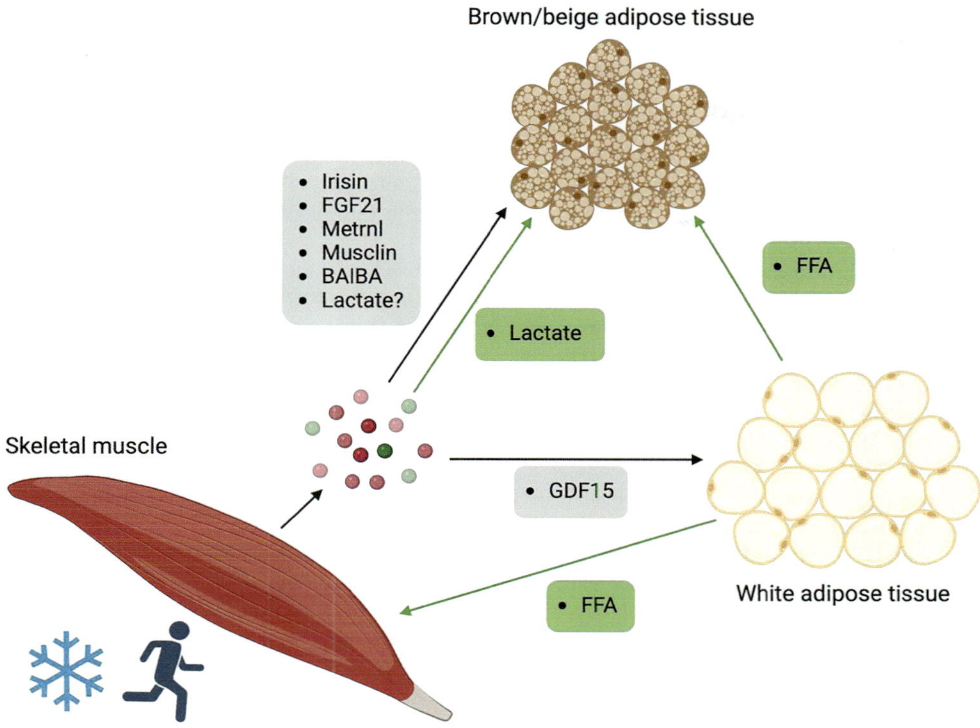

FIGURE 13.1 **Effects of contraction-induced myokines on adipose tissue thermogenesis.** Skeletal muscle releases myokines in response to contractions resulting from cold-induced shivering and exercise. These myokines can affect the adipose tissue in two ways: (1) by direct signaling to the adipose tissue to stimulate thermogenesis (shown in gray) or (2) by providing fuel to supply the energy demands of thermogenesis (shown in green). As described in the text, Irisin, FGF21, Metrnl, Musclin, BAIBA, and potentially lactate directly regulate thermogenic gene expression and function in adipose tissue. GDF15 indirectly modulates thermogenic activity by inducing lipolysis in adipose tissue, which can fuel thermogenesis in adipose tissue and muscle. Lactate can also fuel adipocyte thermogenesis, as it is a preferred energy source for brown adipose tissue. *BAIBA*, β-aminoisobutyric acid; *FGF21*, fibroblast growth factor 21; *GDF15*, growth and differentiation factor 15; *Metrnl*, meteorin-like. Figure created with BioRender.

chronic cold exposure and differing exercise modalities. Care must be taken to establish the specific thermogenic effect of a given myokine by a certain stimulus. In conclusion, muscle-derived factors can significantly affect adipose tissue responses to cold; however, more research is needed to clarify the roles of these factors on thermogenesis and to identify novel candidates.

REFERENCES

1. van Marken Lichtenbelt WD, Vanhommerig JW, Smulders NM, et al. Cold-activated brown adipose tissue in healthy men. *N Engl J Med*. 2009 Apr 9;360(15):1500–8.
2. Virtanen KA, Lidell ME, Orava J, et al. Functional brown adipose tissue in healthy adults. *N Engl J Med*. 2009 Apr 9;360(15):1518–25.
3. Wu J, Bostrom P, Sparks LM, et al. Beige adipocytes are a distinct type of thermogenic fat cell in mouse and human. *Cell*. 2012 Jul 20;150(2):366–76.
4. de Jong JMA, Sun W, Pires ND, et al. Human brown adipose tissue is phenocopied by classical brown adipose tissue in physiologically humanized mice. *Nat Metab*. 2019 Aug;1(8):830–843.
5. van der Lans AA, Hoeks J, Brans B, et al. Cold acclimation recruits human brown fat and increases non-shivering thermogenesis. *J Clin Invest*. 2013 Aug;123(8):3395–403.

6. Cao L, Choi EY, Liu X, et al. White to brown fat phenotypic switch induced by genetic and environmental activation of a hypothalamic-adipocyte axis. *Cell Metab*. 2011 Sep 7;14(3):324–38.
7. Bostrom P, Wu J, Jedrychowski MP, et al. A PGC1-alpha-dependent myokine that drives brown-fat-like development of white fat and thermogenesis. *Nature*. 2012 Jan 11;481(7382):463–8.
8. Rao RR, Long JZ, White JP, et al. Meteorin-like is a hormone that regulates immune-adipose interactions to increase beige fat thermogenesis. *Cell*. 2014 Jun 5;157(6):1279–91.
9. Roberts LD, Bostrom P, O'Sullivan JF, et al. beta-Aminoisobutyric acid induces browning of white fat and hepatic beta-oxidation and is inversely correlated with cardiometabolic risk factors. *Cell Metab*. 2014 Jan 7;19(1):96–108.
10. Stanford KI, Middelbeek RJ, Townsend KL, et al. A novel role for subcutaneous adipose tissue in exercise-induced improvements in glucose homeostasis. *Diabetes*. 2015 Jun;64(6):2002–14.
11. Haman F, Blondin DP. Shivering thermogenesis in humans: Origin, contribution and metabolic requirement. *Temperature (Austin)*. 2017;4(3):217–26.
12. Arnold J, LeBlanc J, Cote J, et al. Exercise suppression of thermoregulatory thermogenesis in warm- and cold-acclimated rats. *Can J Physiol Pharmacol*. 1986 Jul;64(7):922–6.
13. Blondin DP, Haman F. Shivering and nonshivering thermogenesis in skeletal muscles. *Handb Clin Neurol*. 2018;156:153–73.
14. Plotkin DL, Roberts MD, Haun CT, et al. Muscle Fiber Type Transitions with Exercise Training: Shifting Perspectives. Sports (Basel). 2021 Sep 10;9(9).
15. Blondin DP, Labbe SM, Phoenix S, et al. Contributions of white and brown adipose tissues and skeletal muscles to acute cold-induced metabolic responses in healthy men. *J Physiol*. 2015 Feb 1;593(3):701–14.
16. Banerjee P, Clark A, Witte K, et al. Electrical stimulation of unloaded muscles causes cardiovascular exercise by increasing oxygen demand. *Eur J Cardiovasc Prev Rehabil*. 2005 Oct;12(5):503–8.
17. Huh JY, Mougios V, Skraparlis A, et al. Irisin in response to acute and chronic whole-body vibration exercise in humans. *Metabolism*. 2014 Jul;63(7):918–21.
18. Knuth CM, Peppler WT, Townsend LK, et al. Prior exercise training improves cold tolerance independent of indices associated with non-shivering thermogenesis. *J Physiol*. 2018 Sep;596(18):4375–91.
19. Shefer VI, Talan MI. The effect of exercise training in cold on shivering and nonshivering thermogenesis in adult and aged C57BL/6J mice. *Exp Gerontol*. 1998 Aug;33(5):467–76.
20. Lee P, Linderman JD, Smith S, et al. Irisin and FGF21 are cold-induced endocrine activators of brown fat function in humans. *Cell Metab*. 2014 Feb 4;19(2):302–9.
21. He Z, Tian Y, Valenzuela PL, et al. Myokine Response to High-Intensity Interval vs. Resistance Exercise: An Individual Approach. *Front Physiol*. 2018;9:1735.
22. Keipert S, Ost M, Johann K, et al. Skeletal muscle mitochondrial uncoupling drives endocrine cross-talk through the induction of FGF21 as a myokine. *Am J Physiol Endocrinol Metab*. 2014 Mar 1;306(5):E469–82.
23. Hondares E, Iglesias R, Giralt A, et al. Thermogenic activation induces FGF21 expression and release in brown adipose tissue. *J Biol Chem*. 2011 Apr 15;286(15):12983–90.
24. Abu-Odeh M, Zhang Y, Reilly SM, et al. FGF21 promotes thermogenic gene expression as an autocrine factor in adipocytes. *Cell Rep*. 2021 Jun 29;35(13):109331.
25. Laurens C, Parmar A, Murphy E, et al. Growth and differentiation factor 15 is secreted by skeletal muscle during exercise and promotes lipolysis in humans. *JCI Insight*. 2020 Mar 26;5(6).
26. Kleinert M, Clemmensen C, Sjoberg KA, et al. Exercise increases circulating GDF15 in humans. *Mol Metab*. 2018 Mar;9:187–191.
27. Jin L, Han S, Lv X, et al. The muscle-enriched myokine Musclin impairs beige fat thermogenesis and systemic energy homeostasis via Tfr1/PKA signaling in male mice. *Nat Commun*. 2023 Jul 19;14(1):4257.
28. Subbotina E, Sierra A, Zhu Z, et al. Musclin is an activity-stimulated myokine that enhances physical endurance. *Proc Natl Acad Sci U S A*. 2015 Dec 29;112(52):16042–7.
29. Takahashi H, Alves CRR, Stanford KI, et al. TGF-beta2 is an exercise-induced adipokine that regulates glucose and fatty acid metabolism. *Nat Metab*. 2019 Feb;1(2):291–303.
30. Carriere A, Jeanson Y, Berger-Muller S, et al. Browning of white adipose cells by intermediate metabolites: an adaptive mechanism to alleviate redox pressure. *Diabetes*. 2014 Oct;63(10):3253–65.
31. Lund J, Breum AW, Gil C, et al. The anorectic and thermogenic effects of pharmacological lactate in male mice are confounded by treatment osmolarity and co-administered counterions. *Nat Metab*. 2023 Apr;5(4):677–98.

32. Park G, Haley JA, Le J, et al. Quantitative analysis of metabolic fluxes in brown fat and skeletal muscle during thermogenesis. *Nat Metab*. 2023 Jul;5(7):1204–20.
33. Kjaer M, Secher NH, Galbo H. Physical stress and catecholamine release. *Baillieres Clin Endocrinol Metab*. 1987 May;1(2):279–98.
34. Christensen NJ, Galbo H. Sympathetic nervous activity during exercise. *Annu Rev Physiol*. 1983;45: 139–53.

14 Phytochemicals as Potential Candidates to Combat Obesity via Non-Shivering Thermogenesis

Han-bing Li, Yuan-hai Sun, Teng-jiao Zhao, and Sun-ying Cheng
Institute of Pharmacology, Department of Pharmaceutical Sciences, Zhejiang University of Technology, Hangzhou, China

14.1 INTRODUCTION

Obesity is a chronic disease marked by the accumulation of excessive fat due to a number of complicated causes, including genetic susceptibility, a preference for sedentary lifestyles, a high-sugar and high-fat diet (HFD), as well as social and economic circumstances.[1] The idea behind effective management of obesity is to reduce calorie intake while increasing physical activity. Nonetheless, few people can persist and succeed. Therefore, the majority of people should utilize combined interventions to cope with adiposity and adopt appropriate pharmacotherapies (i.e., Orlistat, Liraglutide) and bariatric surgery for losing weight while at the same time considering that there are still side effects.[2–4] At present, the world is confronting severe obesity (BMI ≥ 30) and overweight (BMI ≥ 25) epidemic, which urges us to find efficient therapies and develop bioactive compounds to combat obesity.

Thermogenesis is physiologically defined as any metabolic process that dissipates in the form of heat and is regarded as an alternative therapeutic option against obesity.[5,6] Shivering thermogenesis brought on by repeated muscular contractions and non-shivering thermogenesis in skeletal muscle and adipose tissue constitute the majority of thermogenesis. One of the most heavily recruited thermogenic organs is skeletal muscle, which makes up around 40% of the human mass. Thus, it is the most critical determinant of the basal metabolic rate (BMR).[7] Heat generated by skeletal muscle shivering activity in cold acclimation originates from ATP hydrolysis.[8] However, the prolonged shivering is uncomfortable, exhausting stored muscle glycogen, triggering muscle fatigue, weakening contraction, and eventually leading to hypothermia.[9] During cold acclimation, non-shivering thermogenesis in adipose tissue and skeletal muscle gradually replaces shivering. Adipose tissue thermogenesis generally refers to brown adipose tissue (BAT) activation, conversion of white adipose cells into brown-like cells named white browning, and recruitment of beige adipocytes from adipogenic progenitor cells.[10–12] Adipose tissue–mediated thermogenesis is primarily driven by uncoupling protein 1 (UCP1), which functions to reduce the proton gradient by the electron transport pathway in the mitochondrial intermembranous space. Adipocytes can thereby create proton leakage to decouple ATP generation, which releases intracellular energy as heat.[13,14] Skeletal muscle non-shivering thermogenesis relies predominantly on SLN/SERCA signaling pathways. SLN, as an uncoupler, leads to the futile cycling of the pump, enhances ATP hydrolysis and heat production, and creates energy demand. Also, an increase of cytosolic Ca^{2+} activates the Ca^{2+}-dependent signaling pathway and nuclear transcription factors promoting mitochondrial biogenesis.[8,15,16] Additionally, uncoupling protein 3 (UCP3), in contrast to

DOI: 10.1201/9781003391470-17

UCP1, is primarily expressed in skeletal muscle and the heart.[17] The thermic effect of UCP3 may link with mediating mild uncoupling to protect against obesity and oxidative damage as well as regulate the cellular Ca^{2+} signals by interfering with the activity of SERCA.[18] Here we mainly discuss potential phytochemicals targeting adipose tissue or skeletal muscle non-shivering thermogenesis for future applications in the prevention and therapy of obesity and its related metabolic syndrome.

14.2 PHYTOCHEMICALS TARGETING ADIPOSE TISSUE THERMOGENESIS AGAINST OBESITY

To date, a variety of phytochemicals have been reported to promote $UCP1^+$ adipose differentiation and thermogenesis capacity, thus increasing energy expenditure and curbing obesity and its complications. We summarized the effects of these phytochemicals on obesity and categorized them into phenylpropanoids, flavonoids, terpenoids, alkaloids, glycosides, and phenolic acid, as shown in Table 14.1 and Figure 14.1, in the hope of providing potential lead candidates.

TABLE 14.1
Phytochemicals Targeting Adipose Non-Shivering Thermogenesis Act Against Obesity

	Phytochemicals	Effects of Phytochemicals on Thermogenesis		Models	References
		Upregulation	Downregulation		
Phenylpropanoids	Artepillin C	PPARγ, GLUT1, GLUT4, AP2	/	3T3-L1 adipocytes	19
		PPARγ, PRDM16, UCP1	/	C3H10T1/2 cells C57BL/6J mice	20
	Resveratrol	SIRT1	NF-κB	Rhesus monkeys	21
		Sirt1, Prdm16, Pparγ, Ucp1	/	NMRI-male mice	95
	Pterostilbene	UCP1, GLUT4, PPARα,	/	Male Zucker (*fa/fa*) rats	24
	Curcumin	PGC1α, PPARγ, PRDM16, p-AMPK, UCP1	/	3T3-L1 adipocytes	28
	Chlorogenic acid	PPARα	/	ICR-male mice	96
Flavonoids	Catechin	*Ucp1, Pgc1α, Prdm16*	/	3T3-L1 adipocytes	97
	Quercetin	*Ucp1, Pgc1α, Elovl3*	/	C57Bl/6J mice	37
		UCP1, PGC1α, p-AMPK, SIRT1, *Pparγ*	/	3T3-L1 adipocytes	36
	Nobiletin	C/EBP β, PPAR δ, PPARα, PRDM16, PGC1α, SIRT1, p-AMPK, UCP1	/	3T3-L1 adipocytes HIB1B brown adipocytes	39
	Kaempferol	*Pnpla2, Lipe*	*Cebpa, Pparg, Fans*	3T3-L1 adipocytes	45
	Chrysin	PGC1α, PRDM16, PPARα, PPARγ, PPARδ, p-AMPK, UCP1	C/EBPα	3T3-L1 adipocytes	40
	Cyanidin-3-glucoside	UCP1, SIRT1, PGC1α	/	*db/db* mice	43
		UCP1, *PGC1α, TFAM, Nrf1/2, Ppara, Prdm16*	/	C57BL/6J mice	98
	Genistein	UCP1, Mfn2	/	Human	99
		Ucp1, Cebpb, Sirt1	*ACC, Fasn, Fabp4*	3T3-L1 adipocytes	49
	Licochalcone A	UCP1, PRDM16, PGC1α	/	C57BL/6 mice	100

(continued)

TABLE 14.1 (CONTINUED)

	Phytochemicals	Effects of Phytochemicals on Thermogenesis		Models	References
		Upregulation	Downregulation		
Terpenoids	Ginsenoside	*Pparα, Pgc1α, Ucp1, Ucp3*	*Pparγ, Ap2, leptin*	C57BL/6 mice	51
		Ucp1, Pgc1α, Prdm16	/	3T3-L1 adipocytes	52
		/	*Pparγ, Ap2, Tnfα, Il6*	SD rats	101
		/	*Pparγ, Cebpa*	C57BL/6J mice 3T3-L1 adipocytes	102
	Fucoxanthin	UCP1	/	KK-Ay mouse	54
		UCP1	/	Male Wistar rats and female KK-Ay mice	57
	Phytol	UCP1, PRDM16, PGC1α, PPARγ	/	C57BL/6J male mice	59
		UCP1, *Pgc1α, Prdm16, Pparβ, Pparγ*	/	KK-Ay mouse C57BL/6 mice	60
		RXR, PPARα, UCP1	/	HIB-1B cell	103
	Gypenosides	PGC1α, PRDM16, UCP1	/	C57BL/6J mice	62
	Asperuloside	*Pparγ, Ucp1*	/	Sprague-Dawley rats	64
	Albiflorin	UCP1, PGC1α, p-AMPK, p-AKT, CIDEA	PPARα	hAMS Cells C57 BL/6J mice	65
	Menthol	UCP1, *Pgc1α, Prdm16, Pparγ, Adrb3*		Male C57Bl/6 mice	69
	Thymol	PGC1α, PRDM16, UCP1, β3-AR, PPARγ, PPARα	C/EBPα	3T3-L1 adipocytes	71
Alkaloids	Capsaicin	TRPV1, UCP1, PPARα, SIRT1, PRDM16	PPARγ	$TRPV1^-$ mouse models	76
		UCP1, PRDM16	/	C57BL/6J mice	104
	Caffeine	cAMP	Leptin	ddY mice	105
	Berberine	p-AMPK, PGC1α, UCP1, *Pparα*	/	*db/db* mice	81
	Ephedrine	NA, UCP1	/	CDI mice	84
	Evodiamine	C/EBPβ	Leptin, PPARγ	Ucp^{tm1} knockout (KO) mice 3T3-L1 adipocytes	86
		UCP1, p-AMPK	p-CREB	C57BL/6J mice	87
	Trigonelline	PGC1α, PRDM16, UCP1	/	3T3-L1 adipocytes	90
Glycosides	Oleuropein	UCP1	/	Sprague-Dawley rats	91
Phenolic acid	Gallic Acid	UCP1, PGC1α, SIRT1, PPARγ	/	HepG2 cells	94

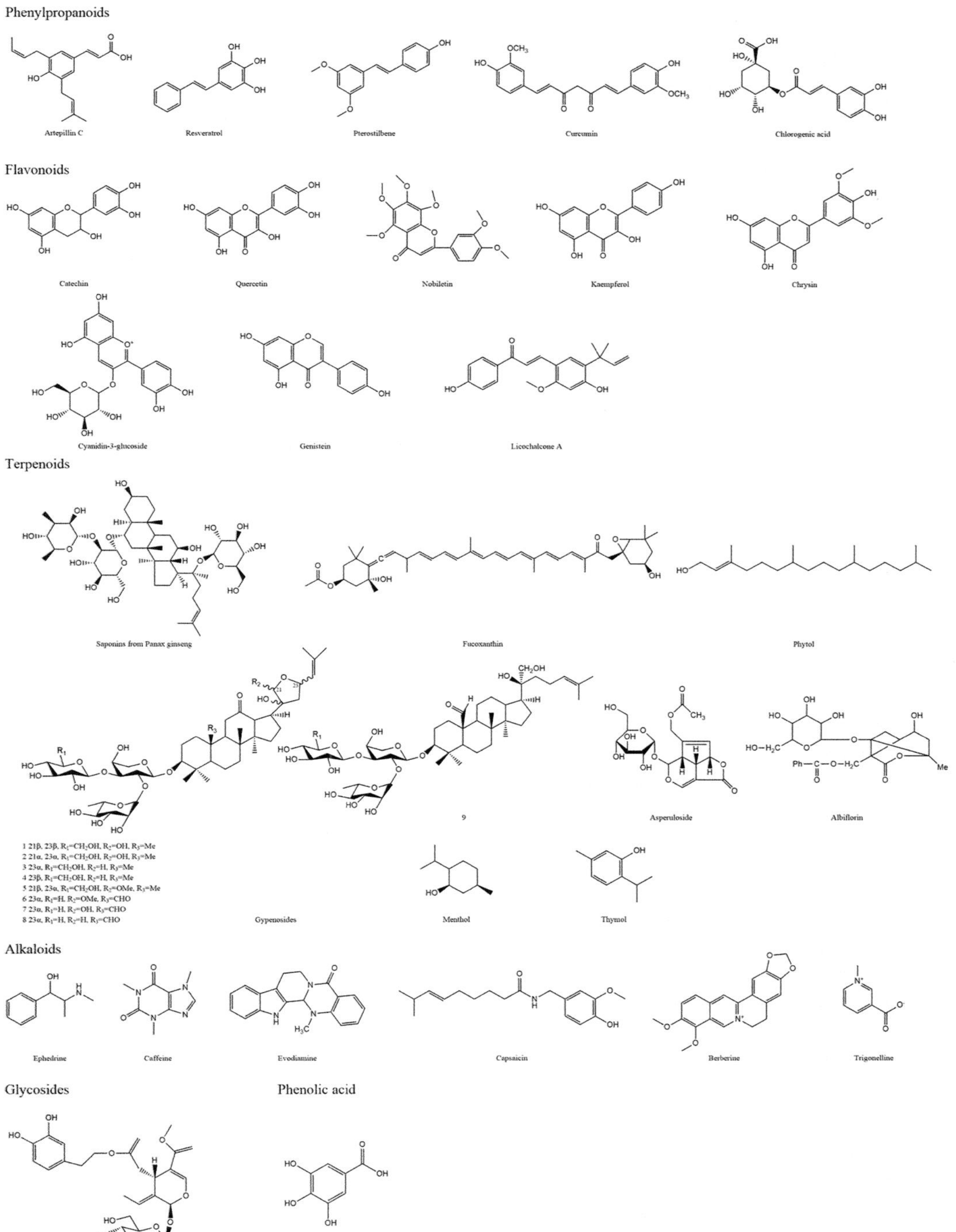

FIGURE 14.1 Chemical structures of phytochemicals targeting adipose tissue–mediated thermogenesis.

14.2.1 Phenylpropanoids

Artepillin C, a typical ingredient derived from the Brazilian green propolis, exhibits properties of antioxidant, anti-inflammation, and anti-tumor. Choi et al. reported that as a peroxisome proliferator-activated receptor gamma (PPARγ) ligand, artepillin C enhanced adipocyte differentiation and glucose uptake in 3T3-L1 adipocytes.[19] Artepillin C significantly induced brown-like adipocytes from inguinal white adipose tissue (WAT) in murine C3H10T1/2 cells through PPARγ activation and stabilization of PR domain containing 16 (PRDM16). Mice experiments showed that it induced brown-like adipocytes in interscapular white adipose tissue (iWAT) of mice.[20]

Resveratrol is a phenylpropanoid existing in various foods and considered to be a promising candidate for the treatment of type 2 diabetes. Long-term administration of resveratrol could increase sirtuin 1 (SIRT1) expression, deacetylate a range of substrates including peroxisome proliferator-activated receptor gamma coactivator 1-alpha (PGC1α), uncoupling protein 2 (UCP2), NF-κB, and forkhead box protein O1 (FOXO1), and thus produce a beneficial metabolic effect on glucose homeostasis and inflammatory adaptations in vWAT in a diet-induced obese nonhuman primate model.[21] However, a previously published paper showed that low-dose resveratrol supplementation was indeed capable of activating the SIRT1-PGC1α pathway in skeletal muscle resulting in improved mitochondrial function in humans, while no effect was observed in BAT.[22] Wang et al. reported that resveratrol could induce brown-like adipocyte formation *in vitro* and beige adipogenesis in mice via adenosine 5'-monophosphate (AMP)-activated protein kinase (AMPK) α1 activation.[23] The discrepancy of resveratrol effects on human and animals might be due to different BAT content between species, and in-depth mechanisms need to be elucidated. Pterostilbene, a resveratrol dimethyl ether derivative, also has a certain effect on adipose tissue thermogenesis. Aguirre et al. reported that the major mechanism of pterostilbene is peroxisome proliferator-activated receptor alpha (PPARα) activation, which is involved in thermogenic and oxidative capacity of BAT in Zucker (*fa/fa*) rats through increased expression of carnitine palmitoyltransferase I (CPT1) and UCP1 with the exception of PGC1α.[24, 25] The increase in PPARα also resulted in the increase of nuclear respiratory factor 1 expression, mitochondriogenesis, and oxidative capacity of iBAT).[24] In short, both stilbenes exert beneficial effects on energy homeostasis through non-shivering thermogenesis against obesity, although further investigations with respect to bioavailability, therapeutic targets, structure-activity relationship, and safety are required.[26]

A substantial bulk of preclinical and clinical investigations has spurred interest in the obesity-fighting potential of curcumin, which is derived from the herbal remedy and dietary spice turmeric. Many studies have shown that curcumin has beneficial effects in reducing weight gain, improving insulin sensitivity, and preventing the development of diabetes in rodent models and prediabetic subjects, mainly by targeting inflammatory pathways.[27] Curcumin appeared to induce fat browning via AMPK-mediated pathway and PGC1α, increased expression of brown adipocyte-specific markers, and increased mitochondrial biogenesis, which sketched out a potential role of curcumin in inducing the beige phenotype.[28] Further *in vitro* and *in vivo* investigations showed that curcumin reduced WAT inflammation and increased BAT UCP1 expression by potentially involving both PPAR-dependent and -independent mechanisms.[29]

Chlorogenic acid is an ester derived from cinnamic acids and quinic acid that is widely available in coffee, plants, fruits, and vegetables. Chlorogenic acid blocked the development of obesity by targeting the liver PPARγ pathway, thereby improving other metabolic parameters such as blood glucose and insulin sensitivity in mice.[30] Interestingly, it has been demonstrated that chlorogenic acid attenuated obesity as a non-stimulant thermogenic substance, which results in upregulation of AMPK, a key sensor of energy metabolism.[31]

14.2.2 Flavonoids

Flavonoids are the most abundant phytochemicals found in various plants with a common structure of C6-C3-C6. These chemicals can be categorized into six groups: flavonols (e.g., quercetin and kaempferol), flavones (e.g., baicalin, nobiletin and chrysin), flavonones (e.g., naringenin), anthocyanins (e.g., cyanidin-3-glucoside), isoflavones (e.g., genistein), and flavanols (e.g., catechin), all of which have been reported to increase UCP1 gene or protein through respective regulatory mechanisms.

Catechin, a category of the most abundant chemical compounds from tea, has a possible role in adipose thermogenesis. Epigallocatechin-3-gallate (EGCG) is one of the catechins present in abundance and is considered to be an effective antioxidant with significant free radical scavenging capacity.[32] The work by Yoneshirc et al. implied that catechin acted directly on transient receptor potential A1 (TRPA1) / transient receptor potential cation channel subfamily V member 1 (TRPV1) in the sensory neurons of the gastrointestinal tract and thus recruited and activated BAT. The recruited BAT by catechin preferentially made use of fatty acids as an energy substrate for thermogenesis.[33] Ahmad et al. have explored catechins and EGCG as promising pharmaceutical agents for ameliorating body weight, blood glucose, and insulin levels in obesity. Treatment with a daily dose of catechins and EGCG for 56 days reduced body weight, cholesterol, and LDL in experimental Sprague-Dawley rats. Interestingly, the EGCG-containing experimental drink showed better results than a catechins-based functional drink in terms of the lipid profile.[34] Catechin in green tea inhibits catechol omethyltransferase, leading to beta-3 adrenergic receptor (ADRB3) signaling, thereby inducing fat oxidation and heat production.[35] Given that assessing the origin of non-shivering thermogenesis is complicated by changes in food intake, heat loss, activity, and weight loss and might be solely secondary to these factors, the definitive role of catechin needs to be elucidated.

Quercetin is one of the most common flavonols found in fruits and vegetables. Lee et al. reported that quercetin could change the profile of white adipocytes toward that of brown-like adipocytes both in mice and 3T3-L1 adipocytes. Levels of the thermogenic genes *Prdm16*, *Pgc1α*, *Ucp1*, fibroblast growth factor-21 (*Fgf21*), and *Cidea* were increased while body weight and fat mass were decreased in supplemented groups, which occurred with concurrent activation of the AMPK/SIRT1/PGC1α pathway.[36] However, it is not clear whether the anti-obesogenic effect of quercetin is a direct consequence of beige and brown adipocyte formation and thermogenic activation. Notably, quercetin promoted sWAT browning and increased TG-derived fatty acid flux to sWAT in C57BL/6J mice, indicating its TG-reducing effect.[37]

Nobiletin is a polymethoxylated flavone present in citrus fruit and possesses an anti-obesity effect. Chou et al. reported that an extraction of immature *Citrus reticulata* increased expression of *Ucp1* and other thermogenic genes in iWAT. This process was regulated by PRDM16 and PGC1α, and thus induced brown-like adipocyte formation and prevented obesity and lipid accumulation.[38] It was demonstrated that nobiletin could induce PRDM16 and UCP1 expression at both the transcriptional and protein levels, thus activating HIB1B brown adipocytes. To get more insight into this issue, they further elucidated that nobiletin enhanced mRNA expression of *Cd137*, *Cidea*, *Tbx1*, *Prdm16*, *Pgc1α*, and *Ucp1* and induced beige phenotype in 3T3-L1 adipocytes. This anti-obesity effect of nobiletin might be mediated by AMPK activation in adipocytes.[39]

Chrysin is 5, 7-dihydroxyflavone naturally present in flowers, honeycombs, and mushrooms. Choi et al. reported that chrysin enhanced brown adipocyte-specific marker expression and increased protein expression of PPARα, PPARγ, PPARδ, CPT1, *p*-AMPK, *p*-ACC, PRDM16, PGC1α, and UCP1, thus inducing fat browning, suggesting its dual regulatory role in thermogenesis and lipid metabolism.[40]

A sizeable body of literature reported that the anti-obesity effects of anthocyanins, a group of flavonoids, might be mediated by various mechanisms including suppression of food intake, inhibition of lipid absorption, regulation of lipid metabolism, and increase of energy expenditure.[41] Cyanidin-3-glucoside, one of the anthocyanins ingested most abundantly, can regulate systemic energy balance and has potential therapeutic implications against obesity. Retrospective studies have shown that the browning effect was due to AMPK activation and the upregulation of CCAAT-enhancer-binding proteins beta (C/EBPβ) mediated by the elevation of cAMP levels.[42] Likewise, Yilin et al. have hinted at a possible role for cyanidin-3-glucoside in *db/db* mice, where they speculated that it promotes BAT activation and beiging in sWAT via the SIRT1-PGC1α-TFAM signaling pathway.[43]

Kaempferol is commonly found in plants used for food and traditional medicine. Gómez-Zorita et al. reported that kaempferol reduced mRNA expressions of *Cebpβ* in human adipocytes differentiated from hMSCs only in the early stage and thus inhibited adipogenesis.[44] Torres-Villarreal et al. confirmed that kaempferol exhibited an anti-obesity effect by downregulating CCAAT-enhancer-binding proteins alpha (*Cebpα*) gene expression and reducing lipid accumulation in mature adipocytes through its positive effects on *Lipe* and *Pnpla2* mRNA levels, thereby regulating adipogenic differentiation in pre-adipocytes.[45] Kaempferol could also potentially increase energy expenditure by thyroid hormone through increased type 2 iodothyronine deiodinase (DIO2) activity and promote non-shivering thermogenesis in skeletal muscle cells *in vitro*.[46]

Genistein is a soy-derived isoflavone and phytoestrogen found in a variety of dietary plants.[47] Rockwood et al. evaluated the tolerability and efficacy of 4-week consumption of genistein-containing diet upon metabolic disturbances in *ob/ob* mice and found that it had beneficial consequences in terms of energy expenditure, T3 production, and corticosterone status.[48] Aziz et al. reported that genistein induced brown-specific markers including *Ucp1*, *Pgc1α*, *C/ebpβ*, and *Sirt1* in 3T3-L1 adipocytes while reducing mRNAs indicative of white adipocytes.[49] However, the browning potential of genistein in humans is yet to be investigated.

Licochalcone A is a flavonoid purified from the ethyl acetate/n-hexane fraction of the ethyl acetate extract of roots of Glycyrrhiza *uralensis*.[50] It enhanced the expression of brown fat markers such as UCP1, PRDM16, and PGC1α and increased the population of brown fat phenotypes in the inguinal WAT of HFD-fed mice.

To sum it up, flavonoids may regulate UCP1-mediated thermogenesis via (1) the activation of AMPK/SIRT1/PGC1α pathway, (2) ADRB3 signaling, (3) the regulation of a PRDM16-containing transcriptional complex with PPARγ, PGC1α, C/EBPα, and β, or (4) TRPA1/TRPV1 activation. Although various categories of flavonoid phytochemicals have been shown to be effective in non-shivering thermogenesis *in vitro* and *in vivo*, we are only halfway to developing a safe and effective medication to treat obesity and related comorbidities. Therefore, future studies should be performed to elucidate structure-activity relationship, pharmacodynamics, and pharmacokinetics of these compounds.

14.2.3 Terpenoids

Ginsenosides are a class of natural steroid glycosides and triterpene saponins found almost exclusively in ginseng and widely used in oriental societies. Chen et al. found that the anti-obesity effect of saponins from *Panax ginseng* (SLG) was related to the regulation of non-shivering thermogenesis, lipogenesis, and lipolysis. Furthermore, the SLG group decreased *leptin*, *Pparγ*, and *Ap2* mRNA levels and increased expression of *Pgc1α*, *Pparα*, *Ucp1*, and *Ucp3* genes compared with the HFD group.[51] Mu et al. further confirmed the browning effects of ginsenoside Rb1 due to significant increases in mRNA expression of *Ucp1*, *Pgc1α*, and *Prdm16* through PPARγ in 3T3-L1 adipocytes.[52] Likewise, Park et al. demonstrated that black ginseng and ginsenoside Rb1 elevated the UCP1, PRDM16, and PGC1α in 3T3-L1 adipocytes and primary white adipocytes.[53]

Fucoxanthin is a xanthophyll with a distinct structure that has attracted wide attention in terms of its anti-obesity and anti-diabetic effects.[54, 55] A fucoxanthin diet increased the mRNA expressions of *Adrb3* in WAT of mice, which might be responsible for adaptive thermogenesis via SNS stimulation and UCP1 expression, thus contributing to its promising anti-obesity effects. The key pharmacophore of the anti-obesity effect has been suggested to be the carotenoid end of polyene chromophore.[56–58]

Phytol is an acyclic diterpene alcohol abundantly found in nature as a constituent of chlorophyll. Administration of phytol reduced body weight gain, reduced iWAT index, and stimulated browning of mice iWAT with increased gene expression of brown adipocytes markers (*Prdm16*, *Pgc1α*, *Ucp1*, *Pdh*, and *Cyto C*) through activating AMPK signaling pathway.[59] Its anti-obesity effect was also confirmed in the investigation of both male wild-type C57BL/6 and male KK-Ay mice by directly and indirectly inducing PPARα target gene expression to improve BAT function.[60]

Gypenosides, triterpenoid saponins isolated from the *Gynostemma pentaphyllum* plant, have been documented to curb obesity in *ob/ob* mice through AMPK activation.[61] Notably, Liu et al. investigated whether and how gypenosides, nine of which were characterized from Jiaogulan tea, ameliorated the development of weight gain in HFD-fed C57BL/6J mice and found that gypenosides increased BAT thermogenic marker proteins including PGC-1α, PRDM16, and UCP1 while also promoting white adipose tissue browning.[62]

Asperuloside, an active constituent isolated from *Eucommia* leaves, was administered to mice on an HFD and exhibited an anti-obesity effect.[63] Further study found that asperuloside significantly increased *Ucp1* mRNA levels in the BAT, contributing to its anti-obesity and anti-metabolic syndrome activity in HFD-fed rats.[64]

Albiflorin, a monoterpene glycoside extract isolated from the *Paeonia albiflora Pall*, could attenuate obesity and enhance the expressions of thermogenic genes both in hAMSCs and HFD-fed obese mice. The anti-obesity effect is partly due to increasing expression of multiple thermogenic genes through AMPK signaling and PI3K/AKT.[65]

Menthol is a food ingredient that serves as the cooling and flavor compound in mint. It is known for its cooling effects due to its action as an agonist of TRPM8, an ion channel that detects cold stimuli in the thermosensory system.[66, 67] Tajino and colleagues reported that topical application of menthol resulted in an increase in TRPM8-dependent core body temperature, which was positively correlated with UCP1 expression in BAT.[68] Previous studies have reported that menthol-induced TRPM8 activation could mimic heat production induced by cold stimulus and can be an intriguing treatment for obesity and related metabolic disorders in humans.[69]

Thymol is a natural monoterpenoid phenol derivative of cymene, which has been found in oil of thyme and extracted from *Thymus vulgaris* and various other kinds of aromatic plants.[70] Regarding the anti-obesity of thymol, it has been recently shown in the literature that thymol enhanced browning of white adipocytes *in vitro* by strongly promoting the biogenesis of mitochondria and the expression of BAT-specific markers that are closely related to stimulation of β3-adernergic receptor (β3-AR), protein kinase A (PKA), mitogen-activated protein kinase (MAPK), and AMPK signaling pathways.[66, 71] However, it remains obscure whether the anti-obesogenic effect of thymol is a direct consequence of adipose thermogenesis, so further *in vivo* studies need to be done to solidify this effect.

14.2.4 Alkaloids

Capsaicin, the most commonly occurring capsaicinoids and a pungent source of the hot sensation in *Capsicum annuum*, [72] is well known as a thermogenic ingredient to promote energy expenditure and reduce body fat.[73] Prolonged intake of capsinoids could recruit active BAT in human individuals and thereby increase non-shivering thermogenesis due to inducible BAT. Further studies indicate that the TRP-BAT axis contributes to the anti-adipogenic effect.[74] The anti-obesity effect of capsaicin in mice was associated with altered gut microbial composition, hypothalamic satiety, induced "browning" in sWAT, and increased expression of genes related to thermogenesis and mitochondrial biogenesis in BAT.[75] Capsaicin triggered WAT browning by promoting SIRT1-dependent deacetylation and facilitating PRDM16-PPARγ interaction via TRPV1-dependent elevation of intracellular Ca^{2+} and phosphorylation of AMPK and CaMKII.[76]

Caffeine is a purine alkaloid in green tea that possesses a thermal effect. A 6-week administration of caffeine extract from green tea slightly ameliorated obesity and markedly reduced interleukin 6 (*Il-6*) and tumor necrosis factor alpha (*Tnfα*) gene expression.[77] Caffeine also stimulates

ADRB3 signaling and has a strong inhibitory effect on the cAMP-degrading enzyme phosphodiesterase.[35] In addition, the thermogenesis mechanism of caffeine may result from the increased release of noradrenaline and dopamine and chemical energy dissipation as heat by stimulating neuronal activity.[78] To our knowledge, caffeine inhibits the phosphodiesterase-induced degradation of cAMP, a critical second messenger for the actions of catecholamines upon thermogenesis.[79]

Berberine is an alkaloid isolated from various plants used in traditional Chinese medicine. Berberine activated AMPK in 3T3-L1 adipocytes and displayed beneficial effects for treating obesity.[80] The Ning group reported that berberine enhanced BAT activity via increasing thermogenic markers including *Cidea*, *Pgc1α*, *Ucp1*, and mitochondrial content, and promoted inguinal brown-like adipocytes development via a mechanism involving AMPK and PGC1α.[81]

Ephedra sinica, a shrub native to Mongolia, Russia, and China, contains sympathomimetic compounds called ephedra alkaloids.[82] Acute administration of ephedrine enhanced sympathetic neuronal release of norepinephrine (NE) and epinephrine, and thus mediated thermogenic effects that are impaired in obese humans.[83] Chronic ephedrine treatment for 28 days reduced body fat content; however, this was not correlated with an increase in BAT activity, suggesting that a key difference exists between rodent and human BAT. Therefore, careful interpretation of these discrepant functions in human and rodent BAT must be assured before it can be considered a potential target against obesity.[84]

Evodiamine is an alkaloid contained in the fruit of *Evodia rutaecarpa* and possesses a promising scaffold targeting energy metabolism against obesity.[85] In pre-adipocytes, evodiamine increased phosphorylation of ERK/MAPK, which suggests that evodiamine potentially prevents the development of obesity partly via the inhibition of adipocyte differentiation through ERK activation.[86] A previous study showed that supplementation with low-dose evodiamine prevented body weight gain in mice via increased AMPK phosphorylation and downregulated mTOR signaling responsible for energy metabolism regulation in WAT.[87]

Trigonelline is a kind of alkaloid found in coffee and has potential anti-diabetic activity. Previous research links trigonelline with reduced adipocyte differentiation and lipid accumulation in cultured adipocytes by downregulating PPARγ.[88] Costa et al. have reported that an intake of trigonelline-enriched yogurt led to a reduction in body weight of HFD-fed mice, suggesting the possibility of using trigonelline as an anti-obesity treatment.[89] Notably, Choi et al. found that trigonelline could increase the expression of BAT signature proteins including PGC-1α, PRDM16, and UCP1 in 3T3-L1 white adipocytes.[90]

14.2.5 Glycosides

Oleuropein is a glycosylated seco-iridoid found in green olives, olive leaves, and argan oil. Oi-Kano et al. reported that oleuropein enhanced thermogenesis via increasing the UCP1 content in iBAT accompanied by the secretion of noradrenaline and adrenaline.[91] Further study showed that oleuropein increased secretion of plasma noradrenaline with concomitant enhancement of UCP1 expression in iBAT and decrease in the visceral fat mass of HFD-induced obese rats. These effects were inhibited by TRPA1 or TRPV1 antagonists, indicating that oleuropein is an agonist of both TRPV1 and TRPA1.[92]

14.2.6 Phenolic Acid

Gallic acid is a type of phenolic acid contained in substantial amounts in red wine and plant-derived foods such as blueberries, strawberries, and mangos. Setayesh et al. reported that gallic acid protected C57BL/6J mice from obesity-induced DNA damage because it possesses hypoglycemic properties and reduces white adipose tissue abundance.[93] Khanh et al. found that gallic acid exerted its beneficial effects on metabolic syndromes by targeting the activation of the AMPK-SIRT1-PGC1α pathway.[94]

14.3 PHYTOCHEMICALS TARGETING SKELETAL MUSCLE THERMOGENESIS AGAINST OBESITY

Phytochemicals also play a pivotal role in promoting skeletal muscle thermogenesis on account of their structural diversity. To date, literature has covered the potential of natural products to fight against obesity through skeletal muscle non-shivering thermogenesis via the SLN/SERCA pathway or UCP3 modulation upon SERCA activity. Here, we summarize small-molecule compounds acting on skeletal muscle non-shivering thermogenesis against obesity, providing lead candidates to fight obesity (Table 14.2 and Figure 14.2).

Salicylate is one of the ancient drugs extracted from plants and applied in clinical practice to treat inflammation and pain caused by rheumatoid arthritis.[106] Aspirin and salsalate are two primary forms of prodrug derived from salicylate, which can be hydrolyzed rapidly to yield salicylate in the small intestine.[107, 108] In recent years, the effects of these compounds on the metabolic syndrome have been extensively studied. Salsalate exerted a positive effect on dyslipidemia and glucose intolerance in T2D and attenuated non-alcoholic steatohepatitis due to its anti-inflammation action.[89] Salicylate directly activated AMPK via interaction with Ser108 residue of the β1 subunit, thus increasing fat utilization and lowering plasma fatty acids *in vivo*,[109] which could be an explanation for the beneficial effects of aspirin and salsalate as well.[110] Salsalate (salicylate) can activate BAT, uncouple mitochondria, improve glucose homeostasis, and reduce liver lipids independent of AMPKβ1 through activation of cAMP-dependent protein kinase (PKA).[111, 112] Furthermore, recent research showed that salsalate enhanced the mitochondria uncoupling and the axis of SLN/SERCA2 of skeletal muscle to facilitate thermogenesis and accelerate energy consumption in mice,[113] which gave us a hint that skeletal muscle is likely to be targeted for pharmacological intervention as a thermogenic organ. Meanwhile, salsalate, as a promising therapeutic drug, has attracted a much more prominent focus in the field of obesity and diabetes.

TABLE 14.2
Phytochemicals Targeting Skeletal Muscle Non-Shivering Thermogenesis Act Against Obesity

Phytochemicals	Effects of Phytochemicals on Thermogenesis		Models	References
	Upregulation	Downregulation		
Salsalate	*Sln, Serca2α*	/	C57BL/6J mice C2C12 cells	113
Capsaicin	SERCA	/	SR vesicles	116
Capsiate	/	*Ucp3*	Sprague-Dawley rats	117
Caffeine	RYR, SERCA	/	Sprague-Dawley rats	122
Resveratrol	TFAM, CPT-1b	/	Wistar rats	125
	TFAM, COX2, UCP3	Acetylated-PGC1α	Sprague-Dawley rats	126
Docosahexaenoic acid	/	SERCA	Wistar-Kyoto rats	129
N-arachidonoyl dopamine	SERCA	/	Rabbits Human skeletal muscle cells	131

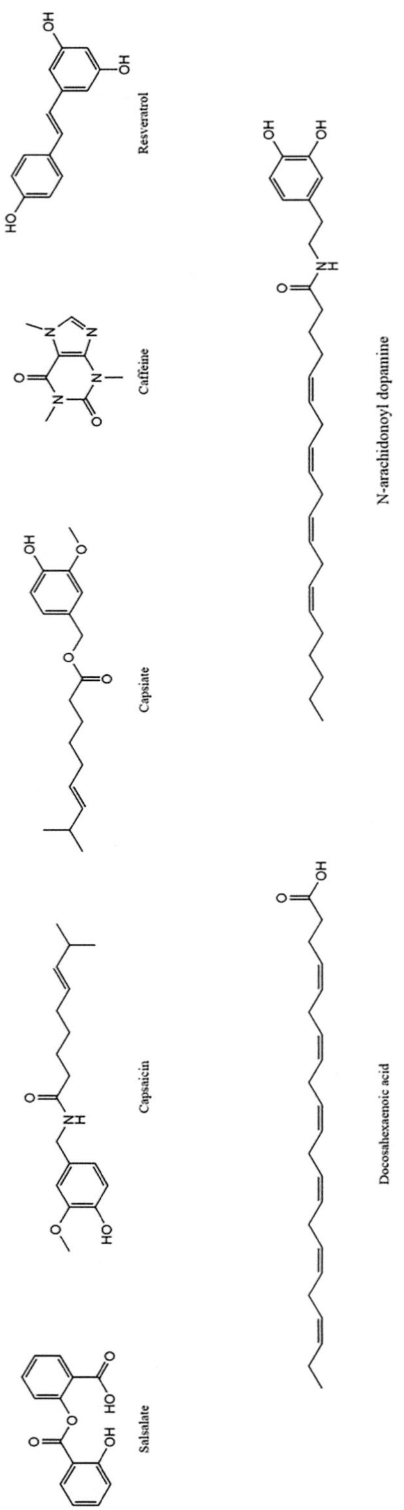

FIGURE 14.2 Chemical structures of phytochemicals targeting skeletal muscle–mediated thermogenesis.

Capsaicin is the principal pungent constituent of chill peppers, responsible for 70% of sensory burn, and reportedly promotes negative energy balance. Capsiate, a non-pungent capsaicin analog, pertains to the capsinoid family and is extracted from the fruit or food of the CH-19 sweet pepper, wherein it is highly concentrated.[114] Both of them activate transient receptor potential vanilloid subtype 1 (TRYV1) receptors, which in turn increase fat tissue lipolysis via SNS, improve insulin sensibility, and enhance thermogenesis via browning of adipocytes,[115] and thus have a decisive role in metabolic syndrome, especially diabetes and obesity. Furthermore, it has been recorded that capsaicin stimulated ATP hydrolysis not coupled with Ca^{2+} transport, resulting in thermogenesis with increased metabolic rate and core body temperature. The preliminary explanation could be that capsaicin can stabilize an ADP-insensitive E2P-related conformation where Ca^{2+} is released from E2-like form to cytoplasm instead of the SR lumen.[116] The precise mechanism of capsaicin as a natural SERCA regulator needs further investigation. Capsitate, an analog to capsaicin, brought about an enhancement of aerobic ATP production, increased energy expenses, and reduced body fat, which might be closely linked with a downregulation of UCP3.[117, 118] Intriguingly, there was a discrepancy report that UCP3 knockout mice failed to illustrate a significant role of UCP3 in regulating the whole energy metabolism.[119] Thus, whether the change in UCP3 expression is a causative factor of weight loss contributing to the effects of capsitate and the detailed mechanism warrants further research.[120]

Caffeine is a purine alkaloid compound found in tea and coffee, which are the most common beverages worldwide, consumed by approximately 55% of adults. It has been extensively studied that caffeine can affect not only endurance and performance but also energy balance. It can alter the activity of SNS and influence peripheral metabolic targets via inhibiting phosphodiesterase enzyme-induced degradation of intracellular cAMP or activating AMPK.[121] Recently, the property of caffeine to enhance skeletal muscle non-shivering thermogenesis was investigated in rats. Energy expenditure and muscle thermogenesis in the presence and absence of caffeine were measured using a graded treadmill walking test, capturing a dynamic range of heat dissipation. The results demonstrated that caffeine administration increased locomotor activity while decreasing locomotor efficiency and elevating SERCA activity and required more ATP to transport Ca^{2+} against the concentration gradient, which is secondary to caffeine's increase of RYR-induced sarcoplasmic reticulum Ca^{2+} release and contributes to caffeine's thermogenic actions. In short, caffeine, regarded as a RyR agonist, directly affects the physiological functions of skeletal muscle by regulating SERCA activity via RyR and uncoupling via SLN.[122] Meanwhile, the results of a double-blind, randomized clinical trial showed that the combination of ephedrine and caffeine did not change muscle UCP3 expression, although it did produce a pronounced thermogenic effect, which may give us a hint that putative thermogenesis stemmed from BAT activation or skeletal muscle non-shivering thermogenesis via mechanisms different from UCP3.[78]

Resveratrol, a natural phenolic compound discovered in various plants and fruits, has garnered much attention due to its reported health benefits, especially antioxidants and phytoestrogenic properties.[123] Milton-Laskibar et al. reported that administration of resveratrol at a dose of 30 mg/kg/day for 6 weeks boosted UCP1 protein expression and thermic effect in iBAT of mice via activating the SIRT1/PGC1α/PPARα axis while promoting TFAM protein expression and CPT-1B activity, and thus increasing oxidative capacity in skeletal muscle.[124, 125] Alberdi et al. showed that resveratrol increased gene expression of TFAM and COX2, decreased acetylated-PGC-1α, and upregulated UCP3 expression in skeletal muscle.[126] In short, resveratrol displays a potential role in the total energy dissipation of skeletal muscle to back up weight loss. Based on the thermogenic mechanism, resveratrol could be further developed as a potential anti-obesity drug candidate.

Docosahexaenoic acid (DHA) is a long-chain polyunsaturated n-3 fatty acid in marine organisms and available to develop the brain and nerve system during pregnancy and breastfeeding.[127, 128] According to the property of DHA, it can participate in cellular energy consumption and mitochondria uncoupling. Fajardo et al. proved that supplementing DHA could reduce SERCA Ca^{2+} transport efficiency. In brief, DHA may promote cell energy metabolism and thermogenesis, which is beneficial for obesity and related complications.[129]

N-arachidonoyl dopamine (NADA) is an endogenous arachidonic acid-based lipid that belongs to the endocannabinoid family and has potency similar to capsaicin. It can stimulate TRYV1, thereby mediating the sense of pain and thermosensation.[130] NADA, as a potential uncoupler for SERCA *in vivo*, uncouples Ca^{2+} cycling with ATP hydrolysis, but the mechanism is distinct from capsaicin. The endocannabinoid family has a regulative effect on Ca^{2+}-handling proteins. Therefore, the uncoupling of SERCA by NADA might be a novel strategy for reducing cellular ATP levels and is closely related to energy metabolism.[18, 131]

14.4 CONCLUSIONS AND PERSPECTIVES

Given the improved living standards worldwide and the increasingly unhealthy lifestyles, the energy imbalance between calorie intake and energy consumption has become a fundamental cause of obesity. The most straightforward way to prevent obesity is to limit caloric intake or to burn more calories through exercise and physical activity; however, the majority of people fail in their attempts to accomplish these goals. With the in-depth understanding of the pathogenesis of obesity, energy consumption through thermogenesis against obesity has become a novel research hotspot, including shivering and non-shivering thermogenesis. Increased non-shivering thermogenesis in adipose tissue and skeletal muscle, which expends calories as heat rather than storing them as fat, is championed as an effective way to combat obesity and related metabolic diseases. Here, we reviewed recent advances in phytochemicals that could be potential candidates for managing obesity through non-shivering thermogenesis in adipose tissue and skeletal muscle (Tables 14.1 and 14.2).

Meanwhile, the promising underlying targets upon which phytochemicals may affect thermogenesis have been proposed: (1) TRP (e.g., oleuropein, catechin, menthol, capsaicin, and curcumin), (2) SNS (e.g., ephedrine and artepillin C), (3) β-AR (e.g., thymol and fucoxanthin), (4) cAMP (e.g., caffeine and chlorogenic acid), (5) AMPK (e.g., anthocyanin, berberine, resveratrol, quercetin, curcumin, gallic acid, evodiamine, nobiletin, phytol, gypenosides, albiflorin), (6) TRPM8 (e.g., menthol), (7) TRPV1 (e.g., catechin, fucoxanthin, and caffeine), (8) SIRT1 (e.g., genistein), (9) PGC1α (e.g., baicalin), (10) PPARα/γ (e.g., pterostilbene, kaempferol, ginsenoside), (11) SLN/SERCA (e.g., salsalate, capsaicin, docosahexaenoic acid, and N-arachidonoyl dopamine), and (12) UCP3 (e.g., capsiate, resveratrol). These phytochemicals can regulate the metabolism and function of adipose tissue and skeletal muscle, especially with regards to non-shivering thermogenesis. Thus, these chemicals might combat obesity and related comorbidities, while the exact mechanism remains to be elucidated.

Despite recent conceptual advances in the etiology of obesity and search for active agents promoting non-shivering thermogenesis, there are still plenty of issues remaining to be addressed. First, due to the inherent complexity of individual differences, our understanding of the pathogenesis of obesity and the corresponding treatment strategies is still incomplete. So far, it is unclear whether BAT dysfunction is a cause or a consequence of obesity. Moreover, the precise mechanisms and contribution of fat browning to overall energy consumption need to be elucidated. Second, to date, the treatment of obesity includes diet, exercise, medical therapies aimed at reducing calorific intake or absorption, and bariatric surgery for extremely obese individuals. Unfortunately, only a fraction of participants in these programs maintain long-term weight loss. Therefore, all current methods to combat obesity are limited in effectiveness and pose significant risk. Meanwhile, no plausible ways are currently available in clinical trials to directly target obesity, although BAT activation and WAT beiging might be among the most promising future ways to prevent and treat obesity. However, safety should always be emphasized to strike a balance between benefit and risk in developing drug candidates. For example, 2,4-dinitrophenol (DNP), a chemical uncoupler, was promoted as a powerful weight loss drug; however, it was quickly withdrawn from the market due to poor tissue selectivity and serious side effects.[92] Thus, systematic research is required to more unequivocally answer questions about the efficacy, side effects, and druggability of these compounds, as some phytochemicals that show potential to stimulate thermogenesis have not been thoroughly studied. Also, randomized controlled clinical trials are needed to definitively understand their effects. Third, rodents have a higher BAT content and rely primarily on BAT for non-shivering thermogenesis and temperature homeostasis, while humans have much less BAT tissue as adults. Therefore, due to different thermal biology, the relevance of BAT in large mammals is controversial compared to small mammals. Fourth, the upstream regulator of SLN-mediated non-shivering thermogenesis remains unclear. Can the SLN-SERCA pathway be pharmacologically regulated through SLN manipulation? If so, will there be any adverse reactions? To date, some small molecule compounds have been illustrated to affect skeletal muscle non-shivering thermogenesis, thereby contributing to whole-body energy metabolism. However, the detailed mechanisms and precise molecular targets need to be elucidated. Fifth, it is difficult to accurately characterize non-shivering thermogenesis; whether and how the neuroendocrine factors (e.g., TH, CNS, NA) activate muscle non-shivering thermogenesis needs further investigation. Sixth, the effect of UCP3 on SERCA activity and muscle non-shivering thermogenesis remains an unresolved issue, as it mainly involves mild uncoupling. Finally, although extensive studies regarding the role of SLN-mediated metabolism in birds and rodents have been reported, it is still relatively unknown what role SLN plays in non-shivering thermogenesis and how much it accounts for overall energy metabolism. Meanwhile, extending the data from animal experiments to human beings requires careful verification.

In a word, deciphering the role of non-shivering thermogenesis of adipose tissue and skeletal muscle in overall energy metabolism and pursuing drug candidates from nature will provide appealing targets for promoting adipose tissue and skeletal muscle metabolism and attractive pharmacological interventions against obesity and its related metabolic syndromes.

REFERENCES

1. Oussaada, S. M.; van Galen, K. A.; Cooiman, M. I.; Kleinendorst, L.; Hazebroek, E. J.; van Haelst, M. M.; Ter Horst, K. W.; Serlie, M. J., The pathogenesis of obesity. *Metabolism* **2019**, *92*, 26–36.
2. Pilitsi, E.; Farr, O. M.; Polyzos, S. A.; Perakakis, N.; Nolen-Doerr, E.; Papathanasiou, A. E.; Mantzoros, C. S., Pharmacotherapy of obesity: Available medications and drugs under investigation. *Metabolism* **2019**, *92*, 170–92.
3. Bray, G. A.; Frühbeck, G.; Ryan, D. H.; Wilding, J. P., Management of obesity. *Lancet (London, England)* **2016**, *387* (10031), 1947–56.
4. Dietz, W. H.; Baur, L. A.; Hall, K.; Puhl, R. M.; Taveras, E. M.; Uauy, R.; Kopelman, P., Management of obesity: improvement of health-care training and systems for prevention and care. *Lancet (London, England)* **2015**, *385* (9986), 2521–33.
5. Nowack, J.; Giroud, S.; Arnold, W.; Ruf, T., Muscle Non-shivering Thermogenesis and Its Role in the Evolution of Endothermy. *Frontiers in Physiology* **2017**, *8*, 889.
6. Tseng, Y. H.; Cypess, A. M.; Kahn, C. R., Cellular bioenergetics as a target for obesity therapy. *Nature Reviews. Drug discovery* **2010**, *9* (6), 465–82.
7. Frontera, W. R.; Ochala, J., Skeletal muscle: a brief review of structure and function. *Calcified Tissue International* **2015**, *96* (3), 183–95.
8. Periasamy, M.; Herrera, J. L.; Reis, F. C. G., Skeletal Muscle Thermogenesis and Its Role in Whole Body Energy Metabolism. *Diabetes & Metabolism Journal* 2017, *41* (5), 327–336.
9. Bal, N. C.; Sahoo, S. K.; Maurya, S. K.; Periasamy, M., The Role of Sarcolipin in Muscle Non-shivering Thermogenesis. *Frontiers in Physiology* **2018**, *9*, 1217.
10. Li, H.; Qi, J.; Li, L., Phytochemicals as potential candidates to combat obesity via adipose non-shivering thermogenesis. *Pharmacological Research* **2019**, *147*, 104393.
11. Montanari, T.; Pošćić, N.; Colitti, M., Factors involved in white-to-brown adipose tissue conversion and in thermogenesis: a review. *Obesity Reviews* **2017**, *18* (5), 495–513.
12. Bartelt, A.; Heeren, J., Adipose tissue browning and metabolic health. *Nature Reviews. Endocrinology* **2014**, *10* (1), 24–36.

13. Sahu, B.; Pani, S.; Swalsingh, G.; Bal, N. C., Non and Epigenetic Mechanisms in Regulation of Adaptive Thermogenesis in Skeletal Muscle. *Front Endocrinol (Lausanne)* **2019**, *10*, 517.
14. Fuller-Jackson, J. P.; Henry, B. A., Adipose and skeletal muscle thermogenesis: studies from large animals. *The Journal of Endocrinology* **2018**, *237* (3), R99–r115.
15. Periasamy, M.; Maurya, S. K.; Sahoo, S. K.; Singh, S.; Sahoo, S. K.; Reis, F. C. G.; Bal, N. C., Role of SERCA Pump in Muscle Thermogenesis and Metabolism. *Comprehensive Physiology* **2017**, *7* (3), 879–890.
16. Palmer, B. F.; Clegg, D. J., Non-shivering thermogenesis as a mechanism to facilitate sustainable weight loss. *Obesity Reviews* **2017**, *18* (8), 819–831.
17. Pohl, E. E.; Rupprecht, A.; Macher, G.; Hilse, K. E., Important Trends in UCP3 Investigation. *Frontiers in Physiology* 2019, *10*, 470.
18. De Marchi, U.; Castelbou, C.; Demaurex, N., Uncoupling protein 3 (UCP3) modulates the activity of Sarco/endoplasmic reticulum Ca2+-ATPase (SERCA) by decreasing mitochondrial ATP production. *The Journal of Biological Chemistry* **2011**, *286* (37), 32533–41.
19. Choi, S. S.; Cha, B. Y.; Iida, K.; Lee, Y. S.; Yonezawa, T.; Teruya, T.; Nagai, K.; Woo, J. T., Artepillin C, as a PPARγ ligand, enhances adipocyte differentiation and glucose uptake in 3T3-L1 cells. *Biochemical Pharmacology* **2011**, *81* (7), 925–33.
20. Nishikawa, S.; Aoyama, H.; Kamiya, M.; Higuchi, J.; Kato, A.; Soga, M.; Kawai, T.; Yoshimura, K.; Kumazawa, S.; Tsuda, T., Artepillin C, a Typical Brazilian Propolis-Derived Component, Induces Brown-Like Adipocyte Formation in C3H10T1/2 Cells, Primary Inguinal White Adipose Tissue-Derived Adipocytes, and Mice. *PLoS One* **2016**, *11* (9), e0162512.
21. Jimenez-Gomez, Y.; Mattison, J. A.; Pearson, K. J.; Martin-Montalvo, A.; Palacios, H. H.; Sossong, A. M.; Ward, T. M.; Younts, C. M.; Lewis, K.; Allard, J. S.; Longo, D. L.; Belman, J. P.; Malagon, M. M.; Navas, P.; Sanghvi, M.; Moaddel, R.; Tilmont, E. M.; Herbert, R. L.; Morrell, C. H.; Egan, J. M.; Baur, J. A.; Ferrucci, L.; Bogan, J. S.; Bernier, M.; de Cabo, R., Resveratrol improves adipose insulin signaling and reduces the inflammatory response in adipose tissue of rhesus monkeys on high-fat, high-sugar diet. *Cell Metabolism* **2013**, *18* (4), 533–45.
22. de Ligt, M.; Bruls, Y. M. H.; Hansen, J.; Habets, M. F.; Havekes, B.; Nascimento, E. B. M.; Moonen-Kornips, E.; Schaart, G.; Schrauwen-Hinderling, V. B.; van Marken Lichtenbelt, W.; Schrauwen, P., Resveratrol improves ex vivo mitochondrial function but does not affect insulin sensitivity or brown adipose tissue in first degree relatives of patients with type 2 diabetes. *Mol Metab* 2018, *12*, 39–47.
23. Wang, S.; Liang, X.; Yang, Q.; Fu, X.; Rogers, C. J.; Zhu, M.; Rodgers, B. D.; Jiang, Q.; Dodson, M. V.; Du, M., Resveratrol induces brown-like adipocyte formation in white fat through activation of AMP-activated protein kinase (AMPK) α1. *International Journal of Obesity* **2015**, *39* (6), 967–76.
24. Aguirre, L.; Milton-Laskibar, I.; Hijona, E.; Bujanda, L.; Rimando, A. M.; Portillo, M. P., Effects of pterostilbene in brown adipose tissue from obese rats. *Journal of Physiology and Biochemistry* **2016**, *73* (3), 457–464.
25. Mosqueda-Solís, A.; Lasa, A.; Gómez-Zorita, S.; Eseberri, I.; Picó, C.; Portillo, M. P., Screening of potential anti-adipogenic effects of phenolic compounds showing different chemical structure in 3T3-L1 preadipocytes. *Food & Function* **2017**, *8* (10), 3576–3586.
26. Pan, M. H.; Wu, J. C.; Ho, C. T.; Lai, C. S., Antiobesity molecular mechanisms of action: Resveratrol and pterostilbene. *BioFactors (Oxford, England)* **2018**, *44* (1), 50–60.
27. Aggarwal, B. B., Targeting inflammation-induced obesity and metabolic diseases by curcumin and other nutraceuticals. *Annual Review of Nutrition* **2010**, *30*, 173–99.
28. Lone, J.; Choi, J. H.; Kim, S. W.; Yun, J. W., Curcumin induces brown fat-like phenotype in 3T3-L1 and primary white adipocytes. *The Journal of Nutritional Biochemistry* **2016**, *27*, 193–202.
29. Song, Z.; Revelo, X.; Shao, W.; Tian, L.; Zeng, K.; Lei, H.; Sun, H. S.; Woo, M.; Winer, D.; Jin, T., Dietary Curcumin Intervention Targets Mouse White Adipose Tissue Inflammation and Brown Adipose Tissue UCP1 Expression. *Obesity (Silver Spring, Md.)* **2018**, *26* (3), 547–558.
30. Ma, Y.; Gao, M.; Liu, D., Chlorogenic acid improves high fat diet-induced hepatic steatosis and insulin resistance in mice. *Pharmaceutical Research* **2015**, *32* (4), 1200–9.
31. Stohs, S. J.; Badmaev, V., A Review of Natural Stimulant and Non-stimulant Thermogenic Agents. *Phytotherapy Research* **2016**, *30* (5), 732–40.
32. Butt, M. S.; Ahmad, R. S.; Sultan, M. T.; Qayyum, M. M.; Naz, A., Green tea and anticancer perspectives: updates from last decade. *Critical Reviews in Food Science and Nutrition* **2015**, *55* (6), 792–805.

33. Yoneshiro, T.; Matsushita, M.; Hibi, M.; Tone, H.; Takeshita, M.; Yasunaga, K.; Katsuragi, Y.; Kameya, T.; Sugie, H.; Saito, M., Tea catechin and caffeine activate brown adipose tissue and increase cold-induced thermogenic capacity in humans. *The American Journal of Clinical Nutrition* **2017**, *105* (4), 873–881.
34. Ahmad, R. S.; Butt, M. S.; Sultan, M. T.; Mushtaq, Z.; Ahmad, S.; Dewanjee, S.; De Feo, V.; Zia-Ul-Haq, M., Preventive role of green tea catechins from obesity and related disorders especially hypercholesterolemia and hyperglycemia. *Journal of Translational Medicine* **2015**, *13*, 79.
35. Okla, M.; Kim, J.; Koehler, K.; Chung, S., Dietary Factors Promoting Brown and Beige Fat Development and Thermogenesis. *Advances in nutrition (Bethesda, Md.)* **2017**, *8* (3), 473–483.
36. Lee, S. G.; Parks, J. S.; Kang, H. W., Quercetin, a functional compound of onion peel, remodels white adipocytes to brown-like adipocytes. *The Journal of Nutritional Biochemistry* **2017**, *42*, 62–71.
37. Kuipers, E. N.; Dam, A. D. V.; Held, N. M.; Mol, I. M.; Houtkooper, R. H.; Rensen, P. C. N.; Boon, M. R., Quercetin Lowers Plasma Triglycerides Accompanied by White Adipose Tissue Browning in Diet-Induced Obese Mice. *International Journal of Molecular Sciences* **2018**, *19* (6).
38. Chou, Y. C.; Ho, C. T.; Pan, M. H., Immature Citrus reticulata Extract Promotes Browning of Beige Adipocytes in High-Fat Diet-Induced C57BL/6 Mice. *Journal of Agricultural and Food Chemistry* **2018**, *66* (37), 9697–9703.
39. Lone, J.; Parray, H. A.; Yun, J. W., Nobiletin induces brown adipocyte-like phenotype and ameliorates stress in 3T3-L1 adipocytes. *Biochimie* **2018**, *146*, 97–104.
40. Choi, J. H.; Yun, J. W., Chrysin induces brown fat-like phenotype and enhances lipid metabolism in 3T3-L1 adipocytes. *Nutrition (Burbank, Los Angeles County, Calif.)* **2016**, *32* (9), 1002–10.
41. Xie, L. H.; Su, H. M.; Sun, C. D.; Zheng, X. D.; Chen, W., Recent advances in understanding the anti-obesity activity of anthocyanins and their biosynthesis in microorganisms. *Trends in Food Science & Technology* **2018**, *72*, 13–24.
42. Matsukawa, T.; Villareal, M. O.; Motojima, H.; Isoda, H., Increasing cAMP levels of preadipocytes by cyanidin-3-glucoside treatment induces the formation of beige phenotypes in 3T3-L1 adipocytes. *The Journal of Nutritional Biochemistry* 2**017**, *40*, 77–85.
43. You, Y.; Yuan, X.; Liu, X.; Liang, C.; Meng, M.; Huang, Y.; Han, X.; Guo, J.; Guo, Y.; Ren, C.; Zhang, Q.; Sun, X.; Ma, T.; Liu, G.; Jin, W.; Huang, W.; Zhan, J., Cyanidin-3-glucoside increases whole body energy metabolism by upregulating brown adipose tissue mitochondrial function. *Molecular Nutrition & Food Research* **2017**, *61* (11), 1–13.
44. Gómez-Zorita, S.; Lasa, A.; Abendaño, N.; Fernández-Quintela, A.; Mosqueda-Solís, A.; Garcia-Sobreviela, M. P.; Arbonés-Mainar, J. M.; Portillo, M. P., Phenolic compounds apigenin, hesperidin and kaempferol reduce in vitro lipid accumulation in human adipocytes. *Journal of Translational Medicine* **2017**, *15* (1), 237.
45. Torres-Villarreal, D.; Camacho, A.; Castro, H.; Ortiz-Lopez, R.; de la Garza, A. L., Anti-obesity effects of kaempferol by inhibiting adipogenesis and increasing lipolysis in 3T3–L1 cells. *Journal of Physiology and Biochemistry* **2019**, *75* (1), 83–88.
46. Kang, H. W.; Lee, S. G.; Otieno, D.; Ha, K., Flavonoids, potential bioactive compounds, and non-shivering thermogenesis. *Nutrients* **2018**, *10* (9), 1168.
47. Kaufman, P. B.; Duke, J. A.; Brielmann, H.; Boik, J.; Hoyt, J. E., A comparative survey of leguminous plants as sources of the isoflavones, genistein and daidzein: implications for human nutrition and health. *Journal of Alternative and Complementary Medicine (New York, N.Y.)* **1997**, *3* (1), 7–12.
48. Rockwood, S.; Broderick, T. L.; Al-Nakkash, L., Feeding Obese Diabetic Mice a Genistein Diet Induces Thermogenic and Metabolic Change. *Journal of Medicinal Food* **2018**, *21* (4), 332–339.
49. Aziz, S. A.; Wakeling, L. A.; Miwa, S.; Alberdi, G.; Hesketh, J. E.; Ford, D., Metabolic programming of a beige adipocyte phenotype by genistein. *Molecular Nutrition & Food Research* **2017**, *61* (2), 1600574.
50. Won, S. R.; Kim, S. K.; Kim, Y. M.; Lee, P. H.; Ryu, J. H.; Kim, J. W.; Rhee, H. I., Licochalcone A: A lipase inhibitor from the roots of Glycyrrhiza uralensis. *Food Research International* **2007**, *40* (8), 1046–1050.
51. Chen, G.; Li, H.; Zhao, Y.; Zhu, H.; Cai, E.; Gao, Y.; Liu, S.; Yang, H.; Zhang, L., Saponins from stems and leaves of Panax ginseng prevent obesity via regulating thermogenesis, lipogenesis and lipolysis in high-fat diet-induced obese C57BL/6 mice. *Food and Chemical Toxicology : An International Journal Published for the British Industrial Biological Research Association* 2017, *106* (Pt A), 393–403.

52. Mu, Q.; Fang, X.; Li, X.; Zhao, D.; Mo, F.; Jiang, G.; Yu, N.; Zhang, Y.; Guo, Y.; Fu, M.; Liu, J. L.; Zhang, D.; Gao, S., Ginsenoside Rb1 promotes browning through regulation of PPARγ in 3T3-L1 adipocytes. *Biochemical and Biophysical Research Communications* 2015, *466* (3), 530–5.
53. Park, S. J.; Park, M.; Sharma, A.; Kim, K.; Lee, H. J., Black Ginseng and Ginsenoside Rb1 Promote Browning by Inducing UCP1 Expression in 3T3-L1 and Primary White Adipocytes. *Nutrients* 2019, *11* (11), 2747.
54. Maeda, H., Nutraceutical effects of fucoxanthin for obesity and diabetes therapy: a review. *Journal of Oleo Science* **2015**, *64* (2), 125–32.
55. Gammone, M. A.; D'Orazio, N., Anti-obesity activity of the marine carotenoid fucoxanthin. *Marine Drugs* **2015**, *13* (4), 2196–14.
56. Maeda, H.; Hosokawa, M.; Sashima, T.; Murakami-Funayama, K.; Miyashita, K., Anti-obesity and anti-diabetic effects of fucoxanthin on diet-induced obesity conditions in a murine model. *Molecular Medicine Reports* **2009**, *2* (6), 897–902.
57. Maeda, H.; Hosokawa, M.; Sashima, T.; Funayama, K.; Miyashita, K., Fucoxanthin from edible seaweed, Undaria pinnatifida, shows antiobesity effect through UCP1 expression in white adipose tissues. *Biochemical and Biophysical Research Communications* **2005**, *332* (2), 392–7.
58. Maeda, H.; Hosokawa, M.; Sashima, T.; Miyashita, K., Dietary combination of fucoxanthin and fish oil attenuates the weight gain of white adipose tissue and decreases blood glucose in obese/diabetic KK-Ay mice. *Journal of Agricultural and Food Chemistry* **2007**, *55* (19), 7701–6.
59. Zhang, F.; Ai, W.; Hu, X.; Meng, Y.; Yuan, C.; Su, H.; Wang, L.; Zhu, X.; Gao, P.; Shu, G.; Jiang, Q.; Wang, S., Phytol stimulates the browning of white adipocytes through the activation of AMP-activated protein kinase (AMPK) α in mice fed high-fat diet. *Food & Function* 2018, *9* (4), 2043–50.
60. An, J. Y.; Jheng, H. F.; Nagai, H.; Sanada, K.; Takahashi, H.; Iwase, M.; Watanabe, N.; Kim, Y. I.; Teraminami, A.; Takahashi, N.; Nakata, R.; Inoue, H.; Seno, S.; Mastuda, H.; Kawada, T.; Goto, T., A Phytol-Enriched Diet Activates PPAR-α in the Liver and Brown Adipose Tissue to Ameliorate Obesity-Induced Metabolic Abnormalities. *Molecular Nutrition & Food Research* 2018, *62* (6), e1700688.
61. Gauhar, R.; Hwang, S. L.; Jeong, S. S.; Kim, J. E.; Song, H.; Park, D. C.; Song, K. S.; Kim, T. Y.; Oh, W. K.; Huh, T. L., Heat-processed Gynostemma pentaphyllum extract improves obesity in ob/ob mice by activating AMP-activated protein kinase. *Biotechnology Letters* 2012, *34* (9), 1607–16.
62. Liu, J.; Li, Y.; Yang, P.; Wan, J.; Chang, Q.; Wang, T. T. Y.; Lu, W.; Zhang, Y.; Wang, Q.; Yu, L. L., Gypenosides Reduced the Risk of Overweight and Insulin Resistance in C57BL/6J Mice through Modulating Adipose Thermogenesis and Gut Microbiota. *Journal of Agricultural and Food Chemistry* 2017, *65* (42), 9237–9246.
63. Hirata, T.; Kobayashi, T.; Wada, A.; Ueda, T.; Fujikawa, T.; Miyashita, H.; Ikeda, T.; Tsukamoto, S.; Nohara, T., Anti-obesity compounds in green leaves of Eucommia ulmoides. *Bioorganic & Medicinal Chemistry Letters* 2011, *21* (6), 1786–91.
64. Fujikawa, T.; Hirata, T.; Hosoo, S.; Nakajima, K.; Wada, A.; Yurugi, Y.; Soya, H.; Matsui, T.; Yamaguchi, A.; Ogata, M.; Nishibe, S., Asperuloside stimulates metabolic function in rats across several organs under high-fat diet conditions, acting like the major ingredient of Eucommia leaves with anti-obesity activity. *Journal of Nutritional Science* 2012, *1*, e10.
65. Jeong, M. Y.; Park, J.; Youn, D. H.; Jung, Y.; Kang, J.; Lim, S.; Kang, M. W.; Kim, H. L.; So, H. S.; Park, R.; Hong, S. H.; Um, J. Y., Albiflorin ameliorates obesity by inducing thermogenic genes via AMPK and PI3K/AKT in vivo and in vitro. *Metabolism* 2017, *73*, 85–99.
66. Silvester, A. J.; Aseer, K. R.; Yun, J. W., Dietary polyphenols and their roles in fat browning. *The Journal of Nutritional Biochemistry* 2019, *64*, 1–12.
67. Khare, P.; Mangal, P.; Baboota, R. K.; Jagtap, S.; Kumar, V.; Singh, D. P.; Boparai, R. K.; Sharma, S. S.; Khardori, R.; Bhadada, S. K.; Kondepudi, K. K.; Chopra, K.; Bishnoi, M., Involvement of Glucagon in Preventive Effect of Menthol Against High Fat Diet Induced Obesity in Mice. *Frontiers in Pharmacology* 2018, *9*, 1244.
68. Tajino, K.; Matsumura, K.; Kosada, K.; Shibakusa, T.; Inoue, K.; Fushiki, T.; Hosokawa, H.; Kobayashi, S., Application of menthol to the skin of whole trunk in mice induces autonomic and behavioral heat-gain responses. *American journal of physiology. Regulatory, Integrative and Comparative Physiology* 2007, *293* (5), R2128–35.

69. Jiang, C.; Zhai, M.; Yan, D.; Li, D.; Li, C.; Zhang, Y.; Xiao, L.; Xiong, D.; Deng, Q.; Sun, W., Dietary menthol-induced TRPM8 activation enhances WAT "browning" and ameliorates diet-induced obesity. *Oncotarget* 2017, *8* (43), 75114–75126.
70. Basch, E.; Ulbricht, C.; Hammerness, P.; Bevins, A.; Sollars, D., Thyme (Thymus vulgaris L.), thymol. *Journal of Herbal Pharmacotherapy* 2004, *4* (1), 49–67.
71. Choi, J. H.; Kim, S. W.; Yu, R.; Yun, J. W., Monoterpene phenolic compound thymol promotes browning of 3T3-L1 adipocytes. *European Journal of Nutrition* **2017**, *56* (7), 2329–41.
72. Bode, A. M.; Dong, Z., The two faces of capsaicin. *Cancer Research* 2011, *71* (8), 2809–14.
73. Fu, C.; Jiang, Y.; Guo, J.; Su, Z., Natural Products with Anti-obesity Effects and Different Mechanisms of Action. *Journal of Agricultural and Food Chemistry* **2016**, *64* (51), 9571–85.
74. Saito, M.; Yoneshiro, T., Capsinoids and related food ingredients activating brown fat thermogenesis and reducing body fat in humans. *Current Opinion in Lipidology* 2013, *24* (1), 71–7.
75. Baboota, R. K.; Murtaza, N.; Jagtap, S.; Singh, D. P.; Karmase, A.; Kaur, J.; Bhutani, K. K.; Boparai, R. K.; Premkumar, L. S.; Kondepudi, K. K.; Bishnoi, M., Capsaicin-induced transcriptional changes in hypothalamus and alterations in gut microbial count in high fat diet fed mice. *The Journal of Nutritional Biochemistry* 2014, *25* (9), 893–902.
76. Baskaran, P.; Krishnan, V.; Ren, J.; Thyagarajan, B., Capsaicin induces browning of white adipose tissue and counters obesity by activating TRPV1 channel-dependent mechanisms. *British Journal of Pharmacology* 2016, *173* (15), 2369–89.
77. Xu, Y.; Zhang, M.; Wu, T.; Dai, S.; Xu, J.; Zhou, Z., The anti-obesity effect of green tea polysaccharides, polyphenols and caffeine in rats fed with a high-fat diet. *Food & Function* 2015, *6* (1), 297–304.
78. Bracale, R.; Petroni, M. L.; Davinelli, S.; Bracale, U.; Scapagnini, G.; Carruba, M. O.; Nisoli, E., Muscle uncoupling protein 3 expression is unchanged by chronic ephedrine/caffeine treatment: results of a double blind, randomised clinical trial in morbidly obese females. *PLoS One* 2014, *9* (6), e98244.
79. Dulloo, A. G.; Duret, C.; Rohrer, D.; Girardier, L.; Mensi, N.; Fathi, M.; Chantre, P.; Vandermander, J., Efficacy of a green tea extract rich in catechin polyphenols and caffeine in increasing 24-h energy expenditure and fat oxidation in humans. *The American Journal of Clinical Nutrition* 1999, *70* (6), 1040–5.
80. Lee, Y. S.; Kim, W. S.; Kim, K. H.; Yoon, M. J.; Cho, H. J.; Shen, Y.; Ye, J. M.; Lee, C. H.; Oh, W. K.; Kim, C. T.; Hohnen-Behrens, C.; Gosby, A.; Kraegen, E. W.; James, D. E.; Kim, J. B., Berberine, a natural plant product, activates AMP-activated protein kinase with beneficial metabolic effects in diabetic and insulin-resistant states. *Diabetes* 2006, *55* (8), 2256–64.
81. Zhang, Z.; Zhang, H.; Li, B.; Meng, X.; Wang, J.; Zhang, Y.; Yao, S.; Ma, Q.; Jin, L.; Yang, J.; Wang, W.; Ning, G., Berberine activates thermogenesis in white and brown adipose tissue. *Nature Communications* 2014, *5*, 5493.
82. Diepvens, K.; Westerterp, K. R.; Westerterp-Plantenga, M. S., Obesity and thermogenesis related to the consumption of caffeine, ephedrine, capsaicin, and green tea. *American journal of physiology. Regulatory, Integrative and Comparative Physiology* 2007, *292* (1), R77–85.
83. Carey, A. L.; Formosa, M. F.; Van Every, B.; Bertovic, D.; Eikelis, N.; Lambert, G. W.; Kalff, V.; Duffy, S. J.; Cherk, M. H.; Kingwell, B. A., Ephedrine activates brown adipose tissue in lean but not obese humans. *Diabetologia* 2013, *56* (1), 147–55.
84. Carey, A. L.; Pajtak, R.; Formosa, M. F.; Van Every, B.; Bertovic, D. A.; Anderson, M. J.; Eikelis, N.; Lambert, G. W.; Kalff, V.; Duffy, S. J.; Cherk, M. H.; Kingwell, B. A., Chronic ephedrine administration decreases brown adipose tissue activity in a randomised controlled human trial: implications for obesity. *Diabetologia* 2015, *58* (5), 1045–54.
85. Gavaraskar, K.; Dhulap, S.; Hirwani, R. R., Therapeutic and cosmetic applications of Evodiamine and its derivatives--A patent review. *Fitoterapia* 2015, *106*, 22–35.
86. Wang, T.; Wang, Y.; Kontani, Y.; Kobayashi, Y.; Sato, Y.; Mori, N.; Yamashita, H., Evodiamine improves diet-induced obesity in a uncoupling protein-1-independent manner: involvement of antiadipogenic mechanism and extracellularly regulated kinase/mitogen-activated protein kinase signaling. *Endocrinology* 2008, *149* (1), 358–66.
87. Yamashita, H.; Kusudo, T.; Takeuchi, T.; Qiao, S.; Tsutsumiuchi, K.; Wang, T.; Wang, Y., Dietary supplementation with evodiamine prevents obesity and improves insulin resistance in ageing mice. *Journal of Functional Foods* 2015, *19*, 320–329.

88. Ilavenil, S.; Arasu, M. V.; Lee, J. C.; Kim, D. H.; Roh, S. G.; Park, H. S.; Choi, G. J.; Mayakrishnan, V.; Choi, K. C., Trigonelline attenuates the adipocyte differentiation and lipid accumulation in 3T3-L1 cells. *Phytomedicine* 2014, *21* (5), 758–65.
89. Costa, M. C.; Lima, T. F. O.; Arcaro, C. A.; Inacio, M. D.; Batista-Duharte, A.; Carlos, I. Z.; Spolidorio, L. C.; Assis, R. P.; Brunetti, I. L.; Baviera, A. M., Trigonelline and curcumin alone, but not in combination, counteract oxidative stress and inflammation and increase glycation product detoxification in the liver and kidney of mice with high-fat diet-induced obesity. *The Journal of Nutritional Biochemistry* 2020, *76*, 108303.
90. Choi, M.; Mukherjee, S.; Yun, J. W., Trigonelline induces browning in 3T3-L1 white adipocytes. *Phytotherapy Research* **2021**, *35* (2), 1113–1124.
91. Oi-Kano, Y.; Kawada, T.; Watanabe, T.; Koyama, F.; Watanabe, K.; Senbongi, R.; Iwai, K., Oleuropein, a phenolic compound in extra virgin olive oil, increases uncoupling protein 1 content in brown adipose tissue and enhances noradrenaline and adrenaline secretions in rats. *Journal of Nutritional Science and Vitaminology* 2008, *54* (5), 363–70.
92. Oi-Kano, Y.; Iwasaki, Y.; Nakamura, T.; Watanabe, T.; Goto, T.; Kawada, T.; Watanabe, K.; Iwai, K., Oleuropein aglycone enhances UCP1 expression in brown adipose tissue in high-fat-diet-induced obese rats by activating β-adrenergic signaling. *The Journal of Nutritional Biochemistry* 2017, *40*, 209–218.
93. Setayesh, T.; Nersesyan, A.; Mišík, M.; Noorizadeh, R.; Haslinger, E.; Javaheri, T.; Lang, E.; Grusch, M.; Huber, W.; Haslberger, A.; Knasmüller, S., Gallic acid, a common dietary phenolic protects against high fat diet induced DNA damage. *European Journal of Nutrition* 2019, *58* (6), 2315–2326.
94. Doan, K. V.; Ko, C. M.; Kinyua, A. W.; Yang, D. J.; Choi, Y. H.; Oh, I. Y.; Nguyen, N. M.; Ko, A.; Choi, J. W.; Jeong, Y.; Jung, M. H.; Cho, W. G.; Xu, S.; Park, K. S.; Park, W. J.; Choi, S. Y.; Kim, H. S.; Moh, S. H.; Kim, K. W., Gallic acid regulates body weight and glucose homeostasis through AMPK activation. *Endocrinology* 2015, *156* (1), 157–68.
95. Serrano, A.; Asnani-Kishnani, M.; Rodríguez, A. M.; Palou, A.; Ribot, J.; Bonet, M. L., Programming of the Beige Phenotype in White Adipose Tissue of Adult Mice by Mild Resveratrol and Nicotinamide Riboside Supplementations in Early Postnatal Life. *Molecular Nutrition & Food Research* 2018, *62* (21), e1800463.
96. Cho, A. S.; Jeon, S. M.; Kim, M. J.; Yeo, J.; Seo, K. I.; Choi, M. S.; Lee, M. K., Chlorogenic acid exhibits anti-obesity property and improves lipid metabolism in high-fat diet-induced-obese mice. *Food and chemical toxicology : an international journal published for the British Industrial Biological Research Association* 2010, *48* (3), 937–43.
97. Zhou, J.; Mao, L.; Xu, P.; Wang, Y., Effects of (-)-Epigallocatechin Gallate (EGCG) on Energy Expenditure and Microglia-Mediated Hypothalamic Inflammation in Mice Fed a High-Fat Diet. *Nutrients* 2018, *10* (11).
98. You, Y.; Han, X.; Guo, J.; Guo, Y.; Yin, M.; Liu, G.; Huang, W.; Zhan, J., Cyanidin-3-glucoside attenuates high-fat and high-fructose diet-induced obesity by promoting the thermogenic capacity of brown adipose tissue. *Journal of Functional Foods* 2018, *41*, 62–71.
99. Grossini, E.; Farruggio, S.; Raina, G.; Mary, D.; Deiro, G.; Gentilli, S., Effects of Genistein on Differentiation and Viability of Human Visceral Adipocytes. *Nutrients* 2018, *10* (8).
100. Lee, H. E.; Yang, G.; Han, S. H.; Lee, J. H.; An, T. J.; Jang, J. K.; Lee, J. Y., Anti-obesity potential of Glycyrrhiza uralensis and licochalcone A through induction of adipocyte browning. *Biochemical and Biophysical Research Communications* 2018, *503* (3), 2117–2123.
101. Jung, S.; Lee, M. S.; Shin, Y.; Kim, C. T.; Kim, I. H.; Kim, Y., High Hydrostatic Pressure Extract of Red Ginseng Attenuates Inflammation in Rats with High-fat Diet Induced Obesity. *Preventive Nutrition and Food Science* 2015, *20* (4), 253–9.
102. Shin, S. S.; Yoon, M., Korean red ginseng (Panax ginseng) inhibits obesity and improves lipid metabolism in high fat diet-fed castrated mice. *Journal of Ethnopharmacology* 2018, *210*, 80–87.
103. Schluter, A.; Giralt, M.; Iglesias, R.; Villarroya, F., Phytanic acid, but not pristanic acid, mediates the positive effects of phytol derivatives on brown adipocyte differentiation. *FEBS Letters* 2002, *517* (1–3), 83–6.
104. Ohyama, K.; Nogusa, Y.; Shinoda, K.; Suzuki, K.; Bannai, M.; Kajimura, S., A Synergistic Antiobesity Effect by a Combination of Capsinoids and Cold Temperature Through Promoting Beige Adipocyte Biogenesis. *Diabetes* 2016, *65* (5), 1410–23.
105. Liu, L.; Sayama, K., The combined administration of EGCG and caffeine induces not only suppression of fat accumulation but also anorexigenic action in mice. *Journal of Functional Foods* 2018, *47*, 156–162.

106. Goldfine, A. B.; Fonseca, V.; Jablonski, K. A.; Chen, Y. D.; Tipton, L.; Staten, M. A.; Shoelson, S. E., Salicylate (salsalate) in patients with type 2 diabetes: a randomized trial. *Annals of Internal Medicine* 2013, *159* (1), 1–12.
107. Dromgoole, S. H.; Cassell, S.; Furst, D. E.; Paulus, H. E., Availability of salicylate from salsalate and aspirin. *Clinical Pharmacology and Therapeutics* 1983, *34* (4), 539–45.
108. Needs, C. J.; Brooks, P. M., Clinical pharmacokinetics of the salicylates. *Clinical Pharmacokinetics* 1985, *10* (2), 164–77.
109. Hawley, S. A.; Fullerton, M. D.; Ross, F. A.; Schertzer, J. D.; Chevtzoff, C.; Walker, K. J.; Peggie, M. W.; Zibrova, D.; Green, K. A.; Mustard, K. J.; Kemp, B. E.; Sakamoto, K.; Steinberg, G. R.; Hardie, D. G., The ancient drug salicylate directly activates AMP-activated protein kinase. *Science (New York, N.Y.)* 2012, *336* (6083), 918–22.
110. Yuan, M.; Konstantopoulos, N.; Lee, J.; Hansen, L.; Li, Z. W.; Karin, M.; Shoelson, S. E., Reversal of obesity- and diet-induced insulin resistance with salicylates or targeted disruption of Ikkbeta. *Science (New York, N.Y.)* 2001, *293* (5535), 1673–7.
111. Liang, W.; Verschuren, L.; Mulder, P.; van der Hoorn, J. W.; Verheij, J.; van Dam, A. D.; Boon, M. R.; Princen, H. M.; Havekes, L. M.; Kleemann, R.; van den Hoek, A. M., Salsalate attenuates diet induced non-alcoholic steatohepatitis in mice by decreasing lipogenic and inflammatory processes. *British Journal of Pharmacology* 2015, *172* (22), 5293–305.
112. Smith, B. K.; Ford, R. J.; Desjardins, E. M.; Green, A. E.; Hughes, M. C.; Houde, V. P.; Day, E. A.; Marcinko, K.; Crane, J. D.; Mottillo, E. P.; Perry, C. G.; Kemp, B. E.; Tarnopolsky, M. A.; Steinberg, G. R., Salsalate (Salicylate) Uncouples Mitochondria, Improves Glucose Homeostasis, and Reduces Liver Lipids Independent of AMPK-β1. *Diabetes* 2016, *65* (11), 3352–61.
113. Nie, L.; Yuan, X. L.; Jiang, K. T.; Jiang, Y. H.; Yuan, J.; Luo, L.; Cui, S. W.; Sun, C., Salsalate Activates Skeletal Muscle Thermogenesis and Protects Mice from High-Fat Diet Induced Metabolic Dysfunction. *eBioMedicine* 2017, *23*, 136–45.
114. Ludy, M. J.; Moore, G. E.; Mattes, R. D., The effects of capsaicin and capsiate on energy balance: critical review and meta-analyses of studies in humans. *Chemical Senses* 2012, *37* (2), 103–21.
115. Panchal, S. K.; Bliss, E.; Brown, L., Capsaicin in Metabolic Syndrome. *Nutrients* 2018, *10* (5).
116. Mahmmoud, Y. A., Capsaicin stimulates uncoupled ATP hydrolysis by the sarcoplasmic reticulum calcium pump. *The Journal of Biological Chemistry* 2008, *283* (31), 21418–26.
117. Faraut, B.; Giannesini, B.; Matarazzo, V.; Le Fur, Y.; Rougon, G.; Cozzone, P. J.; Bendahan, D., Capsiate administration results in an uncoupling protein-3 downregulation, an enhanced muscle oxidative capacity and a decreased abdominal fat content in vivo. *International Journal of Obesity* 2009, *33* (12), 1348–55.
118. Mingrone, G.; Rosa, G.; Greco, A. V.; Manco, M.; Vega, N.; Hesselink, M. K.; Castagneto, M.; Schrauwen, P.; Vidal, H., Decreased uncoupling protein expression and intramyocytic triglyceride depletion in formerly obese subjects. *Obesity Research* 2003, *11* (5), 632–40.
119. Vidal-Puig, A. J.; Grujic, D.; Zhang, C. Y.; Hagen, T.; Boss, O.; Ido, Y.; Szczepanik, A.; Wade, J.; Mootha, V.; Cortright, R.; Muoio, D. M.; Lowell, B. B., Energy metabolism in uncoupling protein 3 gene knockout mice. *The Journal of Biological Chemistry* 2000, *275* (21), 16258–66.
120. Masuda, Y.; Haramizu, S.; Oki, K.; Ohnuki, K.; Watanabe, T.; Yazawa, S.; Kawada, T.; Hashizume, S.; Fushiki, T., Upregulation of uncoupling proteins by oral administration of capsiate, a nonpungent capsaicin analog. *Journal of Applied Physiology (Bethesda, MD: 1985)* 2003, *95* (6), 2408–15.
121. Türközü, D.; Tek, N. A., A minireview of effects of green tea on energy expenditure. *Critical Reviews in Food Science and Nutrition* 2017, *57* (2), 254–258.
122. Clark, K. S.; Coleman, C.; Shelton, R.; Heemstra, L. A.; Novak, C. M., Caffeine enhances activity thermogenesis and energy expenditure in rats. *Clinical and Experimental Pharmacology & Physiology* 2019, *46* (5), 475–482.
123. Fernández-Quintela, A.; Milton-Laskibar, I.; González, M.; Portillo, M. P., Antiobesity effects of resveratrol: which tissues are involved? *Annals of the New York Academy of Sciences* 2017, *1403* (1), 118–131.
124. Lagouge, M.; Argmann, C.; Gerhart-Hines, Z.; Meziane, H.; Lerin, C.; Daussin, F.; Messadeq, N.; Milne, J.; Lambert, P.; Elliott, P.; Geny, B.; Laakso, M.; Puigserver, P.; Auwerx, J., Resveratrol improves mitochondrial function and protects against metabolic disease by activating SIRT1 and PGC-1alpha. *Cell* 2006, *127* (6), 1109–22.

125. Milton-Laskibar, I.; Aguirre, L.; Etxeberria, U.; Milagro, F. I.; Martínez, J. A.; Portillo, M. P., Do the effects of resveratrol on thermogenic and oxidative capacities in IBAT and skeletal muscle depend on feeding conditions? *Nutrients* 2018, *10* (10).
126. Alberdi, G.; Rodríguez, V. M.; Miranda, J.; Macarulla, M. T.; Churruca, I.; Portillo, M. P., Thermogenesis is involved in the body-fat lowering effects of resveratrol in rats. *Food Chemistry* 2013, *141* (2), 1530–5.
127. Calder, P. C., Docosahexaenoic Acid. *Annals of Nutrition & Metabolism* 2016, *69 Suppl 1*, 7–21.
128. Echeverría, F.; Valenzuela, R.; Catalina Hernandez-Rodas, M.; Valenzuela, A., Docosahexaenoic acid (DHA), a fundamental fatty acid for the brain: New dietary sources. *Prostaglandins, Leukotrienes, and Essential Fatty Acids* 2017, *124*, 1–10.
129. Fajardo, V. A.; Bombardier, E.; Irvine, T.; Metherel, A. H.; Stark, K. D.; Duhamel, T.; Rush, J. W.; Green, H. J.; Tupling, A. R., Dietary docosahexaenoic acid supplementation reduces SERCA Ca2+ transport efficiency in rat skeletal muscle. *Chemistry and Physics of Lipids* 2015, *187*, 56–61.
130. Hu, S. S.; Bradshaw, H. B.; Benton, V. M.; Chen, J. S.; Huang, S. M.; Minassi, A.; Bisogno, T.; Masuda, K.; Tan, B.; Roskoski, R., Jr.; Cravatt, B. F.; Di Marzo, V.; Walker, J. M., The biosynthesis of N-arachidonoyl dopamine (NADA), a putative endocannabinoid and endovanilloid, via conjugation of arachidonic acid with dopamine. *Prostaglandins, Leukotrienes, and Essential Fatty Acids* 2009, *81* (4), 291–301.
131. Mahmmoud, Y. A.; Gaster, M., Uncoupling of sarcoplasmic reticulum Ca^{2+}-ATPase by N-arachidonoyl dopamine. Members of the endocannabinoid family as thermogenic drugs. *British Journal of Pharmacology* 2012, *166* (7), 2060–9.

Index

D

J

K

L

M

N

O

P

T

For Product Safety Concerns and Information please contact our
EU representative GPSR@taylorandfrancis.com Taylor & Francis
Verlag GmbH, Kaufingerstraße 24, 80331 München, Germany